Practical Aspects of Computational Chemistry V

Jerzy Leszczynski • Manoj K. Shukla
Editors

Practical Aspects of Computational Chemistry V

Springer

Editors
Jerzy Leszczynski
Department of Chemistry, Physics
and Atmospheric Sciences
Jackson State University
Jackson, Mississippi, USA

Manoj K. Shukla
US Army Engineer Research
and Development Center
Vicksburg, Mississippi, USA

ISBN 978-3-030-83243-8 ISBN 978-3-030-83244-5 (eBook)
https://doi.org/10.1007/978-3-030-83244-5

This Springer imprint is published by the registered company Springer Nature Switzerland AG.
The registered company address is: Gewerbestrasse 11, 6330 Cham, Switzerland

Preface

It is an immense pleasure to bring the next volume of *Practical Aspects of Computational Chemistry* book series (Vol. 5). Our current volume covers an overview of current reactive force-field methodologies, application of coarse-grained modeling in bio- and nanochemistry, application of computational chemistry in the area relevant to fate and transport on soil surfaces, dataset modelability by QSAR, application of alkaline hydrolysis for decontamination, surface-enhanced Raman scattering of organic compounds interacting with silver clusters, modeling of non-covalent interactions, and radiosensitizers. These topics reviewing current state-of-the-science research are distributed in eight chapters in this volume.

Success of any computational methodology largely depends upon the suitability of the applied approach to the problem under investigation. Likewise, the success and failure of molecular dynamics simulations depends upon the suitability of force field for the chemical system under investigation. The first chapter of the current volume, contributed by Bresnahan et al., has reviewed some of the popular reactive force-field approaches and highlighted their respective advantages and potential applications. Coarse-grained approaches are required when size- and timescales are needed to cover systems and phenomena several orders of magnitude larger than those possible using the all-atom simulation. In the second chapter, Liwo et al. have discussed coarse-grained force fields and their application in biological and nanomaterials research.

Non-covalent interactions are prevalent in nature. They play a fundamental role in regulating the structure and function of a variety of materials ranging from biological systems to solid-state materials. Samanta et al. have reviewed first principle approaches to study non-covalent interactions in different systems and their relative accuracy in the third chapter. Radiosensitizers are used in radiation therapy due to hypoxia in solid tumors. Zdrowowicz et al. have reviewed several groups of radiosensitizers geared toward hydrated electrons in the fourth chapter. Fate, transport, and remediation of environmental contaminants are an important area of research. In the next chapter, Jenness et al. have provided a brief description of adsorption of selected munitions compounds on the surface of some components of arid soil, overview of periodic DFT methodologies, and recent work on reducing empiricism in DFTB. In the following chapter, Sviatenko et al. have discussed suitability of computational approaches to analyze alkaline hydrolysis of selected, important energetic compounds. The modelability criteria are used to estimate if predictive QSAR model could be built for a given dataset. Golbraikh et al. have discussed two of such recently introduced modelability criteria in the seventh chapter. In the last chapter of this volume, Trang et al. have discussed a model of silver clusters in order to explain the surface-enhanced Raman scattering of different molecular systems.

We would like to take this opportunity to thank all contributors for devoting their time and hard work to make this project a success. We acknowledge the excellent support from the Presidium of the European Academy of Science as well as editors at Springer Nature. Of course, many thanks go to our family and friends; without their support the realization of the book would not have been possible.

Jackson, MS, USA — Jerzy Leszczynski
Vicksburg, MS, USA — Manoj K. Shukla

Contents

1 **Introductory Roadmap to Current Reactive Force-Field Methodologies** 1
Caitlin G. Bresnahan, Glen R. Jenness, Revati Kumar, and Manoj K. Shukla

2 **Physics-Based Coarse-Grained Modeling in Bio- and Nanochemistry** 31
Adam Liwo, Adam K. Sieradzan, Agnieszka S. Karczyńska, Emilia A. Lubecka, Sergey A. Samsonov, Paweł Krupa, and Magdalena Mozolewska

3 **First-Principles Modeling of Non-covalent Interactions in Molecular Systems and Extended Materials** 71
Pabitra Narayan Samanta, Devashis Majumdar, Szczepan Roszak, and Jerzy Leszczynski

4 **DNA Damage Radiosensitizers Geared Towards Hydrated Electrons** 125
Magdalena Zdrowowicz, Lidia Chomicz-Mańka, Kamila Butowska, Paulina Spisz, Karina Falkiewicz, Anna Czaja, and Janusz Rak

5 **Application of Computational Chemistry for Contaminant Adsorption on the Components of Soil Surfaces** 171
Glen R. Jenness, Levi A. Lystrom, Harley R. McAlexander, and Manoj K. Shukla

6 **Application of Computational Approaches to Analysis of Multistep Chemical Reactions of Energetic Materials: Hydrolysis of Hexahydro-1,3,5-Trinitro-1,3,5-Triazine (RDX) and Octahydro-1,3,5,7-Tetranitro-1,3,5,7-Tetrazocine (HMX)** 215
Liudmyla K. Sviatenko, Leonid Gorb, Danuta Leszczynska, Sergiy I. Okovytyy, Manoj K. Shukla, and Jerzy Leszczynski

7 Dataset Modelability by QSAR: Continuous Response Variable 233
Alexander Golbraikh, Rong Wang, Vinicius M. Alves, Inta Liepina, Eugene Muratov, and Alexander Tropsha

8 A Cluster Model for Interpretation of Surface-Enhanced Raman Scattering of Organic Compounds Interacting with Silver Nanoparticles 255
Nguyen Van Trang, Duy Quang Dao, Pham Vu Nhat, Phan Thi Thuy, and Minh Tho Nguyen

Index 287

Chapter 1
Introductory Roadmap to Current Reactive Force-Field Methodologies

Caitlin G. Bresnahan, Glen R. Jenness, Revati Kumar, and Manoj K. Shukla

Abstract Classical molecular dynamics (MD) simulations have led to many discoveries for a wide variety of reactive chemical systems. Conventional classical potentials are computationally inexpensive and allow for the study of large systems such as the condensed phase with total inclusion of configurational entropy. Unfortunately, the classical MD methodology typically does not allow for bond evolution throughout the simulation. As such, several reactive methods have been developed to study such chemical phenomena using MD simulations. This review provides an overview of current popular reactive potentials and outlines their general framework and current applications.

1.1 Introduction

As the capabilities of computers continue to grow, both in terms of speed and memory, the possible applications into the field of chemistry further expand to include larger system sizes, improved system descriptions, as well as precipitating innovate pathways for approaching chemical problems. Molecular dynamics (MD), a cornerstone in the field of computational chemistry, involves explicitly modeling the evolution of a chemical system as a function of time, based on the numerical integration of the forces in Newton's classical equation of motion [1–4]. The forces

C. G. Bresnahan
Oak Ridge Institute for Science and Education, Oak Ridge, TN, USA

Environmental Laboratory, US Army Engineer Research and Development Center, Vicksburg, MS, USA

G. R. Jenness · M. K. Shukla (✉)
Environmental Laboratory, US Army Engineer Research and Development Center, Vicksburg, MS, USA
e-mail: Manoj.K.Shukla@usace.army.mil

R. Kumar
Department of Chemistry, Louisiana State University, Baton Rouge, LA, USA

J. Leszczynski, M. K. Shukla (eds.), *Practical Aspects of Computational Chemistry V*, https://doi.org/10.1007/978-3-030-83244-5_1

on the particles (atoms in the case of atomistic simulations) are determined from the potential energy surface of the chemical system that, in turn, is described in terms of inter-particle interactions that the particles experience from their local environments.

The accuracy of these simulations is dependent on multiple factors, with the most essential component consisting of the veracity of the underlying potential energy function chosen. The ideal description would result from first principles, also known as an *ab initio* potential. This approach utilizes a fully quantum mechanical (QM) description which inherently includes all the essential physics. These high-level calculations can be used without further parameterization, in principle, to the system of interest, and are transferable across diverse chemical systems. *Ab initio* molecular dynamics (AIMD) can provide the reactivity and high level description required to look at complex systems, however the computational expense limits the technique to system sizes of a couple of hundred atoms for around a hundred picoseconds [5, 6]. Essentially, within the AIMD framework, proper sampling to capture all rare events (e.g., bond breaking/bond forming) and obtain statistically relevant data outside of a cluster system is often intractable due to the spatial and temporal limits. This renders AIMD simulations impractical for condensed phases, where configurational entropy is large and hence sampling for a physically accurate description of many reaction mechanisms is an insurmountable challenge.

In order to circumvent the computational expense of AIMD, a classical model potential can be used, which essentially averages the electronic interactions. This methodology uses classical mechanics in which the potential energy is given as a sum of interatomic interactions including both bonded and non-bonded interactions. The latter can be treated using effective pair potentials that often include Lennard Jones and Coulombic potentials, and may sometimes include the added complexity of many-body functional forms such as a polarizable interaction that can mimic the changes in charge distribution as a function of environment. While empirical two-body potentials provide a rather incomplete representation of interactions, polarizable force-fields improve upon the description but at the cost of added computational expense. Moreover, most widely used classical force-fields, effective pair or many-body models, rely on a fixed bond topology, i.e., molecular connectivity is explicitly defined before the simulations begin and cannot change. An example of a typical force-field is given below in the CHARMM [7] formalism depicted in Eq. (1.1).

$$
\begin{aligned}
V(x) = &\sum_{bonds} \frac{k_i}{2}\left(r_i - r_i^0\right)^2 + \sum_{angles} \frac{k_j}{2}\left(\theta_j - \theta_j^0\right)^2 \\
&+ \sum_{dihedrals} \sum_{m=1}^{m_{max}} k_k^m \cos\left(m\varphi_k - \gamma_k\right) \\
&+ \sum_{impropers} \frac{k_l}{2}\left(\phi - \phi_l^0\right)^2 + \sum_{i<j} \frac{q_i q_j}{4\pi\varepsilon_0 r_{ij}} + \sum_{i<j} 4\varepsilon_{ij}\left[\left(\frac{\sigma_{ij}}{r_{ij}}\right)^{12} - \left(\frac{\sigma_{ij}}{r_{ij}}\right)^{6}\right]
\end{aligned}
\tag{1.1}
$$

The above classical effective pair potential is an elegant approximation of the *ab initio* potential energy surface. The bonded terms are described by the bond length, bond angle, dihedral angle, and improper angle term, penalizing deviations from the ideal structures. The non-bonded terms encompass electrostatic and van der Waals interactions, and in this case, they are described with a Coulomb and Lennard-Jones potential, respectively.

In order to closely approximate the real potential energy surface with a classical potential, parameters are optimized using experimental data, high-level *ab initio* data, or both. However, a universal force-field has not been determined, and as such, force-fields tend to be system specific or tailored to a particular system type. Several popular force-fields have been developed, differing slightly in potential form, for prominent systems of interest including the Chemistry at HARvard Molecular Mechanics (CHARMM) [7–9] force-field, seen above, which was generated with the intention of studying small molecules, and now is used extensively to study both small and macromolecules. The AMBER (Assisted Model Building with Energy Refinement) [10, 11] potential was built to investigate protein systems and DNA. Other widely used force-fields include GROMOS (GROningen Molecular Simulation) [12–14], parameterized for use on biomolecular systems, and the Optimized Potential for Liquid Simulations (OPLS) [15–17] which was developed in order to effectively model ionic liquid systems. While these force-fields are widely available, they are detailed here in order to appreciate the breadth of systems to which classical force-fields are applied. However, this list by no means encompasses the entirety of force-fields which are constantly updated and modified to better describe systems of interest. Of course, due to the fitting of parameters to specific systems, force-fields may not perform well across a variety of systems. Often, in order to utilize a particular force-field, results must be compared and validated to experimental data, or benchmarked to existing high level ab initio data for each molecular system of interest.

These classical potentials perform well on large systems and can model ionic diffusion, protein fluctuation, solvation effects, aggregation events and so on. However, due to the fixed bond nature of a classical potential, reactive events cannot be modeled within this framework. QM/MM methods offer an alternative to this problem by combining the quantum mechanics and molecular mechanics (MM) approach into one simulation [18, 19]. Within the QM/MM framework a majority of the system is treated with classical MM potentials, while a small portion of the system will be treated with high level quantum mechanics (QM). In this ethos, simulations can include full solvation via MM in order to explicitly model the condensed phase, extended surfaces, an entire protein, etc., but still include reactive events (reactive surface, active site, local solvation environment, etc.) via QM at the same time. QM/MM approaches have widely increased the reactive systems we are able to study. However, despite these advances, QM/MM has several drawbacks. In order to effectively use QM/MM, *a priori* knowledge of the reactive site is required. This may lead to inadvertently ignoring side reactions that aid in the reactive process. Additionally, the range of atoms included in the QM sphere is set by the user. Defining a large portion of the simulation to undergo high level calculations defeats the speedup offered by the QM/MM framework, as the limiting factor of the simulation is the QM contribution calculated at each time step. However, utilizing a QM region that is too small may also cause issues with the simulation, such as diffusion of a QM particle into the MM region, which introduces an artificial interaction. In fact, the conditions governing the boundary between the QM and MM region can introduce errors or have a wide influence on the energetics calculated in the system [20]. Adaptive QM/MM methods have been developed which use various mathematical functions based on geometric criteria in order to keep the QM portion limited to the reactive sites and the relevant environment of the reactive process [21–23]. While there has been a significant amount of progress in the QM/MM fields, which will continue its relevance within the computational chemistry sphere, several other approaches have been introduced to bring reactivity into MD for large systems while including the adequate sampling needed for statistical relevance and rare events.

This review is intended to give the reader an idea of the current state of reactive modeling, particularly reactive force-fields. First, two of the widely used bond order methods will be described—REBO and the ReaxFF potential. This will be followed by an introduction into the quantum mechanics-based empirical valence bond method and its extension into a multi-state formalism. Each section will provide the basic outline of how these potentials work, followed by some recent developments and applications. Lastly, there will be a brief discussion of parameterization, followed by a comment on some of the other methods available that are beyond the scope of this review.

1.2 Bond Order Methodologies

Reactive Empirical Bond Order (REBO) Based Potentials One of the first widely used and successful bond order potentials is the Reactive Empirical Bond Order (REBO) potential. This formalism was driven by the idea that a classical description could be used, in which a bond order function would allow for the attractive potential to be modulated throughout the simulation in order to maintain accuracy while reducing computational expense. The energy for a bond between two atoms, i and j, is given as a summation of the attractive (V_{ij}^{A}) and repulsive interaction energies (V_{ij}^{R}), where the attractive component is calibrated to the local environment through the use of a bond order term (b_{ij}) [24–28]. This can be seen in Eq. (1.2).

$$\mathrm{E}_{ij} = V_{ij}^{R} + b_{ij} V_{ij}^{A} \tag{1.2}$$

The bond order term used to describe the effects of the bond between atoms i and j is given in Eq. (1.3).

$$b_{ij} = \frac{1}{2}\left[p_{ij}^{\sigma\pi} + p_{ji}^{\sigma\pi}\right] + \pi_{ij}^{RC} + \pi_{ij}^{DH} \tag{1.3}$$

The first two terms, $p_{ij}^{\sigma\pi}$ and $p_{ji}^{\sigma\pi}$, reflect the atomistic environment for the reference atom (i and j, respectively) by including the impact of the atoms bonded to the reference atom. The last two terms contribute to the bonded interaction through the radical character of the π-bond (π_{ij}^{RC}) as well as the dihedral character (π_{ij}^{DH}).

The first generation of REBO was initially made for large carbon crystal systems [25] and had been extended to parameter sets containing silicon [29, 30]. The original formulation was improved upon in a second iteration which introduced a Coulomb potential. The fitting scheme was enhanced as the training set was comprised of more structures within the REBO2 framework, vs. the initial REBO potential. Due to the modifications, REBO2 can model molecular hydrocarbon systems in addition to the crystals for which the initial potential was built. This framework has been further extended to materials containing oxygen [31], fluorine [32], and silicon [33]. Systems such as graphene and its oxides [34–40], and carbon nanotubes [41, 42], have been successfully modeled by REBO potentials.

The REBO formalism lacks a description of long-range interactions, however, which makes it difficult to use in systems where there is a significant amount of intermolecular interactions. To this end, the approach was extended to the Adaptive Intermolecular REBO (AIREBO) model [28, 43]. In AIREBO, a 12-6 Lennard-Jones potential is integrated into the total potential; however, as a reactive force-field, it is imperative that the non-bonded terms can switch on and off accordingly. Rather than using a simple switching function that could cause discontinuities, the function was built to be continuous and depend on distance, interaction strength, and the bond environment. This introduces adaptivity to the Lennard Jones interactions so that it is only applied if it is unlikely for a bond to occur, but is within a certain distance of the atom. Additionally, a torsional potential was added, however it took the form seen in Eq. (1.4), rather than the typical form of Eq. (1.1).

$$V_{torsional}(\omega) = \epsilon\left[\frac{256}{405}\cos^{10}\left(\frac{\omega}{2}\right) - \frac{1}{10}\right] \tag{1.4}$$

In general, the AIREBO potential takes the form seen in Eq. (1.5) with all details laid out in the original paper by Stuart et al. [43]

$$E^{AIREBO} = E^{REBO} + E^{LJ} + E^{torsional} \tag{1.5}$$

The AIREBO method was further improved so that the bond order takes into account the immediate environment of the bond, instead of averaging the environment around the two bonded atoms [26]. This method has been widely used on amorphous carbon [44], graphene systems [40, 45–47], carbon nanotubes [48], and fullerenes [49], and has recently seen code improvements to increase the speed of these simulations [50]. Additionally, the potential has been used to model gas collisions with a graphite surface, where the gas to surface interactions took the form of a Lennard Jones interaction [51]. Other modifications to the REBO potential include, mod-LJ AIREBO [52], AIREBO-M [53], REBO-S [54], and qAIREBO [55], which are outside the scope of this review, but an excellent discussion of them may be found in the review by Harrison et al. [28]

ReaxFF Methodologies The ReaxFF [56] force-field is the most well-known bond order method available due to its sophistication and successful implementation for a variety of systems, as well as its integration into several MD packages. In this formalism, the typical bonded and non-bonded terms of a traditional force-field were kept. Meanwhile, an elegant bond order term is used to modulate the bond topologies, allowing it to evolve while the system undergoes a reaction. The energy is calculated as a sum over the various interaction components as can be seen in Eq. (1.6) [56].

$$E_{system} = E_{bond} + E_{lp} + E_{over} + E_{under} + E_{val} + E_{pen} + E_{coa} + E_{C2} \\ + E_{triple} + E_{tors} + E_{conj} + E_{H-bond} + E_{vdW} + E_{Coulomb} \tag{1.6}$$

E_{bond} is the is the corrected bond energy after the bond order calculation, E_{lp} is the energy associated with a lone pair, E_{over} is an energy penalty associated with an over coordinated atom, E_{under} is related to under coordinated atoms, E_{val} includes the valence angle energy and an associated penalty energy E_{pen} is introduced for the situation when an angle contains two double bonds. E_{coa} was introduced in the initial formalism to describe three-body conjugation. E_{C2} was integrated into the formalism to correct the original models inability to reproduce C_2. E_{triple} only influences C-O systems and is an energy correction that was introduced for carbon monoxide. E_{tors} is related to the torsional energy, while E_{conj} reflects the energy related to a conjugated system. $E_{H\text{-}Bond}$ was incorporated for hydrogen bonding. The non-bonded interactions are given by E_{vdW} and $E_{Coulomb}$ which are calculated for all atomic pairs in the system. An electronegativity equalization method (EEM) charge scheme is utilized so that charges can fluctuate throughout the simulations allowing an accurate description of both covalent and electrostatic effects. As of a 2016 review on the state of ReaxFF [57], it was specified that the above equation is the current formalism released by the van Duin group, and an in depth description of each term can be found in the 2008 paper by Chenoweth et al. [56] Recent advances in 2020 have led to the introduction of the atom-condensed Kohn-Sham DFT approximated to second order (ACKS2) charge model [58]. The introduction of this charge description requires reparameterization of several parameters and was used to study Li_2O systems.

This potential energy description maintains the accuracy of a typical atomic description throughout a simulation while also capturing the fluctuations required to model a reactive event through the bond order term. The bond order term for an atom pair is given by Eqs. (1.7) and (1.8) [57].

$$BO_{ij} = BO_{ij}^{\sigma} + BO_{ij}^{\pi} + BO_{ij}^{\pi\pi} \tag{1.7}$$

$$BO_{ij} = \exp\left[p_{bo1}\left(\frac{r_{ij}}{r_0^{\sigma}}\right)^{p_{bo2}}\right] + \exp\left[p_{bo3}\left(\frac{r_{ij}}{r_0^{\pi}}\right)^{p_{bo4}}\right] + \exp\left[p_{bo5}\left(\frac{r_{ij}}{r_0^{\pi\pi}}\right)^{p_{bo6}}\right] \tag{1.8}$$

The bond orders are dependent on the distance between atoms, given by r_{ij}. Each r_0 is an equilibrium bond length, and $p_{bo\#}$ are fit parameters. The above equations demonstrate the BO inclusion of single, double, and triple bonds for each possible atomic pair given in a continuous function, so that it may be successfully used within the workhorse of molecular dynamic simulations.

ReaxFF treats each element the same, regardless of the environment it is in, or rather, one element has the same description regardless of the species to which it is connected. Therefore, each pair-wise interaction must be taken into account when generating parameter sets. Currently three distinct branches exist; the combustion branch, an aqueous branch, and an independent branch. These branches contain parameter sets which are not necessarily transferable between different systems [57]. ReaxFF was initially built for high temperature systems (combustion branch) in which H_2O would only be present in gas phase. Consequently, the combustion branch did not work well on aqueous descriptions, as it could not mimic liquid water. This has since been addressed [59], hence the formation of the aqueous branch. Many groups have modified parameters or adjusted the potential description for their own uses, (for instance, a recent publication altered the potential description in order to better describe coordinate bonds [60]), however what has been described thus far in this review is the formalism pioneered by the van Duin group [56]. The other parameter sets and potential descriptions may not work correctly with the original parameter set and should always be validated before use. ReaxFF is available in platforms such as PuReMD, and LAMMPS [61].

A recent publication has parameterized ReaxFF for Cu/C/H/O for investigating catalysis on a Cu-metal surface [62]. This paper is intriguing as it used a new transition state search algorithm to add transition states into the parameter fitting set. ReaxFF has also been utilized in pathway sampling studies which included effect of temperature on these pathways, before refining with DFT, showing how versatile the applications of the force-field have become [63].

Recent studies within the ReaxFF framework include supercritical conditions such as acetic acid water mixtures at high temperatures and pressures [64], sol-gel condensation of silica (where the effect of parameterization set had a large influence) [65], applicability to subcritical and supercritical water [66], phase separation in relation to explosive events [67, 68], energetic materials [69], biological systems [70], liquid metals [71], MOFs [72], lithium cation in electrolyte solution [73], and polymer crosslinking using accelerated ReaxFF [74, 75]. This list, while not exhaustive, highlights the versatility of chemical investigations to which ReaxFF may be successfully applied.

However, many systems require reparameterization before use to capture system specific interactions. For instance, in 2019 there have been several reparameterizations in order to handle C/H/O/N polymers and its carbonization process [76], AlF_x compounds [77] and liquid metals [71]. This list is not complete but illustrates examples in which ReaxFF parameters may require refinement. Indeed, Bertels et al. [78] recently tested several ReaxFF parameter sets to benchmark them against hydrogen combustion systems, and offer a close examination of the performance of several parameter sets. Additionally, another recent study compared ReaxFF parameter sets to the previously described AIREBO model highlighting the problems and advantages of each method [79]. These studies are included to show how the performance of the model depends on which parameter set is used, and to highlight the importance of selecting a parameter set that can accurately model the system of interest.

A recent paper looked at the numerical stability of the force-field and suggested utilizing more tapering functions, particularly on the bond order terms, to allow for smoother transitions. They found that the newly tapered functions allowed for better chemical minimization without effecting the overall accuracy of the energetic terms [80].

A note on some specialized ReaxFF examples:

ReaxFF-lg In order to properly simulate energetic materials such as explosives, the ReaxFF-lg model was created. These materials tend to be crystalline solids; thus, the ReaxFF-lg integrated a van der Waals like term in the form of a low-gradient model to describe the long range dispersion interactions that are commonplace within these solids [81]. The energy calculated in the ReaxFF-lg model is shown in Eq. (1.9), where the E_{reax} term is the ReaxFF description and E_{lg} term is described in Eq. (1.10).

$$E_{Reax-\mathrm{lg}} = E_{reax} + E_{lg} \tag{1.9}$$

$$E_{lg} = -\sum_{ij,i<j}^{N} \frac{C_{\mathrm{lg},ij}}{r_{ij}^{6} + dR_{eij}^{6}} \tag{1.10}$$

In Eq. (1.10), $C_{lg,ij}$ is a parameter for the dispersion correction, r_{ij} refers to the interatomic distance between the atoms *i* and *j*, d is a scaling factor, and R_{eij} is the equilibrium van der Waals distance. This addition required a new parameterization, which restricts is transferability to other ReaxFF parameter sets. The ReaxFF-lg has seen several parameterizations and uses on nitroamine systems [82–89], and pyrolysis of carbon oxide systems [90].

eReaxFF While force-fields essentially coarse grain the electronic degrees of freedom, eReaxFF was created in order to include an explicit electron when necessary, so that the force-field could be used for redox reactions [91]. In this way, the electron is treated as a particle where it can either be an electron (–1 charge) or a hole (+1 charge) and several modifications are included to properly model the electron behavior and its affects. Additionally, the EEM charge scheme used in the original ReaxFF force-field was altered to ACKS2 to have a better description of the charge throughout the simulations. This method has been applied to study electron transfer from lithium [92]. Recently, the method was expanded to look at Ag nanoclusters [93]. This system is significantly more complicated as several electron states are possible, and as such, a function was added to determine the number of electrons that are associated with the Ag.

1.3 Valence Bond Models

Empirical Valence Bond (EVB) Framework Thus far the current review has focused on the popular bond order methodologies in the field. As discussed in detail above, these calculations tend to involve seamlessly turning the bond strength up or down based on a geometric criteria, and tuning the intramolecular terms accordingly. An alternative to this method is the more quantum inspired method of the empirical valence bond (EVB) formalism.

This sophisticated method was initially introduced by Arieh Warshel in 1980 [94]. Warshel was interested in combining the ionic and covalent resonance forms so that the environmental effects on bond breaking could be seen. In this case, a reactant species of XY would become the product system of $X^- + Y^+$. This is a standard elementary decomposition reaction; however, these reactions are abundant in biological systems where a full QM treatment is still intractable. Using the idea of valence bond theory, Warshel envisioned a methodology where reactivity could be realized with a classical potential description, via writing the total energy as a linear combination of states.

An example of the two state method is given from the water/HCl system in the gas phase and is depicted in Fig. 1.1. The system can be described by two states: state |1> consisting of one water and HCl, while state |2> consists of H_3O^+ and Cl^-. The real system has both covalent and ionic character and can be described as a linear combination of these two states. In this description, the coordinates are the same in both states, but the bonding topology differs by one bond. The potential for state |1> is simply the classical force-field which comprises all intermolecular and intramolecular interactions given in Eq. (1.11) as H_{11}. State two is described by a force-field which describes hydroniums interaction with chloride given as H_{22} given in Eq. (1.12).

$$H_{11} = \sum V_{water}^{intramolecular} + \sum V_{HCl}^{intramolecular} + \sum V_{HCl-water}^{intermolecular} \tag{1.11}$$

$$H_{22} = \sum V_{H_3O^+}^{intramolecular} + \sum V_{H_3O^+-Cl^-}^{intermolecular} \tag{1.12}$$

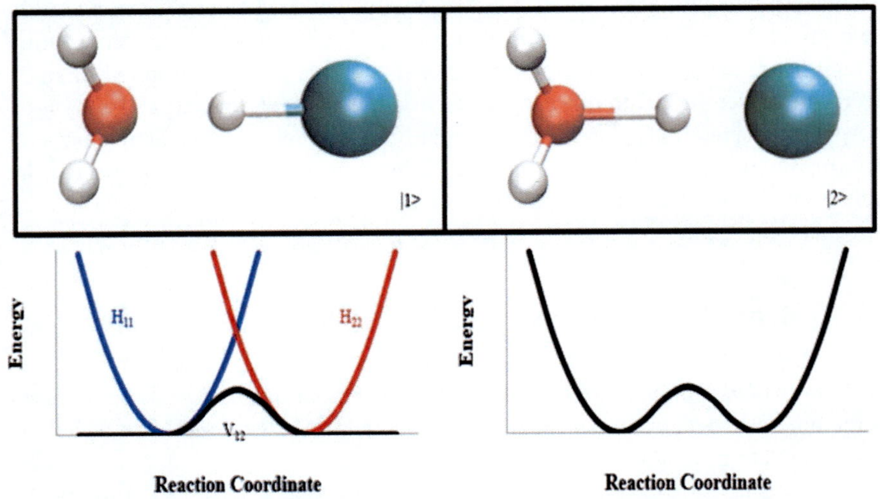

Fig. 1.1 The top half shows two states with the same coordinates but differing by connectivity. The lower left quadrant illustrates the potential energy well of both bonding types before coupling. The lower right quadrant depicts the coupled EVB potential energy surface

The total Hamiltonian is written as a linear combination of the two states, which can be seen in Eq. (1.13). The states are now coupled through the off-diagonal element.

$$H = \begin{bmatrix} H_{11} & H_{12} \\ H_{21} & H_{22} \end{bmatrix} \tag{1.13}$$

The matrix is Hermitian so $H_{12}=H_{21}$. Upon diagonalization of the matrix, the lowest eigenvalue will correspond to the energy of the system and the square of the elements of the corresponding eigenvector will correspond to the weight of each state, or the probability of the system existing in that state. The potential energy surface of the EVB model for the HCl-water system is given in the lower right quadrant of Fig. 1.1 where the system can propagate through both states with a seamless transition. This methodology, coined by Warshel [94, 95], allowed for countless investigations into two-state, reactant-product systems in the condensed phase.

The off-diagonal term is essential for the EVB formalism to accurately reproduce the potential energy surface. The simplest method is to add a constant value, however this is not always a sufficient method. Warshel and Weiss used a Morse function to describe V_{12} in the original formalism [94]. Vuilleumier and Borgis modeled the excess proton in water using another physically motivated and straight forward coupling term which includes an exponential function dependent on the distance between oxygens sharing the excess proton, and the position of the proton compared to the O-O bond [96–98]. Chang and Miller adopted a distributed Gaussian function to describe $V_{12}(\mathbf{q})^2$ where the vector $\mathbf{q}$ depends on the configuration [99]. This method was extended so that the distributed Gaussians were multiplied by a quadratic polynomial that did improve the accuracy of the off-diagonal potential [100, 101]. The current MS-EVB3.2 formulation (discussed in the next section) uses a complex function which includes the Zundel ion interaction with the other species in the system [102]. Currently, common forms include a simple Gaussian based on a reaction coordinate descriptive of the reactivity used (i.e., bond distance) [103, 104] or a spline interpolation [105].

EVB methodologies have seen successful usage across a multitude of systems/investigations. Many have focused on proton transfer such as that of 2-pyridone to 2-hydroxypyridine [106] and charge delocalization in water wires and water clusters [96, 107]. Other systems include Claisen rearrangement in allyl vinyl ethers and a methane and chlorine system [106, 108]. Perhaps one of the greatest applications can be seen in biological systems where simple reactivity can be brought into a large condensed system and solvent effects can be studied [95, 103, 109–113]. The Amber package has been successfully used to run EVB simulations [111, 114].

Multi-state Empirical Valence Model The EVB model was further extended to more comprehensively model Grotthuss shuttling in aqueous systems. The Grotthuss shuttling mechanism describes the ability of an excess proton to hop or shuttle over several water molecules in a very brief window of time through bond breakage/formation within the hydrogen bonded network [115–117]. In these systems, the excess charge is effectively delocalized over its solvation shell and hence the two state model was extended to a multi-state approach by the Voth group [102, 117–129], to the so called multi-state empirical valence bond method (MS-EVB). In this expansion, rather than writing the Hamiltonian as a linear combination of just two states, the system is described as a combination of several states involving waters in the solvation shell of the excess charge. A simple example is shown in Fig. 1.2 for the $HCl(H_2O)_3$ system. In order to properly describe all possible bonding topologies, four states are included. The coordinates in each state are the same, however the bonding topology is different. Only one HCl state is considered, as the other two water molecules are too far to see proton transfer with the chlorine atom (state |1>). Three possible hydronium states are also depicted (states |2> through |4>).

As before, the total Hamiltonian is written as a linear combination of states, however states that do not share a bonded atom are not coupled and therefore have no off-diagonal term. This leaves a sparse matrix, which is quite quick to calculate and is depicted in Eq. (1.14).

$$H = \begin{bmatrix} H_{11} & H_{12} & 0 & 0 \\ H_{12} & H_{22} & H_{23} & H_{24} \\ 0 & H_{23} & H_{33} & 0 \\ 0 & H_{24} & 0 & H_{44} \end{bmatrix} \tag{1.14}$$

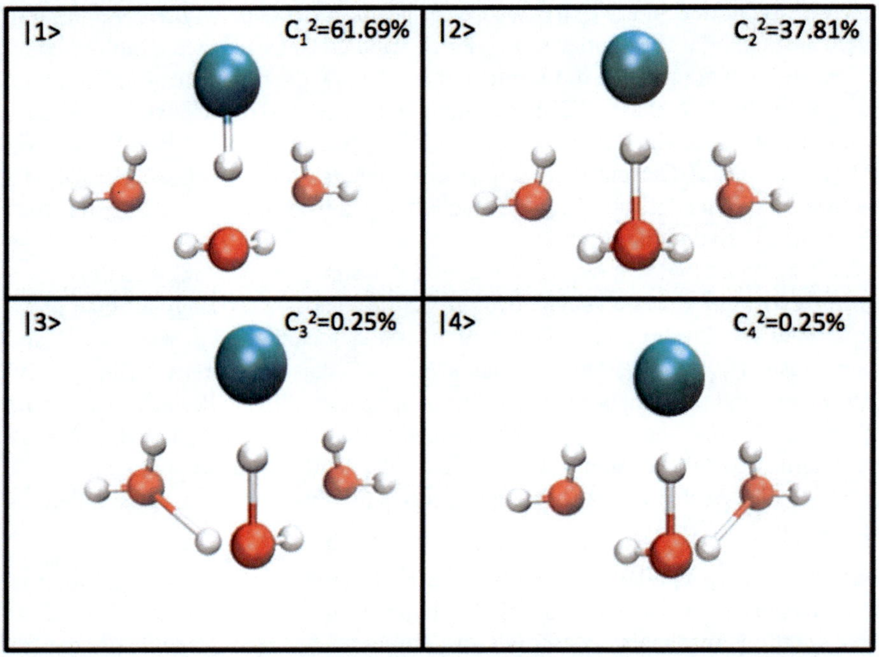

Fig. 1.2 Configuration of the $HCl(H_2O)_3$ system. Each state has the same coordinates, but differs in bonding topology. Percent contribution of each state is included

The square of the elements of the eigenvector corresponding to the lowest energy eigenvalue depicts the weight of each state. Figure 1.2 has the contribution labeled for each state showing that while the HCl is the predominant state, there is a significant amount of hydronium character to that configuration as well.

In order to accurately model systems with rapid proton diffusion, a search algorithm must identify all states to include. Wu et al. [125] generated an extremely comprehensive search algorithm for the MS-EVB3 model for proton transport in neat water, which has been re-parameterized in the latest MS-EVB3.2 model [102]. The algorithm includes every possible hydronium state within three solvation shells of the pivot hydronium (otherwise known as the hydronium with the highest contribution calculated from the eigenvector).

The MS-EVB formalism has been widely used to study proton transfer in many different aqueous systems, especially to understand the influence of the environment on the transport process. For instance, it has been used extensively to study the effect of the excess proton in bulk water and small water clusters [102, 117, 120–125, 130–132]. It is also quite useful for studying interfaces, ion behavior around an interface, the excess proton in a water channel, acidic groups in proteins, membranes, concentrated acid solutions, and other systems of biological interest [102, 104, 105, 118, 119, 122, 124, 126, 128, 129, 133–139]. As the formalism depends on the underlying classical description, water anharmonicity can be included through the use of an anharmonic force field [140]. In fact, the anharmonic SPC/FW/ice model was used as the underlying potential energy surface in order to study the surface of ice and properties of the quasi-liquid layer [141]. Not only can the MS-EVB formalism model the initial dissociation, it can determine how the presence of other groups affects the proton transport, and give an accurate picture of the chemistry at an active site. Additionally it has been used in systems like liquid imidazole [142, 143] and hydroxy groups [138]. In the above examples proton shuttling to the ionic Cl^- counterion is not considered, however this can become important in water clusters with HCl as well as at air-solution interface of concentrated HCl solutions. Two models exist in the literature which model explicit shuttling of the acidic proton to the chloride in water including the work of Wick for HCl accommodation to a bulk water surface [105] and Bresnahan et al. [104] on HCl water clusters. It was found that in the cluster systems, explicit incorporation of the covalent HCl states improves the energetics for contact ion pairs, and the MS-EVB-HCl model truly captures the covalent HCl interactions.

The MS-EVB formalism has further been expanded by Voth and co-workers [121, 123, 125] for the movement of large molecules, rather than modeling chemical reactions. Here a multi-configurational method is used to simulate the folding of dodeca-alanine, using a coarse-grained approach [144]. Additionally, the Voth group [145] has recently published a method called RCG, for reactive coarse-grained, which allows for chemical reactions of coarse-grained molecules which was validated on an S_N2 reaction mechanism.

1.4 Parameterization

One common feature in all of these methods is the requirement of parameters to closely approximate the potential energy surface. This can be quite difficult, particularly for methods like ReaxFF, in which a system containing only C/H/O atoms has over 600 parameters that must be optimized [146]. The EVB methods require less parameters but does necessitate an appropriate function to describe the coupling between states.

Several steps are required for parameterization. The first step is to generate data representative of the problem you are trying to solve. Generally, that includes structural data on configurational motifs for the system of interest. Typically, one chooses global minimum structures and may also require a detailed potential energy surface in which to match the force-field to. In the case of the MS-EVB-HCl model, however, the purpose was to describe both covalent and ionic forms of HCl water-clusters of various sizes [104]. Given that its intended purpose was for atmospheric aerosols which can experience configurations that are quite high in energy, it was essential that parameters were fit to local minima as well as the global minima. Furthermore, the parameterization set included various configurations with covalent HCl, contact ion pairs of hydronium and Cl^- as well as solvent separation ion pairs (where the ions are separated by one water molecule) of various hydration levels .

Once a parameterization set has been identified, the next step involves a parameter search and a fit criterion. The fit criterion is a measure of the force-fields ability to reproduce the higher level *ab initio* calculations to a desired accuracy. The parameter search is used to generate parameters which are tested via the fit criterion. Unfortunately, parameter space is expansive. In theory, one could determine the parameters by starting with a best guess, and refining one parameter at a time, iteratively. However, this approach is time consuming and so several expedient and automated processes have been developed.

One extensively used parameterization search method for reactive force-fields is the genetic algorithm. In this method, the user inputs bounds for each parameter and genetic operations occur over a set amount of iterations in order to determine the best fit based on the fit criteria. Genetic operations include mating and mutations of the parameter set throughout a given amount of iterations.

While the genetic algorithm is a popular parameter optimization scheme, an abundance of other schemes exist in which parameter space is traversed. In fact, Shchygol et al. [147] compared the performance two other schemes, a Monte-Carlo FF (MCFF) optimizer, and a covariance matrix adaption evolutionary strategy (CMA-ES) with the genetic algorithm. The MCFF scheme borrows from the idea of simulated annealing where a system at a very high temperature is cooled down; all while introducing random changes to the parameter set. By cooling very slowly, the optimizer is able to sample the parameter space. In the CMA-ES method parameters are allowed to evolve in regards to a multivariate normal distribution. The authors found that a single optimization run may not be sufficient, as the reproducibility between runs were low or convergence was not ideal. While CMA-ES performed well in local minimums, the genetic algorithm by nature, is less likely to be caught in a local minima. The list of parameter optimization schemes is not exhaustive but rather is included in order to illustrate how different methods can traverse parameter space. While seemingly straightforward, there are many issues with parameterization, such as getting stuck within a local minima or poor convergence. The matter of generating parameters for a force-field is far from trivial.

1.5 Other Reactive Methods

RexPoN Several other reactive methods have also been developed. For instance RexPoN is being developed for both reactive and nonreactive simulations [148, 149]. Its functional form is depicted in Eq. (1.15) [148]. E_{bond} includes the bonding terms, E_{angle} includes bond angles, E_{elect} is the electrostatic energy, E_{vdw} is the pure van der Waals term, and E_{HB} is included for hydrogen bond energy.

$$E_{total} = E_{bond} + E_{angle} + E_{elect} + E_{vdW} + E_{HB} \tag{1.15}$$

RexPoN uses a polarizable charge equilibration method, allowing dynamic evolution of the charge, and has the van der Waals terms trained to single element crystals. In this manner, the non-bonded terms have been separated and trained to the quantum data. This is a developing bond order formalism that is aiming for improved transferability.

LEWIS Reactive Force Field The Herzfeld group [150–155] has published several papers on a reactive force-field that is inspired by the breakdown of a molecule into Lewis structures. In the initial formalism a water molecule is decomposed into its Lewis features, which can be described by six pairwise interactions [150–152]. The oxygen of the water molecules exists as O^{-2}. That species would be broken down into separate pieces as seen in a Lewis dot structure; The core of the oxygen would be describe by a core particle (O^{+6}), and then four valence electron pairs would be explicitly modeled while the hydrogen of water is then described by H^{+} particles. While the specific requirements of the model are described in more detail in the original publications [150, 151] a quick overview of the charge with respect to the Lewis model is presented. The charge description also reflects the Lewis structure, where the core oxygen particle maintains a charge of 6, the valence electron pairs are described by a –2 charge, and the hydrogen particles round out the molecular neutrality with a charge of +1.

In itself, this formalism is pseudo-classical. Electrons are being explicitly treated as core electrons, and valence electrons. Unlike the reactive-force fields previously described, there is no standard potential in which to describe these interactions, and as such, the physically motivated potentials seen before are set aside for a more heuristic approach where the potential that performs the best on the system is searched for [150, 151]. Around 300 combinations of potential forms were analyzed with the final functional form described in Kale et al. [151] The potentials describe the pair-wise interactions between the particles, inherently allowing for polarization and reactivity as the valence electrons are explicitly modeled [150, 151]. Training sets included neutral water, as well as variations with protonated or deprotonated species allowing the model to accurately describe both hydronium and hydroxide [152, 154, 155].

The original model offered significant flexibility, however was limited by the valence electrons being described as an electron pair. In order to address this, the LEWIS· method was introduced, where each valence electron was treated individually [153]. A spin description was added via a pair interaction to describe interactions between particles of the same and opposing spin. Additionally, another coordinate, which describes the spread of the electron, or rather its cloud diameter, was integrated into the formalism. In theory, this addition will create a more transferable model. These interactions are described in more detail in Ekesan et al. [153] Furthermore, the model was extended to include carbon interactions [155].

Chebyshev Interaction Model for Efficient Simulation (ChIMES) The initial motivation behind the ChIMES force field was to apply force matching to generate a force-field to model reactivity in water in which the thermodynamic conditions were extreme [156]. The original method used a pairwise interactions in conjunction with an over coordination term. Future iterations, however, would focus on a pairwise interaction with the addition of a three-body term as seen in Eq. (1.16) [157–160].

$$E_{ChIMES} = \sum_{i}^{N}\sum_{j>i}^{N} E_{ij} + \sum_{i}^{N}\sum_{j>i}^{N}\sum_{k>j}^{N} E_{ijk} \tag{1.16}$$

There is no topology dependence in this method, making it a reactive force-field where the total energy of ChIMES comes from the pairwise interaction of the total N particles between atoms *i* and *j* as well as the three body interaction between atoms *i*, *j* and *k*. The two body potential is described in Eq. (1.17), and is dependent on the use of Chebyshev Polynomials.

$$E_{ij} = f_p^{e_i,e_j}\left(r_{ij}\right) + f_s^{e_i,e_j}\left(r_{ij}\right)\sum_{n=1}^{\mathcal{O}_{2B}} c_n^{e_i,e_j} T_n\left(s_{ij}^{e_i,e_j}\right) \tag{1.17}$$

$f_p(r_{ij})$ is a penalty term which keeps the bond distance from falling under a minimum value. $f_s(r_{ij})$ allows for a smooth transition to zero for a given cutoff distance. The order of the polynomial is given by $\mathcal{O}_{2B}$=n, where the T_n is the Chebyshev polynomial, modulated by its lead coefficient, $c_n^{e_i,e_j}$. The symbols e_i is representative of the element of atom *i*, and e_j describes the element of atom *j*. A more detailed discussion of these terms is given in the work of Lindsey et al. [157–160]

The three body term is an extension of the two body term, without the penalty function listed in Eq. (1.17), because this penalty is already included in the two body term. Equation (1.18) shows the three body potential with analogous terms and coefficients. Lindsey et al. [158] mention that this model can easily expand to higher many body effects.

$$E_{ijk} = f_s^{e_i,e_j}\left(r_{ij}\right) f_s^{e_i,e_k}\left(r_{ik}\right) f_s^{e_j,e_k}\left(r_{jk}\right)\sum_{m=0}^{\mathcal{O}_{3B}}\sum_{p=0}^{\mathcal{O}_{3B}}\sum_{q=0}^{\mathcal{O}_{3B}} c_{m,p,q}^{e_{ij},e_{ik},e_{jk}} T_m\left(s_{ij}^{e_i,e_j}\right) T_p\left(s_{ik}^{e_i,e_k}\right) T_q\left(s_{jk}^{e_j,e_k}\right) \tag{1.18}$$

This three body method has been used for nonreactive dynamics of water at ambient conditions [158], studying molten carbon [157, 159], and was recently extended for carbon-oxygen systems to look at condensation [160].

An advantage to the ChIMES model is its linearity, which allows for fitting to different systems to be executed quickly [157–160]. While several parameters are part of the fitting scheme, the user of the model is given flexibility in inputting certain parameters such as the order of the polynomial, cut off distances, and a Morse variable [159]. While parameters would likely need to be generated for the system of interest, the quick fitting scheme and flexibility afforded the model by the input parameters allows for the model to quickly be tuned to a particular system of interest. The breadth of environments studied with this force-field show its unique ability to the application of a wide variety of problems, particularly in extreme conditions.

RGATS Another method under development is RGATS, which stands for restrained geometries and topology switching [161]. In this strategy, standard restraints were used, and a scan of the topologies for potential reactive states occurs. In this scan the reactive species is not connected to either the reactant or the product, rather a harmonic restraint is utilized to describe the transition. The publication outlines the systematic approach but mentions a user patch is required to have the topology switch occur. By construct, all dynamical properties of the system are lost, such as the activation barrier, however it gives an idea of the system response to the reactivity.

Density Functional Theory Tight Bonding (DFTB) For the sake of completeness, the DFTB method will be discussed. The formalism found its origins in semi-empirical tight-binding electronic structure methods that treated a crystal as a series of isolated atoms, with the crystal density subsequently being the summation of these isolated atomic densities. However, the parameters were highly empirical, and suffered from a lack of transferability. The pioneering work of Foulkes and Haydock [162], however, demonstrated that the parameters required for tight-binding calculations can be derived directly from DFT; giving rise to what is now known as density functional tight-binding (DFTB).

The DFTB formalism assigns a set of basis functions to each atom. The traditional approach uses Slater functions [162–168], however recent work has also used grid-based basis functions [169–171]. Once done, the atomic parameters are determined by solving Eq. (1.19) self-consistently to generate what is known as the pseudo-atom.

$$\left[T + V^{pseudo-atom}(r)\right]\phi_v(\boldsymbol{r}) = \varepsilon_v^{pseudo-atom}\phi_v(\boldsymbol{r}) \tag{1.19}$$

While the terms in the above equation have their usual meaning (for a detailed discussion on Eq. (1.19), we refer the reader to Chap. 5 by Jenness et al. contained in this volume), the potential of the pseudo-atom is defined in Eq. (1.20).

$$V^{pseudo-atom}(r) = V_{nuclear} + V_{Hartree}[n(r)] + V_{XC}^{LDA}[n(r)] + V_{confinement}(r) \tag{1.20}$$

The confinement term transforms the atomic Schrodinger equation into the pseudo-atom one, and becomes parameter dependent. Following the construction of the pseudo-atoms, a zeroth-order molecular Hamiltonian is constructed, as depicted in Eq. (1.21).

$$H_{\mu v}^{0} = \begin{cases} \varepsilon_{\mu}^{free-atom}, \text{ if } \mu = \upsilon \\ \left\langle \phi_{\mu}^{A} \middle| \hat{T} + V_0^A + V_0^B \middle| \phi_{v}^{B} \right\rangle, \text{ if } A \neq B \\ 0, \text{Otherwise.} \end{cases} \tag{1.21}$$

A repulsive potential is then added to the solution of the Hamiltonian. Within a fitting scheme, this repulsive potential is defined as the difference between the DFT energy of the molecule and the DFTB Hamiltonian [163–168]. A second-order Hamiltonian can be formed by consideration of atomic charges [164] as shown in Eq. (1.22) in which the charges (Δq_α) are determined in a self-consistent manner via Mulliken charge analysis.

$$E^{SCC-DFTB} = \sum_{i}^{occupied} \left\langle \phi_i | \hat{H}_0 | \phi_i \right\rangle + \frac{1}{2} \sum_{\alpha,\beta}^{N} \gamma_{\alpha\beta} \Delta q_\alpha \Delta q_\beta + E_{repulsion} \tag{1.22}$$

The self-consistent charge (SCC) approach allows for fluctuations in the atomic environment arising from the molecular environment to be accounted for. More recently, third-order terms have been included (SCC-DFTB3) [172–175].

The Hamiltonian and overlap terms are pre-computed and stored into Slater-Koster tables for each atomic pair as a function of atom-atom distance, which are read in at the start of the DFTB calculation. Through interpolation the values at any distance can be generated rapidly allowing DFTB to be orders of magnitude faster than DFT; however interactions for each atom pair needs to be calculated *a priori*. Consequently, this limits applicability as the majority of DFTB codes do not allow for the calculation of the relevant matrix elements. Additionally, the repulsion terms are also dependent on both atom-atom pairs and having a ready data-base of DFT repulsive curves. This makes for a rather arduous parameterization process, and there has been a lot of work into automatic generation of these potentials [176–180]. Recently, three of the authors on the current report have utilized combination rules in conjunction with exponential repulsion potentials in an effort to allow for heterogeneous repulsive terms to be generated from homogeneous repulsive curves [171].

Machine Learning Methods Lastly, the authors of the review would like to mention machine learning (ML) based methods. ML techniques have seen a surge in the field of chemistry. They can be trained to quantum data to act as a reactive potential, and have been known to reproduce *ab initio* calculations far better than a classical force-field [181–183]. There are several steps required in order to gain accuracy within a ML model [184]. First is the creation of reference data; unfortunately, the ML-potentials require a plethora of data to accurately train to a system, more so than the traditional force-field fitting methods. One must take care that the data set include all relevant configurations and interactions to a given system. Additionally, it is essential that the training set consist of unique configurations so that one area of the potential energy surface is not over sampled, and subsequently over trained to that area. From the dataset, one needs to be able to describe the data going in through a fingerprint—a description of the configurational space. Lastly, a learning algorithm is required to reach from the input (molecular configurations) to the desired output.

ML potentials have yet to exhibit real transferability outside the system upon which they were trained. To circumvent this issue, some groups have looked into

ML with the atoms in molecules approach for property prediction [185]. Other researchers have implemented ML into the QM/MM formalism, the so called QM/ML model [186].

The Accurate NeurAl networK engINe for Molecular Energies (ANAKIN-ME, or ANI) method was created in order to develop highly accurate and transferable neural network potentials (NNP) [187]. Utilizing this method, the AN1-1 potential was generated for systems containing carbon, hydrogen, nitrogen, and oxygen [187]. Rather than using an AIMD simulation to generate data, Normal Mode Sampling was utilized to avoid biasing to a specific trajectory. In order to test the transferability of the model, the fitting set included molecules with up to eight heavy atoms, and an extensive amount of data, which is discussed in more detail by Smith et al. [187] The model performed well on molecules that had more than eight heavy atoms. ANI was improved in the AN1-1x iteration which used an automated data generation scheme with active learning [188]. This reduced the datasets by searching for and removing redundant data followed by subsequent data generation via normal mode sampling as well as via molecular dynamics. This method increased the amount of configurational space explored in the dataset. As can be seen from the work by Smith et al. [187–189], ML potentials are a highly data driven process and significant care has to be taken when generating datasets. While the previous models had used DFT methods to describe the energetics, higher accuracy was desired. To this end, Smith et al. [190] used transfer learning to improve the accuracy to the AN1-ccx potential. Briefly, data is generated using DFT and is then used to train the neural network. Next, a smaller data set was created using the CCSD(T)/CBS level. By retraining the neural network from the large data set at a lower level of theory followed by a small data set at a higher level of theory, the retrained AN1-ccx potential performs at a higher level of accuracy than the original DFT calculations in their benchmarking.

1.6 Conclusion

The current methods reviewed herein bring reactivity into MD simulations in an efficient manner so that the simulations can properly sample rare events, and reach large system sizes to include condensed phase, extended surfaces, proteins, and so on. The intent is not to prioritize or argue for one method over another, but rather highlight the advantages and disadvantages of each. The REBO/AIREBO methods are extremely fast, however have limited use outside of the solid systems for which they were developed. ReaxFF has the capability to model several different types of system environments, and has extended their parameter sets to an extensive portion of the periodic table. However, the method is considerably slower than a typical conventional force-field calculation. MS-EVB methodologies have fewer parameters and have shown great success in modeling proton transport. That being said, *a priori* knowledge of the reaction mechanism is required in the EVB methodologies, unlike in the bond order framework.

Several methods were discussed in detail including REBO, AIREBO, ReaxFF, ReaxFF-lg, eReaxFF, EVB, MS-EVB, ChIMES, LEWIS, and even DFTB. Applications of all models were presented, as well as their general framework. A brief statement on parameterization and genetic algorithms was included in order to drive home the extensive parameterization that goes into the reactive methodology. Lastly, several other methods outside the scope of this review were briefly mentioned with the intent to impart the reader with awareness and further resources regarding the methodologies that may be most applicable to their research goals.

Acknowledgments The use of trade, product, or firm names in this report is for descriptive purposes only and does not imply endorsement by the U.S. Government. The tests described and the resulting data presented herein, unless otherwise noted, were obtained from research conducted under the Environmental Quality Technology Program of the United States Army Corps of Engineers and the Environmental Security Technology Certification Program of the Department of Defense by the USAERDC. Permission was granted by the Chief of Engineers to publish this information. The findings of this report are not to be construed as an official Department of the Army position unless so designated by other authorized documents. This work was also supported by a grant of computer time from the DOD High Performance Computing Modernization Program at ERDC, Vicksburg. This document has been approved for public release (Distribution Statement A).

References

1. Farah K, Müller-Plathe F, Böhm MC (2012) Classical reactive molecular dynamics implementations: state of the art. ChemPhysChem 13(5):1127–1151
2. Berendsen HJC, Postma JPM, van Gunsteren WF, DiNola A, Haak JR (1984) Molecular dynamics with coupling to an external bath. J Chem Phys 81(8):3684–3690
3. van Gunsteren WF, Berendsen HJC (1990) Computer simulation of molecular dynamics: methodology, applications, and perspectives in chemistry. Angewandte Chemie Int Ed Engl 29(9):992–1023
4. Andersen HC (1980) Molecular dynamics simulations at constant pressure and/or temperature. J Chem Phys 72(4):2384–2393
5. Nakata H, Bai S (2019) Development of a new parameter optimization scheme for a reactive force field based on a machine learning approach. J Comput Chem 40:2000–2012
6. Dittner M, Müller J, Aktulga HM, Hartke B (2015) Efficient global optimization of reactive force-field parameters. J Comput Chem 36(20):1550–1561
7. Vanommeslaeghe K, Hatcher E, Acharya C, Kundu S, Zhong S, Shim J, Darian E, Guvench O, Lopes P, Vorobyov I, Mackerell Jr AD (2010) CHARMM general force field: a force field for drug-like molecules compatible with the CHARMM all-atom additive biological force fields. J Comput Chem 31(4):671–690
8. Soteras Gutiérrez I, Lin F-Y, Vanommeslaeghe K, Lemkul JA, Armacost KA, Brooks CL, MacKerell AD (2016) Parametrization of halogen bonds in the CHARMM general force field: improved treatment of ligand–protein interactions. Bioorg Med Chem 24(20):4812–4825
9. Mackerell Jr AD (2004) Empirical force fields for biological macromolecules: overview and issues. J Comput Chem 25(13):1584–1604
10. Maier JA, Martinez C, Kasavajhala K, Wickstrom L, Hauser KE, Simmerling C (2015) ff14SB: improving the accuracy of protein side chain and backbone parameters from ff99SB. J Chem Theory Comput 11(8):3696–3713

11. Cheatham Iii TE, Case DA (2013) Twenty-five years of nucleic acid simulations. Biopolymers 99(12):969–977
12. Reif MM, Hünenberger PH, Oostenbrink C (2012) New interaction parameters for charged amino acid side chains in the GROMOS force field. J Chem Theory Comput 8(10):3705–3723
13. Schuler LD, Daura X, van Gunsteren WF (2001) An improved GROMOS96 force field for aliphatic hydrocarbons in the condensed phase. J Comput Chem 22(11):1205–1218
14. Soares TA, Hünenberger PH, Kastenholz MA, Kräutler V, Lenz T, Lins RD, Oostenbrink C, van Gunsteren WF (2005) An improved nucleic acid parameter set for the GROMOS force field. J Comput Chem 26(7):725–737
15. Jorgensen WL, Tirado-Rives J (1988) The OPLS [optimized potentials for liquid simulations] potential functions for proteins, energy minimizations for crystals of cyclic peptides and crambin. J Am Chem Soc 110(6):1657–1666
16. Doherty B, Zhong X, Gathiaka S, Li B, Acevedo O (2017) Revisiting OPLS force field parameters for ionic liquid simulations. J Chem Theory Comput 13(12):6131–6145
17. Sambasivarao SV, Acevedo O (2009) Development of OPLS-AA force field parameters for 68 unique ionic liquids. J Chem Theory Comput 5(4):1038–1050
18. Lin H, Truhlar DG (2006) QM/MM: what have we learned, where are we, and where do we go from here? Theor Chem Acc 117(2):185
19. Warshel A, Levitt M (1976) Theoretical studies of enzymic reactions: dielectric, electrostatic and steric stabilization of the carbonium ion in the reaction of lysozyme. J Mol Biol 103(2):227–249
20. Jiang T, Boereboom JM, Michel C, Fleurat-Lessard P, Bulo RE (2015) Proton transfer in aqueous solution: exploring the boundaries of adaptive QM/MM. In: Rivail J-L, Ruiz-Lopez M, Assfeld X (eds) Quantum modeling of complex molecular systems. Springer, Cham, pp 51–91
21. Pezeshki S, Lin H (2015) Adaptive-partitioning QM/MM for molecular dynamics simulations: 4. proton hopping in bulk water. J Chem Theory Comput 11(6):2398–2411
22. Zheng M, Waller MP (2016) Adaptive quantum mechanics/molecular mechanics methods. Wiley Interdiscip Rev Comput Mol Sci 6(4):369–385
23. Duster AW, Garza CM, Aydintug BO, Negussie MB, Lin H (2019) Adaptive partitioning QM/MM for molecular dynamics simulations: 6. Proton transport through a biological channel. J Chem Theory Comput 15(2):892–905
24. Brenner DW, Shenderova OA, Harrison JA, Stuart SJ, Ni B, Sinnott SB (2002) A second-generation reactive empirical bond order (REBO) potential energy expression for hydrocarbons. J Phys Condens Matter 14(4):783–802
25. Brenner DW (1990) Empirical potential for hydrocarbons for use in simulating the chemical vapor deposition of diamond films. Phys Rev B 42(15):9458–9471
26. Hur J, Stuart SJ (2012) Modified reactive empirical bond-order potential for heterogeneous bonding environments. J Chem Phys 137(5):054102
27. Brenner DW (1992) Erratum: Empirical potential for hydrocarbons for use in simulating the chemical vapor deposition of diamond films. Phys Rev B 46(3):1948–1948
28. Harrison JA, Schall JD, Maskey S, Mikulski PT, Knippenberg MT, Morrow BH (2018) Review of force fields and intermolecular potentials used in atomistic computational materials research. Appl Phys Rev 5(3):031104
29. Dyson AJ, Smith PV (1996) Extension of the Brenner empirical interatomic potential to C-Si-H systems. Surf Sci 355(1):140–150
30. Sbraccia C, Silvestrelli PL, Ancilotto F (2002) Modified XB potential for simulating interactions of organic molecules with Si surfaces. Surf Sci 516(1):147–158
31. Ni B, Lee K-H, Sinnott SB (2004) A reactive empirical bond order (REBO) potential for hydrocarbon–oxygen interactions. J Phys Condens Matter 16(41):7261–7275
32. Jang I, Sinnott SB (2004) Molecular dynamics simulations of the chemical modification of polystyrene through CxFy+ beam deposition. J Phys Chem B 108(49):18993–19001

33. Schall JD, Gao G, Harrison JA (2008) Elastic constants of silicon materials calculated as a function of temperature using a parametrization of the second-generation reactive empirical bond-order potential. Phys Rev B 77(11):115209
34. Zhang P, Ma L, Fan F, Zeng Z, Peng C, Loya PE, Liu Z, Gong Y, Zhang J, Zhang X, Ajayan PM, Zhu T, Lou J (2014) Fracture toughness of graphene. Nat Commun 5(1):3782
35. Shakouri A, Ng TY, Lin RM (2011) A new REBO potential based atomistic structural model for graphene sheets. Nanotechnology 22(29):295711
36. Fonseca AF, Lee G, Borders TL, Zhang H, Kemper TW, Shan T-R, Sinnott SB, Cho K (2011) Reparameterization of the REBO-CHO potential for graphene oxide molecular dynamics simulations. Phys Rev B 84(7):075460
37. Shakouri A, Ng TY, Lin RM (2013) A study of the scale effects on the flexural vibration of graphene sheets using REBO potential based atomistic structural and nonlocal couple stress thin plate models. Physica E 50:22–28
38. Tavakol M, Montazeri A, Aboutalebi SH, Asgari R (2020) Mechanical properties of graphene oxide: the impact of functional groups. Appl Surf Sci 525:146554
39. Damasceno DA, Mesquita E, Rajapakse RKND, Pavanello R (2019) Atomic-scale finite element modelling of mechanical behaviour of graphene nanoribbons. Int J Mech Mater Des 15(1):145–157
40. Lebedeva IV, Minkin AS, Popov AM, Knizhnik AA (2019) Elastic constants of graphene: comparison of empirical potentials and DFT calculations. Physica E 108:326–338
41. Eberhardt O, Wallmersperger T (2019) Advanced molecular structural mechanics model for carbon nanotubes incorporating the 2nd generation REBO potential. IntJ Eng Sci 144:103137
42. Vijayaraghavan V, Garg A, Wong CH, Tai K (2014) Estimation of mechanical properties of nanomaterials using artificial intelligence methods. Appl Phys A 116(3):1099–1107
43. Stuart SJ, Tutein AB, Harrison JA (2000) A reactive potential for hydrocarbons with intermolecular interactions. J Chem Phys 112(14):6472–6486
44. Rahman SM, Song J, Yeo C-D (2019) Computational study on surface energy of amorphous DLC with respect to hybridization state of carbon and potential functions. Diamond Relat Mater 95:127–134
45. Amiri H, Aghazadeh Mohandesi J, Marashi P (2020) Tensile properties of pillared graphene block. Mater Sci Eng B 257:114557
46. Dhaliwal G, Nair PB, Singh CV (2019) Uncertainty analysis and estimation of robust AIREBO parameters for graphene. Carbon 142:300–310
47. Li R, Liu B, Gao F (2019) Molecular dynamics study of the material property changes induced by accumulated point defects in graphite. Nucl Instrum Methods Phys Res Sect B 455:52–56
48. Umeno Y, Yachi Y, Sato M, Shima H (2019) On the atomistic energetics of carbon nanotube collapse from AIREBO potential. Physica E 106:319–325
49. Nejat Pishkenari H, Golzari A (2020) A temperature-calibrated continuum model for vibrational analysis of the fullerene family using molecular dynamics simulations. App Math Model 80:115–125
50. Höhnerbach M, Bientinesi P (2019) Accelerating AIREBO: navigating the journey from legacy to high-performance code. J Comput Chem 40(14):1471–1482
51. Andric N, Jenny P (2018) Molecular dynamics investigation of energy transfer during gas-surface collisions. Phys Fluids 30(7):077104
52. Liu A, Stuart SJ (2008) Empirical bond-order potential for hydrocarbons: adaptive treatment of van der Waals interactions. J Comput Chem 29(4):601–611
53. O'Connor TC, Andzelm J, Robbins MO (2015) AIREBO-M: a reactive model for hydrocarbons at extreme pressures. J Chem Phys 142(2):024903
54. Pastewka L, Klemenz A, Gumbsch P, Moseler M (2013) Screened empirical bond-order potentials for Si-C. Phys Rev B 87(20):205410
55. Knippenberg MT, Mikulski PT, Ryan KE, Stuart SJ, Gao G, Harrison JA (2012) Bond-order potentials with split-charge equilibration: application to C-, H-, and O-containing systems. J Chem Phys 136(16):164701

56. Chenoweth K, van Duin ACT, Goddard WA (2008) ReaxFF reactive force field for molecular dynamics simulations of hydrocarbon oxidation. Chem A Eur J 112(5):1040–1053
57. Senftle TP, Hong S, Islam MM, Kylasa SB, Zheng Y, Shin YK, Junkermeier C, Engel-Herbert R, Janik MJ, Aktulga HM, Verstraelen T, Grama A, van Duin ACT (2016) The ReaxFF reactive force-field: development, applications and future directions. NPJ Comput Mater 2:15011
58. O'Hearn KA, Swift MW, Liu J, Magoulas I, Piecuch P, van Duin ACT, Aktulga HM, Qi Y (2020) Optimization of the Reax force field for the lithium–oxygen system using a high fidelity charge model. J Chem Phys 153(8):084107
59. Zhang W, van Duin ACT (2017) Second-generation ReaxFF water force field: improvements in the description of water density and OH-anion diffusion. J Phys Chem B 121(24):6021–6032
60. Pai SJ, Lee HW, Han SS (2019) Improved description of a coordinate bond in the ReaxFF reactive force field. J Phys Chem Lett 10(22):7293–7299
61. Kylasa SB, Aktulga HM, Grama AY (2014) PuReMD-GPU: a reactive molecular dynamics simulation package for GPUs. J Comput Phys 272:343–359
62. Zhu W, Gong H, Han Y, Zhang M, van Duin ACT (2020) Development of a ReaxFF reactive force field for simulations on the catalytic conversion of C/H/O molecules on Cu-metal and Cu-oxide surfaces and application to Cu/CuO based chemical looping. J Phys Chem C 124:12512–12520
63. Gao P, Huang Z, Yu H (2020) Exploration of the dehydrogenation pathways of ammonia diborane and diammoniate of diborane by molecular dynamics simulations using reactive force fields. Chem A Eur J 124(9):1698–1704
64. Sengul MY, Randall CA, van Duin ACT (2018) ReaxFF molecular dynamics simulation of intermolecular structure formation in acetic acid-water mixtures at elevated temperatures and pressures. J Chem Phys 148(16):164506
65. Du T, Li H, Sant G, Bauchy M (2018) New insights into the sol–gel condensation of silica by reactive molecular dynamics simulations. J Chem Phys 148(23):234504
66. Manzano H, Zhang W, Raju M, Dolado JS, López-Arbeloa I, van Duin ACT (2018) Benchmark of ReaxFF force field for subcritical and supercritical water. J Chem Phys 148(23):234503
67. Lv L, Zhang L, Yang M (2018) Understanding the phase separation of N2/H2O and CO2/H2O binary systems through reactive force fields-based molecular dynamics simulations. J Appl Phys 124(23):235901
68. Islam MM, Strachan A (2020) Role of dynamical compressive and shear loading on hotspot criticality in RDX via reactive molecular dynamics. J Appl Phys 128(6):065101
69. Hao W, Niu L, Gou R, Zhang C (2019) Influence of Al and Al2O3 nanoparticles on the thermal decay of 1,3,5-trinitro-1,3,5-triazinane (RDX): reactive molecular dynamics simulations. J Phys Chem C 123(22):14067–14080
70. Yang S, Zhao T, Zou L, Wang X, Zhang Y (2019) ReaxFF-based molecular dynamics simulation of DNA molecules destruction in cancer cells by plasma ROS. Phys Plasma 26(8):083504
71. Huang HS, Ai LQ, van Duin ACT, Chen M, Lü YJ (2019) ReaxFF reactive force field for molecular dynamics simulations of liquid Cu and Zr metals. J Chem Phys 151(9):094503
72. Mohamed SA, Chong S, Kim J (2019) Thermal stability of methyl-functionalized MOF-5. J Phys Chem C 123(49):29686–29692
73. Hossain MJ, Pawar G, Liaw B, Gering KL, Dufek EJ, van Duin ACT (2020) Lithium-electrolyte solvation and reaction in the electrolyte of a lithium ion battery: a ReaxFF reactive force field study. J Chem Phys 152(18):184301
74. Vashisth A, Ashraf C, Zhang W, Bakis CE, van Duin ACT (2018) Accelerated ReaxFF simulations for describing the reactive cross-linking of polymers. Chem A Eur J 122(32):6633–6642
75. Ashraf C, Vashisth A, Bakis CE, van Duin ACT (2019) Reactive molecular dynamics simulations of the atomic oxygen impact on epoxies with different chemistries. J Phys Chem C 123(24):15145–15156

76. Kowalik M, Ashraf C, Damirchi B, Akbarian D, Rajabpour S, van Duin ACT (2019) Atomistic scale analysis of the carbonization process for C/H/O/N-based polymers with the ReaxFF reactive force field. J Phys Chem B 123(25):5357–5367
77. Liu Y, Qi Y, Hu X, van Duin ACT (2019) Formation of AlFx gaseous phases during high temperature etching: a reactive force field based molecular dynamics study. J Phys Chem C 123(27):16823–16835
78. Bertels LW, Newcomb LB, Alaghemandi M, Green JR, Head-Gordon M (2020) Benchmarking the performance of the ReaxFF reactive force field on hydrogen combustion systems. Chem A Eur J 124(27):5631–5645
79. Orekhov N, Ostroumova G, Stegailov V (2020) High temperature pure carbon nanoparticle formation: validation of AIREBO and ReaxFF reactive molecular dynamics. Carbon 170:606–620
80. Furman D, Wales DJ (2019) Transforming the accuracy and numerical stability of ReaxFF reactive force fields. J Phys Chem Lett 10(22):7215–7223
81. Liu L, Liu Y, Zybin SV, Sun H, Goddard WA (2011) ReaxFF-lg: correction of the ReaxFF reactive force field for london dispersion, with applications to the equations of state for energetic materials. Chem A Eur J 115(40):11016–11022
82. Larentzos JP, Rice BM (2017) Transferable reactive force fields: extensions of ReaxFF-lg to nitromethane. Chem A Eur J 121(9):2001–2013
83. Huang X, Qiao Z, Dai X, Zhang K, Li M, Pei G, Wen Y (2019) Effects of different types of defects on ignition mechanisms in shocked β-cyclotetramethylene tetranitramine crystals: a molecular dynamics study based on ReaxFF-lg force field. J Appl Phys 125(19):195101
84. Zhou T, Zybin SV, Liu Y, Huang F, Goddard WA (2012) Anisotropic shock sensitivity for β-octahydro-1,3,5,7-tetranitro-1,3,5,7-tetrazocine energetic material under compressive-shear loading from ReaxFF-lg reactive dynamics simulations. J Appl Phys 111(12):124904
85. Wang F, Chen L, Geng D, Lu J, Wu J (2018) Effect of density on the thermal decomposition mechanism of ε-CL-20: a ReaxFF reactive molecular dynamics simulation study. Phys Chem Chem Phys 20(35):22600–22609
86. Zeng T, Yang R, Li J, Tang W, Li D (2019) Thermal decomposition mechanism of nitroglycerin by ReaxFF reactive molecular dynamics simulations. Combust Sci Technol 1–15
87. Zhao Y, Zhao F-Q, Xu S-Y, Ju X-H (2020) Molecular reaction dynamics simulation of thermal decomposition for aluminiferous RDX composites. Comput Mater Sci 177:109556
88. Wang F, Chen L, Geng D, Lu J, Wu J (2019) Molecular dynamics simulations of an initial chemical reaction mechanism of shocked CL-20 crystals containing nanovoids. J Phys Chem C 123(39):23845–23852
89. Zhao Y, Mei Z, Zhao F-Q, Xu S-Y, Ju X-H (2020) Thermal decomposition mechanism of 1,3,5,7-tetranitro-1,3,5,7-tetrazocane accelerated by nano-aluminum hydride (AlH3): ReaxFF-Lg molecular dynamics simulation. ACS Omega 5(36):23193–23200
90. Bidault X, Pineau N (2018) Dynamic formation of nanodiamond precursors from the decomposition of carbon suboxide (C3O2) under extreme conditions—A ReaxFF study. J Chem Phys 149(11):114301
91. Islam MM, Kolesov G, Verstraelen T, Kaxiras E, van Duin ACT (2016) eReaxFF: a pseudoclassical treatment of explicit electrons within reactive force field simulations. J Chem Theory Comput 12(8):3463–3472
92. Islam MM, van Duin ACT (2016) Reductive decomposition reactions of ethylene carbonate by explicit electron transfer from lithium: an eReaxFF molecular dynamics study. J Phys Chem C 120(48):27128–27134
93. Evangelisti B, Fichthorn KA, van Duin ACT (2020) Development and initial applications of an e-ReaxFF description of Ag nanoclusters. J Chem Phys 153(10):104106
94. Warshel A, Weiss RM (1980) An empirical valence bond approach for comparing reactions in solutions and in enzymes. J Am Chem Soc 102(20):6218–6226
95. Warshel A, Florián J (2004). The Empirical Valence Bond (EVB) Method. In Encyclopedia of Computational Chemistry (eds P. Ragué Schleyer, N.L. Allinger, T. Clark, J. Gasteiger, P.A.

Kollman, H.F. Schaefer, P.R. Schreiner, W. Thiel, W.L. Jorgensen and R.C. Glen). https://doi.org/10.1002/0470845015.cu0002
96. Vuilleumier R, Borgis D (1997) Molecular dynamics of an excess proton in water using a non-additive valence bond force field. J Mol Struct 436-437:555–565
97. Vuilleumier R, Borgis D (1998) Quantum dynamics of an excess proton in water using an extended empirical valence-bond hamiltonian. J Phys Chem B 102(22):4261–4264
98. Vuilleumier R, Borgis D (2000) Wavefunction quantization of the proton motion in a H5O2+ dimer solvated in liquid water. J Mol Struct 552(1):117–136
99. Chang YT, Miller WH (1990) An empirical valence bond model for constructing global potential energy surfaces for chemical reactions of polyatomic molecular systems. J Phys Chem 94(15):5884–5888
100. Sonnenberg JL, Wong KF, Voth GA, Schlegel HB (2009) Distributed Gaussian valence bond surface derived from ab initio calculations. J Chem Theory Comput 5(4):949–961
101. Schlegel HB, Sonnenberg JL (2006) Empirical valence-bond models for reactive potential energy surfaces using distributed Gaussians. J Chem Theory Comput 2(4):905–911
102. Biswas R, Tse Y-LS, Tokmakoff A, Voth GA (2016) Role of presolvation and anharmonicity in aqueous phase hydrated proton solvation and transport. J Phys Chem B 120(8):1793–1804
103. Mones L, Kulhánek P, Florián J, Simon I, Fuxreiter M (2007) Probing the two-metal ion mechanism in the restriction endonuclease BamHI. Biochemistry 46(50):14514–14523
104. Bresnahan CG, David R, Milet A, Kumar R (2019) Ion pairing in HCl–water clusters: from electronic structure investigations to multiconfigurational force-field development. Chem A Eur J 123(43):9371–9381
105. Wick CD (2013) HCl accommodation, dissociation, and propensity for the surface of water. Chem A Eur J 117(47):12459–12467
106. Sonnenberg JL, Schlegel HB (2007) Empirical valence bond models for reactive potential energy surfaces. II. Intramolecular proton transfer in pyridone and the Claisen reaction of allyl vinyl ether. Mol Phys 105(19-22):2719–2729
107. Mei HS, Tuckerman ME, Sagnella DE, Klein ML (1998) Quantum nuclear ab initio molecular dynamics study of water wires. J Phys Chem B 102(50):10446–10458
108. Hornung B, Harvey JN, Preston TJ, Dunning GT, Orr-Ewing AJ (2015) Empirical valence bond theory studies of the CH4 + Cl → CH3 + HCl reaction. Chem A Eur J 119(37):9590–9598
109. Kamerlin SC, Warshel A (2010) The EVB as a quantitative tool for formulating simulations and analyzing biological and chemical reactions. Faraday Discuss 145:71–106
110. Kamerlin SCL, Warshel A (2011) The empirical valence bond model: theory and applications. WIREs Comput Mol Sci 1(1):30–45
111. Kamath G, Howell EE, Agarwal PK (2010) The tail wagging the dog: insights into catalysis in R67 dihydrofolate reductase. Biochemistry 49(42):9078–9088
112. Blaha-Nelson D, Krüger DM, Szeler K, Ben-David M, Kamerlin SCL (2017) Active site hydrophobicity and the convergent evolution of paraoxonase activity in structurally divergent enzymes: the case of serum paraoxonase 1. J Am Chem Soc 139(3):1155–1167
113. Calixto AR, Moreira C, Pabis A, Kötting C, Gerwert K, Rudack T, Kamerlin SCL (2019) GTP hydrolysis without an active site base: a unifying mechanism for Ras and related GTPases. J Am Chem Soc 141(27):10684–10701
114. Duboué-Dijon E, Pluhařová E, Domin D, Sen K, Fogarty AC, Chéron N, Laage D (2017) Coupled valence-bond state molecular dynamics description of an enzyme-catalyzed reaction in a non-aqueous organic solvent. J Phys Chem B 121(29):7027–7041
115. Agmon N (1995) The Grotthuss mechanism. Chem Phys Lett 244(5):456–462
116. Cukierman S (2006) Et tu, Grotthuss! and other unfinished stories. Biochimica et Biophysica Acta (BBA) Bioenerg 1757(8):876–885
117. Knight C, Voth GA (2012) The curious case of the hydrated proton. Acc Chem Res 45(1):101–109
118. Wang F, Izvekov S, Voth GA (2008) Unusual "amphiphilic" association of hydrated protons in strong acid solution. J Am Chem Soc 130(10):3120–3126

119. Taraphder S, Maupin CM, Swanson JMJ, Voth GA (2016) Coupling protein dynamics with proton transport in human carbonic anhydrase II. J Phys Chem B 120(33):8389–8404
120. Biswas R, Carpenter W, Voth GA, Tokmakoff A (2016) Molecular modeling and assignment of IR spectra of the hydrated excess proton in isotopically dilute water. J Chem Phys 145(15):154504
121. Day TJF, Soudackov AV, Čuma M, Schmitt UW, Voth GA (2002) A second generation multistate empirical valence bond model for proton transport in aqueous systems. J Chem Phys 117(12):5839–5849
122. Swanson JMJ, Maupin CM, Chen H, Petersen MK, Xu J, Wu Y, Voth GA (2007) Proton solvation and transport in aqueous and biomolecular systems: insights from computer simulations. J Phys Chem B 111(17):4300–4314
123. Schmitt UW, Voth GA (1998) Multistate empirical valence bond model for proton transport in water. J Phys Chem B 102(29):5547–5551
124. Voth GA (2006) Computer simulation of proton solvation and transport in aqueous and biomolecular systems. Acc Chem Res 39(2):143–150
125. Wu Y, Chen H, Wang F, Paesani F, Voth GA (2008) An improved multistate empirical valence bond model for aqueous proton solvation and transport. J Phys Chem B 112(2):467–482
126. Lee S, Liang R, Voth GA, Swanson JMJ (2016) Computationally efficient multiscale reactive molecular dynamics to describe amino acid deprotonation in proteins. J Chem Theory Comput 12(2):879–891
127. Savage J, Voth GA (2016) Proton solvation and transport in realistic proton exchange membrane morphologies. J Phys Chem C 120(6):3176–3186
128. Iuchi S, Chen H, Paesani F, Voth GA (2009) Hydrated excess proton at water−hydrophobic interfaces. J Phys Chem B 113(13):4017–4030
129. Wang F, Voth GA (2005) A linear-scaling self-consistent generalization of the multistate empirical valence bond method for multiple excess protons in aqueous systems. J Chem Phys 122(14):144105
130. Brancato G, Tuckerman ME (2005) A polarizable multistate empirical valence bond model for proton transport in aqueous solution. J Chem Phys 122(22):224507
131. Kumar R, Christie RA, Jordan KD (2009) A modified MSEVB force field for protonated water clusters. J Phys Chem B 113(13):4111–4118
132. Zeng Y, Li A, Yan T (2020) Hydrogen bond dynamics in the solvation shell on proton transfer in aqueous solution. J Phys Chem B 124(9):1817–1823
133. Lee HJ, Svahn E, Swanson JMJ, Lepp H, Voth GA, Brzezinski P, Gennis RB (2010) Intricate role of water in proton transport through cytochrome c oxidase. J Am Chem Soc 132(45):16225–16239
134. Sumner I, Voth GA (2012) Proton transport pathways in [NiFe]-hydrogenase. J Phys Chem B 116(9):2917–2926
135. Qin Z, Tepper HL, Voth GA (2007) Effect of membrane environment on proton permeation through gramicidin A channels. J Phys Chem B 111(33):9931–9939
136. Wick CD (2012) Hydronium behavior at the air–water interface with a polarizable multistate empirical valence bond model. J Phys Chem C 116(6):4026–4038
137. Kumar R, Knight C, Wick CD, Chen B (2015) Bringing reactivity to the aggregation-volume-bias Monte Carlo based simulation framework: water nucleation induced by a reactive proton. J Phys Chem B 119(29):9068–9075
138. Wick CD (2017) Comparing hydroxide and hydronium at the instantaneous air-water interface using polarizable multi-state empirical valence bond models. Comput Theor Chem 1116:64–72
139. Li Z, Li C, Wang Z, Voth GA (2020) What coordinate best describes the affinity of the hydrated excess proton for the air-water interface? J Phys Chem B 124:5039–5046
140. Park K, Lin W, Paesani F (2012) A refined MS-EVB model for proton transport in aqueous environments. J Phys Chem B 116(1):343–352
141. Park K, Lin W, Paesani F (2014) Fast and slow proton transfer in ice: the role of the quasi-liquid layer and hydrogen-bond network. J Phys Chem B 118(28):8081–8089

142. Chen H, Yan T, Voth GA (2009) A computer simulation model for proton transport in liquid imidazole. Chem A Eur J 113(16):4507–4517
143. Li A, Yan T (2020) Proton propensity and orientation of imidazolium cation at liquid imidazole–vacuum interface: a molecular dynamics simulation. J Phys Chem B 124(19):4010–4016
144. Sharp ME, Vázquez FX, Wagner JW, Dannenhoffer-Lafage T, Voth GA (2019) Multiconfigurational coarse-grained molecular dynamics. J Chem Theory Comput 15(5):3306–3315
145. Dannenhoffer-Lafage T, Voth GA (2020) Reactive coarse-grained molecular dynamics. J Chem Theory Comput 16(4):2541–2549
146. Pahari P, Chaturvedi S (2012) Determination of best-fit potential parameters for a reactive force field using a genetic algorithm. J Mol Model 18(3):1049–1061
147. Shchygol G, Yakovlev A, Trnka T, van Duin ACT, Verstraelen T (2019) ReaxFF parameter optimization with Monte-Carlo and evolutionary algorithms: guidelines and insights. J Chem Theory Comput 15(12):6799–6812
148. Naserifar S, Goddard WA (2018) The quantum mechanics-based polarizable force field for water simulations. J Chem Phys 149(17):174502
149. Naserifar S, Oppenheim JJ, Yang H, Zhou T, Zybin S, Rizk M, Goddard WA (2019) Accurate non-bonded potentials based on periodic quantum mechanics calculations for use in molecular simulations of materials and systems. J Chem Phys 151(15):154111
150. Kale S, Herzfeld J, Dai S, Blank M (2012) Lewis-inspired representation of dissociable water in clusters and Grotthuss chains. J Biol Phys 38(1):49–59
151. Kale S, Herzfeld J (2012) Natural polarizability and flexibility via explicit valency: the case of water. J Chem Phys 136(8):084109
152. Kale S, Herzfeld J (2012) Proton defect solvation and dynamics in aqueous acid and base. Angew Chem Int Ed Engl 51(44):11029–11032
153. Ekesan S, Kale S, Herzfeld J (2014) Transferable pseudoclassical electrons for aufbau of atomic ions. J Comput Chem 35(15):1159–1164
154. Bai C, Herzfeld J (2016) Surface propensities of the self-ions of water. ACS Central Sci 2(4):225–231
155. Bai C, Kale S, Herzfeld J (2017) Chemistry with semi-classical electrons: reaction trajectories auto-generated by sub-atomistic force fields. Chem Sci 8(6):4203–4210
156. Koziol L, Fried LE, Goldman N (2017) Using force matching to determine reactive force fields for water under extreme thermodynamic conditions. J Chem Theory Comput 13(1):135–146
157. Lindsey RK, Fried LE, Goldman N (2017) ChIMES: a force matched potential with explicit three-body interactions for molten carbon. J Chem Theory Comput 13(12):6222–6229
158. Lindsey RK, Fried LE, Goldman N (2019) Application of the ChIMES force field to nonreactive molecular systems: water at ambient conditions. J Chem Theory Comput 15(1):436–447
159. Lindsey R, Kroonblawd M, Fried L, Goldman N (2019) Force matching approaches to extend density functional theory to large time and length scales. In: Computational approaches for chemistry under extreme conditions. Challenges and advances in computational chemistry and physics, vol 28, pp 71–93
160. Lindsey RK, Goldman N, Fried LE, Bastea S (2020) Many-body reactive force field development for carbon condensation in C/O systems under extreme conditions. J Chem Phys 153(5):054103
161. Manjari SR, Banavali NK (2018) Structural articulation of biochemical reactions using restrained geometries and topology switching. J Chem Inf Model 58(2):453–463
162. Foulkes MC, Haydock R (1989) Tight-binding models and density-functional theory. Phys Rev B 39:12520
163. Porezag D, Frauenheim T, Köhler T, Seifert G, Kaschner R (1995) Construction of tight-binding-like potentials on the basis of density-functional theory: application to carbon. Phys Rev B 51:947–957

164. Elstner M, Porezag D, Jungnickel G, Elsner J, Haugk M, Frauenheim T, Suhai S, Seifert G (1998) Self-consistent charge density-functional tight-binding method for simulations of complex materials properties. Phys Rev B 58:7260–7268
165. Elstner M (2006) The SCC-DFTB method and its application to biological systems. Theor Chem Acc 116:316–325
166. Aradi B, Hourahine B, Frauenheim T (2007) DFTB+, a sparse matrix-based implementation of the DFTB method. J Phys Chem A 111:5678–5684
167. Seifert G, Joswig J-O (2012) Density-functional tight binding-an approximate density-functional theory method. WIREs Comput Mol Sci 2:456–465
168. Hourahine B, Aradi B, Blum V, Bonafé F, Buccheri A, Camacho C, Cevallos C, Deshaye MY, Dumitric T, Dominguez A, Ehlert S, Elstner M, Van Der Heide T, Hermann J, Irle S, Kranz JJ, Köhler C, Kowalczyk T, Kubař T, Lee IS, Lutsker V, Maurer RJ, Min SK, Mitchell I, Negre C, Niehaus TA, Niklasson AMN, Page AJ, Pecchia A, Penazzi G, Persson MP, Řezáč J, Sánchez CG, Sternberg M, Stöhr M, Stuckenberg F, Tkatchenko A, Yu VWZ, Frauenheim T (2020) DFTB+, a software package for efficient approximate density functional theory based atomistic simulations. J Chem Phys 152:124101
169. Koskinen P, Mäkinen V (2009) Density-functional tight-binding for beginners. Comput Mater Sci 47:237–253
170. Mäkinen V, Koskinen P, Häkkinen H (2013) Modeling thiolate-protected gold clusters with density-functional tight-binding. Eur Phys J D 67:1–6
171. Jenness GR, Bresnahan CG, Shukla MK (2020) Adventures in DFTB: towards an automatic parameterization scheme. J Chem Theory Comp 16(11):6894–6903
172. Gaus M, Cui Q, Elstner M (2011) DFTB3: extension of the self-consistent-charge density-functional tight-binding method (SCC-DFTB). J Chem Theory Comput 7:931–948
173. Gaus M, Goez A, Elstner M (2012) Parametrization and benchmark of DFTB3 for organic molecules. J Chem Theory Comput 9:338–354
174. Gaus M, Lu X, Elstner M, Cui Q (2014) Parameterization of DFTB3/3OB for sulfur and phosphorus for chemical and biological applications. J Chem Theory Comput 10:1518–1537
175. Kubillus M, Kubař T, Gaus M, Řezáč J, Elstner M (2015) Parameterization of the DFTB3 method for Br, Ca, Cl, F, I, K, and Na in organic and biological systems. J Chem Theory Comput 11:332–342
176. Gaus M, Chou CP, Witek H, Elstner M (2009) Automatized parametrization of SCC-DFTB repulsive potentials: application to hydrocarbons. J Phys Chem A 113:11866–11881
177. Krishnapriyan A, Yang P, Niklasson AMN, Cawkwell MJ (2017) Numerical optimization of density functional tight binding models: application to molecules containing carbon, hydrogen, nitrogen, and oxygen. J Chem Theory Comput 13:6191–6200
178. Kranz JJ, Kubillus M, Ramakrishnan R, Von Lilienfeld OA, Elstner M (2018) Generalized density-functional tight-binding repulsive potentials from unsupervised machine learning. J Chem Theory Comput 14:2341–2352
179. Goldman N, Fried LE, Koziol L (2015) Using force-matched potentials to improve the accuracy of density functional tight binding for reactive conditions. J Chem Theory Comput 11:4530–4535
180. Doemer M, Liberatore E, Knaup JM, Tavernelli I, Rothlisberger U (2013) In situ parameterisation of SCC-DFTB repulsive potentials by iterative Boltzmann inversion. Mol Phys 111:3595–3607
181. Behler J (2015) Constructing high-dimensional neural network potentials: a tutorial review. Int J Quant Chem 115(16):1032–1050
182. Jose KVJ, Artrith N, Behler J (2012) Construction of high-dimensional neural network potentials using environment-dependent atom pairs. J Chem Phys 136(19):194111
183. Kondati Natarajan S, Morawietz T, Behler J (2015) Representing the potential-energy surface of protonated water clusters by high-dimensional neural network potentials. Phys Chem Chem Phys 17(13):8356–8371
184. Huan TD, Batra R, Chapman J, Krishnan S, Chen L, Ramprasad R (2017) A universal strategy for the creation of machine learning-based atomistic force fields. NPJ Comput Mater 3(1):37

185. Zubatyuk R, Smith JS, Leszczynski J, Isayev O (2019) Accurate and transferable multitask prediction of chemical properties with an atoms-in-molecules neural network. Sci Adv 5(8):eaav6490
186. Shen L, Yang W (2018) Molecular dynamics simulations with quantum mechanics/molecular mechanics and adaptive neural networks. J Chem Theory Comput 14(3):1442–1455
187. Smith JS, Isayev O, Roitberg AE (2017) ANI-1: an extensible neural network potential with DFT accuracy at force field computational cost. Chem Sci 8(4):3192–3203
188. Smith JS, Nebgen B, Lubbers N, Isayev O, Roitberg AE (2018) Less is more: sampling chemical space with active learning. J Chem Phys 148(24):241733
189. Smith JS, Roitberg AE, Isayev O (2018) Transforming computational drug discovery with machine learning and AI. ACS Med Chem Lett 9(11):1065–1069
190. Smith JS, Nebgen BT, Zubatyuk R, Lubbers N, Devereux C, Barros K, Tretiak S, Isayev O, Roitberg AE (2019) Approaching coupled cluster accuracy with a general-purpose neural network potential through transfer learning. Nat Commun 10(1):2903

Chapter 2
Physics-Based Coarse-Grained Modeling in Bio- and Nanochemistry

Adam Liwo, Adam K. Sieradzan, Agnieszka S. Karczyńska, Emilia A. Lubecka, Sergey A. Samsonov, Cezary Czaplewski, Paweł Krupa, and Magdalena Mozolewska

Abstract Coarse-grained approaches, in which groups of atoms are represented by single interaction sites, are very important in biological and materials sciences because they enable us to cover the size- and time-scales by several orders of magnitude larger than those available all-atom simulations, while largely keeping the details of the systems studied. The coarse-grained approaches differ by the scheme of reduction and by the origin and parameterization of the respective force fields. Both statistical (database-derived) and physics-based potentials are used, the physics-based potentials enabling us to bridge the coarse-grained level with the all-atom level, which is necessary when carrying out the simulations at multiple resolutions (multiscale simulations). The physics-based potentials originate from the potential of mean force (PMF) of a system under study, in which the degrees of freedom that are not considered in the model are averaged out. For tractability and transferability the PMF has to be expressed as a sum of contributions that constitute the effective energy terms. These terms are often assigned analytical

A. Liwo (✉)
Faculty of Chemistry, University of Gdańsk, Gdańsk, Poland

School of Computational Sciences, Korea Institute for Advanced Study, Seoul, Republic of Korea
e-mail: adam.liwo@ug.edu.pl

A. K. Sieradzan · S. A. Samsonov · C. Czaplewski
Faculty of Chemistry, University of Gdańsk, Gdańsk, Poland

A. S. Karczyńska
Faculty of Chemistry, University of Gdańsk, Gdańsk, Poland

Univ. Grenoble Alpes, Inria, CNRS, Grenoble INP, LJK, Grenoble, France

E. A. Lubecka
Faculty of Electronics, Telecommunications and Informatics, Gdansk University of Technology, Gdańsk, Poland

P. Krupa
Institute of Physics, Polish Academy of Sciences, Warsaw, Poland

M. Mozolewska
Institute of Computer Science, Polish Academy of Sciences, Warsaw, Poland

J. Leszczynski, M. K. Shukla (eds.), *Practical Aspects of Computational Chemistry V*, https://doi.org/10.1007/978-3-030-83244-5_2

expressions imported from all-atom force fields or engineered to reproduce certain structural patterns (e.g., the secondary structures of proteins or nucleic acids). Tabulated (model-free) potentials are also applied. Approaches also exist in which the effective energy terms are derived systematically by splitting the potential of mean force into transferable terms, e.g., by expressing the PMF by the Kubo cluster-cumulant functions. Two approaches, or a combination thereof, are applied in the parameterization of the coarse-grained force fields: the bottom-up one, in which the potentials of mean force are determined from atomistically-detailed calculations and then used to parameterize the respective expressions, and the top-down approach, in which the force field is tuned to fit the experimental data. In this chapter, the theory and parameterization of the physics-based coarse-grained force fields, along with the corresponding methods of conformational search are reviewed. Examples of physics-based coarse-grained force fields applied to study biomolecules and their assemblies and nanosystems are discussed.

2.1 Introduction

Computer simulations of soft matter are nowadays of unquestionable importance in biology, biophysics, medical sciences and materials sciences [1–7]. This way of research enables us to understand the mechanisms of functioning of elements of living cells, cancer and hereditary diseases, pathogen attack and, ultimately, to design new drugs and therapies. Materials (including nanomaterials) modeling is important in materials science (see, e.g., https://emmc.info), while the modeling of the interactions of man-engineered nanostructures (nanotubes, fullerenes, etc.) with biological macromolecules contributes to finding the best way of drug delivery and enables us to study the impact of man-introduced nanostructures on the living organisms (in particular, nanotoxicity) and the environment.

The simulation community makes a great effort to evaluate the existing modeling approaches, the best example being the Community Wide Experiments on the Critical Assessment of Techniques for Protein Structure Prediction (CASP; http://www.predictioncenter.org) [8], which are conducted, since 1994, every other year. Following this initiative, the Critical Assessment of PRediction of Interactions (CAPRI; https://www.ebi.ac.uk/msd-srv/capri/) [9], RNA-Puzzles (http://www.rnapuzzles.org/) [10], and crystal-structure prediction [11] experiments have been established.

Depending on the size of a system, time scale, and required accuracy, different modeling approaches are used [12]. The most accurate modeling is carried out at the quantum mechanics (QM) level, with the use of Schrödinger-equation [13] or Density Functional Theory (DFT) [14] based approaches. This level covers systems with up to nanometer size and up to picosecond time scales and is the most accurate and general one, enabling us to study chemical reactions that occur in a system (e.g., enzymatic catalysis or polymerization processes). Another advantage of the QM approach is its high transferability, which is 100% for the *ab initio* methods and still high for the semiempirical QM methods [15].

In order to handle larger systems at longer time scales, the QM Hamiltonian needs to be replaced with a classical one, the classical Hamiltonians being termed *force fields* [2]. The force fields can be considered as parameterized QM potential-energy surfaces (PESs) and, consequently, the electron degrees of freedom are averaged out, the degrees of freedom that are kept being the nuclear coordinates. Switching to the classical approach enables us to extend the time- and size-scales to micrometers and microseconds, respectively. However, the accuracy, the scope of applications, and transferability are reduced compared to those of the QM approaches. Chemical reactions cannot be studied, except for using specially designed reactive force fields, and force fields become specific for the systems studied with many, often engineered, energy terms and many parameters, the physical meaning of which is often not clear.

Owing to the use of graphical processing units (GPUs) [16], worldwide distributed computing [17] and the construction of dedicated machines, such as the ANTON supercomputer developed by the D.E. Shaw company [18, 19], the size- and timescale of all-atom simulations become gradually extended. Nevertheless, for larger systems such as proteins, nucleic acids and their complexes, as well as fragments of cellular organelles, the use of the atomistically-detailed approaches is still infeasible and higher degree of reduction must be sought. Such an approach is termed *coarse graining* (CG) and involves merging several atoms or even whole entities (e.g., entire proteins) into extended interaction sites [4, 12, 20–38]. This approach can be understood as averaging over the positions and momenta of individual atoms, leaving only those of the entire sites [12, 38]. Consequently, the physical origin of a CG effective energy function is the potential of mean force (PMF) of a system [24, 35, 39]. When carried out to the limit, coarse graining becomes the mesoscale or even continuous modeling.

With coarse graining, the time scale can be extended to seconds and systems with even millimeter size can be treated, depending on the degree of reduction [12]. The extension of the time scale in coarse-grained molecular dynamics simulations is a consequence of averaging out the degrees of freedom that are not present in the reduced representation. On one hand, it enables us to cover much larger time scales compared to all-atom simulations but, on the other hand, the relationship between the time unit in all-atom and CG dynamics can only be estimated by comparing the simulation times in which well-defined events (e.g., helix formation) occur [40, 41].

It should be noted that a given system needs to be treated at different resolution and accuracy at the same time. For example, when studying a chemical reaction that takes place in the enzyme's active center, QM is needed, in principle. However, the QM description needs to be applied only to the center, while the rest of the proteins can be treated at the classical level; this is known as the QM/MM approach [42]. In general, such an approach is termed *multiscale modeling* [5, 24] and has been originated by the year 2013 Nobel Prize laureates in Chemistry [43–45].

The design of coarse-grained approaches has two dimensions. The first of them is the top-down vs. the bottom-up approach. In the top-down approach, the model is constructed and parameterized to reproduce the structures and properties of training systems. In the bottom-up approaches, one starts from the atomistically-detailed representation of a system to construct the effective site-site interaction potentials by averaging over secondary degrees of freedom. The second dimension is the source of parameterization, which divides the coarse-grained force fields into knowledge-based (statistical) and physics-based ones, depending on whether the potentials are obtained by Boltzmann inversion of the distribution or correlation functions extracted from structural databases or by integrating all-atom PESs or using thermodynamic data. The physical meaning of the statistical potentials is not clear [46]; however, they are quite successful in *ab initio* protein structure prediction and protein-folding simulations [12]. The most advanced CG force fields combine the bottom-up approach, regarding the design and parameterization of specific energy terms, with the top-down approach, executed as the last step of parameterization, as well as physics-based and statistical data for parameterization.

This chapter is devoted to the physics-based CG approaches. In what follows, we describe the construction of CG energy functions, including the scale-consistent approach developed in our laboratory [35, 39], the methods of force-field parameterization, and the methods for the search of the conformational space, including the use of information from structural databases and experimental data. Lastly, we briefly describe some of the physics-based CG force fields and their applications. We conclude with summarizing the present status and outlining the perspectives of the field.

2.2 Designing Coarse-Grained Models and Force Fields

2.2.1 *Potential of Mean Force as the Origin of CG Force Fields*

A coarse-grained effective energy function of a system originates from its potential of mean force (PMF), with all degrees of freedom that are not present in the coarse-grained representation averaged out [24, 35, 39]. The variables describing the geometry of the system, including the surrounding solvent molecules if applicable, are divided into the *primary* variables (**X**), which describe the coarse-grained degrees of freedom, and the *secondary* variables (**Y**) that are averaged over. The secondary variables often include the solvent degrees of freedom; with such a choice, the solvent is implicit in an effective CG energy function. For polymer chains, it is natural to identify the dominant secondary degrees of freedom with the dihedral angles λ for the rotation of the atoms comprising the extended sites about the respective virtual-bond axes [35] (Fig. 2.1). Consequently, the PMF $[F(\mathbf{X}; T)]$ is expressed by Eq. 2.1.

$$F(\mathbf{X}; T) = -\frac{1}{\beta} \ln \left\{ \frac{1}{V_{\mathbf{Y}}} \int\limits_{\Omega_{\mathbf{Y}}} \exp\left[-\beta E(\mathbf{X}; \mathbf{Y})\right] dV_{\mathbf{Y}} \right\} \tag{2.1}$$

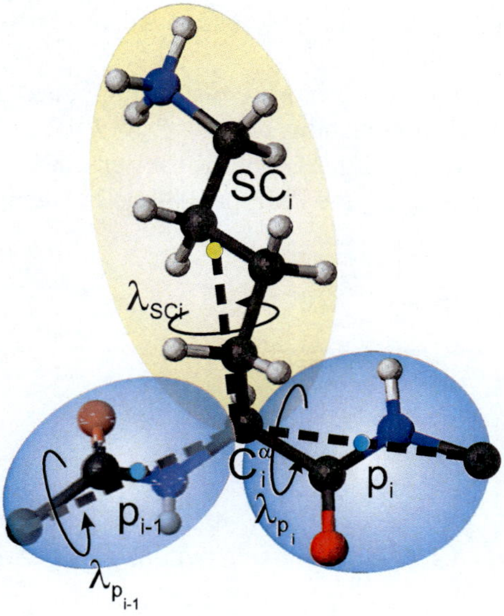

Fig. 2.1 Illustration of the primary (kept in the CG representation) and secondary (averaged out) degrees of freedom with a fragment of a polypeptide chain containing an L-lysine residue (which has position i in the sequence) in the UNited RESidue (UNRES) [48] CG representation. The only interaction sites are the united peptide groups (p), the centers of which are located in the middle of the $C^\alpha \cdots C^\alpha$ virtual-bond axes and marked as small light-blue circles, and the united side chains (SC), the lysine side-chain center being located on the top of the $C^\alpha \cdots$ SC virtual bond and marked with a small yellow circle. The virtual bonds are marked with black dashed lines and the atoms that belong to the centers are contained in light-blue (peptide groups) and light-orange (side chain) spheroids, respectively. The primary degrees of freedom are the positions of the C^α and the side-chain centers (SC); these constitute the variables **X** in Eq. 2.1. The degrees of freedom averaged over (variables **Y** in Eq. 2.1) are, primarily, the collective rotation angles λ of the atoms of the united groups about the $C^\alpha \cdots C^\alpha$ ($\lambda_{p_{i-1}}$ and λ_{p_i} in the Figure) and $C^\alpha \cdots$ SC (λ_{SC_i} in the Figure) virtual-bond axes, respectively; they are marked with curved arrows perpendicular to the virtual-bond axes, and the coordinates of the solvent molecules (not shown in the Figure). These secondary variables undergo the greatest variation. Additionally, the atoms of the united groups and those of the water molecules oscillate about the equilibrium values determined by valence geometry. It should be noted that averaging implies that the shape and interaction potentials of the united groups have the axial and not the spherical symmetry

where $E(\mathbf{X};\mathbf{Y})$ is the original (all-atom) energy function, equivalent to the Born-Oppenheimer potential-energy surface, in which the electron degrees of freedom are averaged out, $\Omega_\mathbf{Y}$ is the region of the $\mathbf{Y}$ subspace of variables over which the integration is carried out, $V_\mathbf{Y} = \int_{\Omega_\mathbf{Y}} dV_\mathbf{Y}$ is the volume of that region, and $\beta = 1/RT$, R being the universal gas constant and T the absolute temperature. From Eq. 2.1 it also follows that a CG energy function depends on temperature (in other words, it contains both the energy and the entropy components); however, this dependence is rarely taken into account in the CG force fields [47].

It can be noted at this point that the PMF (Eq. 2.1) defines the prototype of coarse-grained energy functions in the same way as the Born-Oppenheimer potential energy defines all-atom energy functions; in the second case averaging is done over electron degrees of freedom.

The choice of the PMF as a prototype of coarse-grained energy functions is motivated by the fact that the probability density, $P(\mathbf{X};T)$ of a given coarse-grained structure is directly related to it (Eq. 2.2) and also any ensemble and thermodynamic averages can be computed using the PMF or PMF and its temperature derivatives [35].

$$P(\mathbf{X};T) = \frac{\int\limits_{\Omega_\mathbf{Y}} \exp[-\beta E(\mathbf{X};\mathbf{Y})]\,dV_\mathbf{Y}}{\int\limits_{\Omega_\mathbf{X}}\int\limits_{\Omega_\mathbf{Y}} \exp[-\beta E(\mathbf{X};\mathbf{Y})]\,dV_\mathbf{Y}dV_\mathbf{X}} = \frac{\exp[-\beta F(\mathbf{X};\mathbf{Y})]}{\int\limits_{\Omega_\mathbf{X}} \exp[-\beta F(\mathbf{X};\mathbf{Y})]\,dV_\mathbf{X}} \tag{2.2}$$

2.2.2 *Derivation of CG Force Fields from PMF Surfaces*

The PMF defined by Eq. 2.1 cannot be applied as it is because, except for very simple systems, its evaluation would be much more expensive than doing the atomistically-detailed simulations of a system and, moreover, it would be system-specific. Therefore, as in the case of the construction of all-atom force fields from the Born-Oppenheimer potential-energy surfaces, a simplified form must be sought. Generally, the PMF can be split into local, pairwise, and multibody terms, as expressed by Eq. 2.3.

$$F(\mathbf{X}) = \sum_i U_i + \sum_{i<j} U_{ij} + \sum_{i<j<k} U_{ijk} + \ldots \tag{2.3}$$

The local terms (U_i) correspond to virtual-bond stretching, virtual-bond bending, etc., the pairwise (U_{ij}) terms are site-site interaction potentials. The multibody terms (U_{ijk} and higher) are the most difficult to derive and, therefore, are present in few force fields, although their presence is mandatory in principle, because the potential of mean force is not expressible in terms of pairwise terms except for systems composed of isolated pairs of sites. A systematic method of splitting the PMF into components, including the multibody terms, will be outlined later in this section [35, 39].

The terms in Eq. 2.3 can be introduced as tabulated potentials or as parameterizable analytical expressions. The tabulated potentials are effectively restricted to functions in one variable, e.g., a site-site distance. Such potentials can be applied directly in Monte Carlo simulations, although spline functions are used to make them suitable for CG molecular dynamics [49].

The analytical expressions for the local and pairwise terms are usually imported from all-atom force fields, which include the harmonic bond-, bond-angle, and torsional terms for local interactions and the Coulomb-electrostatic and van der

Waals terms for long-range interactions; this is the case of the most widely used MARTINI force field [50]. However, such neoclassical expressions are not necessarily relevant for coarse-grained models, because the long-range potentials depend not only on distance but also on orientation. An example of introducing the dependence on site orientation are the Gay-Berne potentials [51], which were initially developed for liquid-crystal simulations.

In our laboratory, we have developed a systematic method for the derivation of effective coarse-grained energy terms [35, 38, 39]. First, the atoms are assigned to the CG interaction sites and the interaction energies are divided into the contributions ascribed to single sites, pairs of sites, and clusters of sites, the latter corresponding to all-atom multibody terms. The respective groups of site, site-site, and site-cluster energies are termed the *component energies*. Subsequently, the PMF is rewritten in terms of Kubo cluster cumulant functions [52], termed *factors*, which correspond to Boltzmann integrals over combinations of the component energies. Each factor is a potential of mean force or a combination of potentials of mean forces corresponding to a respective section of the system and can be treated as a force-field term. Because the factors correspond to repeated units (e.g., amino-acid side chains and their pairs or higher clusters), they are transferable between systems just as all-atom force-field terms are. The order of a factor is the number of component energies that enter into the respective Boltzmann integrals and summing all the factors restores the original PMF. For practical reasons, the factor expansion needs to be truncated. Factors of order 1 correspond to site-local (virtual-bond potentials), site-pair local (virtual-bond-angle potentials) and long-range pairwise potentials, while higher-order factors correspond to multibody terms. Specifically, the second order-factors composed of the local interactions of three consecutively bonded sites, but excluding the interactions between the first and the third site, correspond to the torsional potentials [35, 39]. We found that expansion of up to order 3 is sufficient for a force field to work [35, 38]. The neoclassical CG force fields contain mostly the first-order factors, the only second-order factors being the torsional potentials.

To get approximate analytical expressions for the effective CG energy terms (PMF factors), we expand them into Kubo's generalized-cumulant series [52]. The lowest-order cumulants serve as parameterizable analytical energy-term expressions [35, 39]. Further, by expressing the energy of a system at the all-atom level in terms of interatomic distances, expressing the interatomic distances in terms of the distances between the coarse-grained-site centers, the orientation of the virtual-bond axes, and the angles for collective rotation of the atoms of the coarse-grained groups about the virtual-bond axes λ (cf. Fig. 2.1), and averaging over λs to compute the cumulants, we derived scale-consistent expressions for the effective energy terms, which embed the details of the all-atom geometry that are not present explicitly when passing to a coarse-grained representation [35]. The above approach enables us to obtain the energy expressions that exhibit correct dependence on orientation and the multibody energy terms that account for the coupling between the coarse-grained degrees of freedom through the fine-grained degrees of freedom, which are omitted from a coarse-grained model. In particular, we found that the torsional potentials depend not only on the virtual-bond-dihedral angles but also on the virtual-bond angles [35].

2.2.3 *Solvent Treatment in CG Force Fields*

Water is the most common solvent in biomolecular simulations. Its role in protein folding is not limited to that of a medium. Water molecules often become part of the biomolecule's structure [53] and can be found in the interior of the proteins [54]. They are known to play an essential role in protein folding [55]. In particular, when studying protein-ligand interactions, it is essential to simulate the nearby water molecules[56], because they participate in protein-ligand hydrogen-bond-network formation.

Explicit water molecules or water clusters are considered in many CG force fields such as, e.g., MARTINI [31, 50], SIRAH (South American Initiative for a Rapid and Accurate Hamiltonian) [57, 58], and in the family of force fields developed by Voth and coworkers [24, 49]. In the latter, a single water molecule is considered as an extended interaction site [49]. Several (4 or 8) water molecules are merged into a single CG bead in the MARTINI force field [50, 59]. More sophisticated water models with polarizable beads have also been proposed [60]. A good review about explicit representation of the water molecules is presented in ref [61].

In most of the CG force fields, including a variant of MARTINI termed "dry Martini" [62], the interactions with water are absorbed into the CG potentials, which is formally accomplished by treating the water degrees of freedom as the degrees of freedom averaged out in the expression for the PMF (Eq. 2.1). This simplification significantly reduces the number of degrees of freedom of the system and speeds up the simulations; however, its disadvantage is that structural water cannot be considered. In dry MARTINI [62], only the well-depths (ϵ) and radii (σ) in the Lennard-Jones potential are modified to account for the interactions with water molecules implicitly. A similar approach has been implemented in the early version of UNited RESidue (UNRES) CG force field for proteins [63], in which the interactions with water are implicit in the side-chain-interaction potentials. A more sophisticated model, in which the effective interactions with the solvent are handled through a cavity term and a Generalized-Born term, similar to that of all-atom implicit-solvent models [64] was later implemented in UNRES [65]. This model also takes into account side-chain anisotropy. An analytical formula, which is based on differential Gaussian overlap for the cavity term was proposed [66], which has been subsequently generalized to sites with axial symmetry [67]. This formula is easier to handle and more stable numerically compared to those based on the molecular surface area. The model was successfully applied in protein simulations [68].

2.2.4 *Force-Field Parameterization*

The data for parameterization are the PMFs of model systems and the experimental data, which include thermodynamical and structural properties and the structures themselves. Additionally, the distribution functions derived from structural databases are used to parameterize statistical potentials. In this section we describe briefly the methods of parameterization.

2.2.4.1 Direct Computation of PMF Surfaces

This approach is a bottom-up one and is particularly suitable to determine the PMF factors in the factor-expansion approach (see Sect. 2.2.4). The respective model systems correspond exactly to the evaluated PMF factors, e.g., single or connected sites (to evaluate the local potentials), pairs of disconnected sites (to evaluate pairwise potentials), or clusters of sites or interacting fragments of polymer chains (to evaluate multibody terms). This approach has been originated by Levitt [44] and also used later by Pincus and Scheraga [69] to determine the CG potentials from atomistically-detailed potentials.

For small systems such as, e.g., fragments of polypeptide backbone, the PMF can be evaluated by numerical integration of the PESs calculated at the QM level. For example, in the computation of the virtual-bond-stretching PMF of the peptide group, taking into account drastic virtual-bond-length change upon cis-trans isomerization, the PES was evaluated at the MP2/6-31G** *ab initio* QM level, (300 grid points only) [70]. The calculations were carried out at the same high level to derive the virtual-bond-angle bending and the virtual-bond-dihedral-angle torsional potentials of polypeptide chains for the UNRES force field [71]. However, to derive the side-chain-rotamer potentials [72], which require about 700,000 grid points for the largest side chains, the AM1 semi-empirical QM method [15] was used.

For systems with more degrees of freedom such as, e.g., the pairs of interacting sites in water or the fragments of polysaccharide or saccharide derivative chains in water, where the solvent degrees of freedom need to be considered, the PMFs must be evaluated by umbrella-sampling MD simulations. In these simulations, biasing potentials are imposed on the reaction coordinate (e.g., the distance between site centers) to cover the whole range, which are removed by using the Weighted Histogram Analysis Method (WHAM) [73] in the post-processing step to obtain the PMF. This approach has been implemented in determining the potentials of side-chain–side-chain interactions [65], the potentials of interaction of heparin units [74] and the torsional potentials of polysaccharides [75] and heparin [74].

2.2.4.2 Force Matching

Force matching (FM) is a bottom-up approach, in which the forces acting on CG sites are determined to fit the average forces computed from the respective all-atom forces, which are usually obtained by carrying all-atom MD simulations. This method was first applied to develop coarse-grained force fields by the Voth group [24, 49, 76] and termed Multiscale Coarse Graining (MSCG) method. Originally the MSCG/FM method was designed for radial pairwise potentials, the site centers being their centers of mass. This is accomplished by linear least-squares fitting, following Eq. 2.4.

$$\sum_{i=1}^{M}\sum_{j=1}^{N}||\mathbf{F}_{ij}^{CG} - \mathbf{F}_{ij}^{MD}||^2 = \min \tag{2.4}$$

where $\mathbf{F}_{ij}^{CG}$ is the vector of CG forces acting on the center of the mass of site j in MD snapshot i, $\mathbf{F}_{ij}^{MD}$ is the respective vector of CG forces computed from all-atom MD simulations, which is defined by Eq. 2.5.

$$\mathbf{F}_{ij}^{MD} = \sum_{k\in\{j\}} \mathbf{f}_{ijk} \tag{2.5}$$

where $\mathbf{f}_{ijk}$ is the all-atom force acting at atom k of site j in snapshot i. The CG forces are further expressed in terms of radial pairwise contributions (Eq. 2.6).

$$\mathbf{F}_{ij}^{CG} = \sum_{k\neq j} \boldsymbol{\varphi}_{jk}(r_{ijk}) \tag{2.6}$$

where $\boldsymbol{\varphi}_{jk}$ is the effective CG force acting between sites j and k and r_{ijk} is the distance between the centers of the masses of sites j and k for snapshot i. The $\boldsymbol{\varphi}$s are binned, the values at the bins being determinable parameters, hence minimization of the target function given be Eq. 2.4 is a linear least-squares problem. Even though the multibody terms are not considered directly, they are implicit in the pairwise potentials [76].

Once the $\boldsymbol{\varphi}_{jk}$ profiles have been determined, the respective PMFs are obtained by numerical integration over site-site distance, thus giving the force-field components, which can be used in CG simulations [77]. However, the potentials are specific to the systems they have been derived with and, thus, weakly transferable.

Recently, we developed another variant of the FM approach [78], in which the analytical formulas for the CG forces derived by using the scale-consistent methodology [35, 38] are fitted to the MD forces, which improves transferability. Our method also covers the CG models in which the interaction potentials are axially- and not radially-symmetric, this enabling more extensive coarse graining.

Application of this approach to the UNRES force field resulted in a transferable force field that is able to predict protein structures [78].

2.2.4.3 Iterative Boltzmann Inversion and Inverse Monte Carlo Iteration

These two approaches are similar to FM in that they extract the effective potentials based on Monte Carlo or MD simulations of multicomponent systems. In the iterative Boltzmann inversion (IBI) method [79, 80], starting from an initial CG energy function (which can have a tabulated or analytical form), its parameters are refined given the probability distribution function obtained from all-atom simulations, as expressed by Eq. 2.7.

$$U^{(i+1)} = U^{(i)} + RT \ln \frac{P^{(i)}}{P_{ref}} \tag{2.7}$$

where $U^{(i)}$ and $P^{(i)}$ are the CG energy function and the respective probability distribution obtained in the ith iteration and P_{ref} is the reference probability distribution (in CG variables) obtained from all-atom simulations.

In the inverse Monte Carlo (IMC) method [81, 82], the CG Hamiltonian of a system, $H(\{q\})$, is defined in terms of basis functions $S_\alpha(\{q\})$, q being a selected CG coordinate and $\{q\}$ denoting the set of coordinates used in parameterization, as given by Eq. 2.8.

$$H(\{q\}) = \sum_{\alpha=1}^{N} V_\alpha S_\alpha(\{q\}) \tag{2.8}$$

where V_α is the coefficient corresponding to the basis function S_α and N is the number of the basis functions. Usually the basis functions are associated with the bins of a reaction coordinate (e.g., the site-site distance), being 1 inside and 0 outside a bin. From all-atom simulations, the averages over the basis functions $\langle S_\alpha \rangle$ (proportional to the counts of points in the bins if the basis functions are bin-based) can be obtained, which can also be computed using the V_αs (Eq. 2.8). Thus, the V_αs can be iteratively updated using the Newton iteration, as expressed by Eq. 2.9.

$$\sum_{\gamma=1}^{N} \frac{\partial \langle S_\alpha \rangle}{\partial V_\gamma} \Delta V_\gamma = \langle S_\alpha^{ref} - S_\alpha^{CG} \rangle, \; \alpha = 1, 2, \ldots, N \tag{2.9}$$

where the derivatives are defined by Eq. 2.10.

$$\frac{\partial \langle S_\alpha \rangle}{\partial V_\gamma} = \frac{1}{RT} \left(\langle S_\alpha \rangle \langle S_\gamma \rangle - \langle S_\alpha S_\gamma \rangle \right) \tag{2.10}$$

Both the IBI and IMC methods have been implemented in the MAGIC package for automatic parameterization of CG force fields [37, 83].

2.2.4.4 Fitting Force-Field Parameters to Reproduce the Experimental Data

In this top-down approach, the force-field parameters are adjusted to reproduce the thermodynamical and mechanical properties such as hydration free energies of model compounds, partition coefficients or interfacial tensions. This approach has been used to parameterize the MARTINI force field [27, 31]. Other properties to use are the free energies of DNA hybridization or protein folding-unfolding transition and transition heat capacities; these quantities were used as part of parameterization of the UNRES [47] and NARES-2P [84] coarse-grained force fields for proteins and nucleic acids, respectively.

2.2.4.5 Calibration with Experimental Structures

This top-down approach is used to parameterize the force fields aimed at reproducing the structures of the systems studied, e.g., physics-based prediction of protein structures, protein-protein docking, protein dynamics, prediction of RNA structure and dynamics, etc. The underlying principle is to make the native-like structures the most probable in simulations.

Wolynes and coworkers [85–88] developed a method based on maximizing the ratio of the folding-transition and the glass-transition temperature (T_f/T_g). With this condition satisfied, a system achieves the native structure at a temperature higher than the glass-transition temperature, this enabling efficient folding. Maximization of the T_f/T_g ratio can be approximated by the requirement for obtaining the most negative value of the Z-score of the energy distribution, which is the difference between the native-structure energy and the mean energy scaled by the standard deviation of the energy, as given by Eq. 2.11.

$$Z = \frac{E_{nat} - \langle E \rangle}{\sqrt{\langle E^2 \rangle - \langle E \rangle^2}} \tag{2.11}$$

Another approach is the maximization of the energy gap between the native structure and the lowest-energy alternative structure, which was first developed by Crippen and coworkers [89]. We generalized this approach to develop a hierarchical optimization method [90, 91], in which free-energy gaps, determined from unfolding experiments if possible, are set between subensembles with various degree of folding. We applied this method to the UNRES force field, obtaining its predictive variants [47, 90, 91].

An undesirable feature of the approaches described above is that they heavily rely on a small (and, thereby, volatile) subset of simulated conformations, which are classified as native-like or belonging to a certain level of the folding hierarchy. Since the procedure consists of iterations, in which the energy-function parameters are determined by optimization of the target function of choice computed by using the ensembles of conformations simulated in a previous iteration, the procedure is unstable. To overcome this instability, recently we developed the maximum-likelihood approach [92], in which the simulated conformational ensembles of the training systems are fitted to the corresponding experimental ensembles (usually determined by NMR), at best at a range of temperatures bracketing the folding-unfolding transition. Because optimization is based on whole ensembles, the instability is removed. This method has been used with success to produce predictive variants of the UNRES force field [93, 94] for proteins and of the NARES-2P force field for nucleic acids [84].

2.3 Methods of Conformational Search

Because the multidimensional energy surfaces of complex systems possess zillions of local energy minima, even if the coarse-grained models are used, exploring the configurational space in order to locate stable structures is a challenge [95, 96]. Local energy minimization is not suitable as a search tool except for very simple systems. Moreover, low energy does not guarantee that a structure is stable; it needs to be located in a wide enough energy basin. Furthermore, computing structural and thermodynamic averages requires a wider search.

Regarding their purpose, the methods of conformational search can be divided into those aimed at finding canonical averages; these are canonical Monte Carlo (MC) [97] and canonical molecular dynamics (MD) methods [1], and methods for finding the lowest-energy minima in the potential-energy surface (global minimization methods) [95]. Methods of the first class also enable the researchers to study the time evolution of a system. Multi-canonical extensions of these methods, such as those based on replica exchange (REMC and REMD, respectively) enable us to perform even a wider search at many temperatures simultaneously, and make it possible to find structures with the lowest energy. Although global energy minima are not necessarily the most stable structures because of neglecting the entropy component, the global minimization methods [95] are useful because they usually lead to low-energy structures faster than the canonical sampling even with the multicanonical extension. They are often used in data-assisted structure determination.

In Sects. 2.3.1–2.3.4 the basic conformational-search techniques used with CG force fields are discussed, while the use of experimental and database-derived information is described in Sect. 2.3.5.

2.3.1 Canonical Monte Carlo

The basic canonical MC algorithm [97] consists of a sequence of steps, in each of which the current conformation is randomly perturbed and the energy of the perturbed conformation is computed. If the energy is lower, the old conformation is replaced with the new one. If not, the new conformation is accepted with probability $\exp(-\Delta E/RT)$, where ΔE is the difference between the energy of the new and the old conformation, R is the universal gas constant, and T is the absolute temperature. The process is iterated until the ensemble averages converge. The ensemble averages are updated every given number of steps. The MC approach requires only energy evaluation but its efficiency is reduced by small perturbation-step size that must be taken to achieve a sufficient acceptance rate.

Except for simulations of model Lennard-Jones systems, Cartesian coordinates are rarely used as the variables of choice. For polymers, it is common to sample in the dihedral-angle space or even to introduce collective many-bond motions [98]. However, care must be taken, because the use of angular variables makes it only too easy to introduce a bias in the sampling [1]. Also, the use of lattice representation of polymer chains, which enables us to precompute many energy terms is common [98].

2.3.2 Molecular Dynamics

Molecular dynamics is based on numerical integration of Newton's equations of motion [1, 6]. For the all-atom representation, the construction of these equations is straightforward because the atomic nuclei can be treated as point objects with masses, whose coordinates are the variables. This scheme is carried over to most of the coarse-grained MD approaches by assuming that the CG beads are spherical. Consequently, each bead can be represented as a point mass positioned at the center of the mass of the bead [24]. However, this assumption precludes extensive coarse graining, in which case it is impossible to keep the spherical shape of the CG beads.

In our work [99], we proposed a general MD scheme, in which each site is characterized by the position of its center and the direction of its axis, identified with the virtual-bond axis. Assuming such axial symmetry is natural for polymer chains. Consequently, the system is composed of stretchable rods, with the mass of each rod smeared over the rod, according to the distribution of the atoms along the direction of the virtual-bond axis. The resulting equations of motion do contain a non-diagonal matrix of inertia ($\mathbf{G}$) but this matrix is constant, which enables us to compute its inversion only once [99]. Moreover, with an appropriate choice of CG coordinates ($\mathbf{q}$), the matrix of inertia becomes a five-diagonal matrix, which makes the memory requirements and compute costs liner in system size. Thus, in general, the equations of motion for a CG system can be expressed by Eq. 2.12.

$$\mathbf{G}\ddot{\mathbf{q}} = -\nabla_{\mathbf{q}} U(\mathbf{q}) + \mathbf{f}_{fric} + \mathbf{f}_{rand} \tag{2.12}$$

where $\ddot{\mathbf{q}}$ are the accelerations, U is the effective CG potential energy, and $\mathbf{f}_{fric}$ and $\mathbf{f}_{rand}$ are the friction and stochastic forces, respectively which, apart from accounting for part of the interactions with the usually implicit solvent, serve to maintain a constant average temperature, a condition necessary for canonical simulations. The system of differential equations given by Eq. 2.12 corresponds to Langevin dynamics; without the friction and stochastic term the system maintains constant total energy (microcanonical mode).

The Langevin thermostat (inclusion of the friction and stochastic forces) is recommended even with explicit solvent [100]; however, the weak-coupling (Berendsen) [101], Nosé-Hoover [102] or Nosé-Poincaré [103] thermostats can also be used. It should be noted, though, that the Berendsen thermostat, although the simplest to implement and exhibiting stable behavior, results in too narrow a temperature distribution.

A variety of algorithms have been developed for numerical integration of the equations of motion, the most popular belonging to the Verlet algorithm family; these are the Verlet, the leap-frog, and the velocity-Verlet algorithm [6]. Except for the velocity-Verlet algorithm, they require the estimation of the position or velocity, respectively, prior to the initial step, which can be accomplished by doing a backward step of a simple Euler integration. All three algorithms are symplectic algorithms, which means that the total energy undergoes only small oscillations (the extent of which depends on time-step size) about the average value. The equations of motion with the Nosé-Hoover and Nosé-Poincaré thermostats, on the other hand, are symplectic regarding the kinetic energy (kinetic temperature) [102, 103].

Compared to MC methods, MD usually results in a faster (in terms of compute time) evolution of the system. The size of the time step in MD cannot be too large and a value 1 or 2 fs is taken. With a larger time step, the simulations often exhibit unstable behavior due to rapid change of acceleration and, consequently, the calculated structures and properties become unreliable. The time step can be extended even up to 20 fs when the multiple-time-step (MTS) algorithms are used [6, 104–107], in which the forces are divided into the short-range and fast-varying ones and the slow-varying forces. The fast-varying forces, which are less expensive to compute, are integrated with a time step which is an integer fraction of the original time step, while the slow forces, which are more expensive to compute, are integrated with the original time step. The number of divisions of the original time step is termed the *split number*. With a fixed split number, this algorithm is time reversible, however, setting a large split number (which might be needed for bumpy energy landscapes) reduces the efficiency. We developed [107] the Adaptive Multiple Time Step (A-MTS) algorithm, in which the split number is adjusted depending on acceleration change. This algorithm is not time reversible but more efficient than the original MTS algorithms.

As mentioned in the Introduction, the time scale of CG simulations is dilated compared to the all-atom time scale, owing to pre-averaging over the secondary degrees of freedom. This dilatation amounts to 3–4 orders of magnitude when the solvent degrees of freedom are also averaged out in the CG model [40]. Therefore, 1 time unit of CG simulations corresponds to about 1000 units of all-atom simulations.

2.3.3 Extensions of MC and MD

To enhance the search of the conformational space, the replica-exchange variants of the canonical Monte Carlo and molecular dynamics approaches, also known as parallel tempering, have been developed [108, 109]. These extensions are called replica-exchange MC (REMC) or replica-exchange MD (REMD), respectively. In these methods, a number of parallel MC or MD simulations (replicas) are carried out at different temperatures ($T_0, T_1, \cdots, T_{M_T}$). The replicas evolve independently and, after a certain number of MC or MD steps, an exchange of temperatures between neighboring replicas (with indices i and $i+1$, respectively) is attempted, the acceptance of the exchange following the Metropolis criterion with the exchange probability ω expressed by Eq. 2.13, which reflects the fact that the effective CG energy function can depend on temperature [47, 110]:

$$\omega(\mathbf{q}_i \rightarrow \mathbf{q}_j) = \min[1, \exp(-\Delta)] \tag{2.13}$$

with

$$\Delta = \left[\beta_j U(\mathbf{q}_j; \beta_j) - \beta_i U(\mathbf{q}_j; \beta_i)\right] - \left[\beta_j U(\mathbf{q}_i; \beta_j) - \beta_i U(\mathbf{q}_i; \beta_i)\right] \tag{2.14}$$

where $\beta_i = 1/RT_i$, T_i being the absolute temperature corresponding to the ith trajectory, and $\mathbf{q}_i$ denotes the variables of the conformation of the ith trajectory at the attempted exchange point. The multiplexed variants of the REMC/REMD methods (MREMC/MREMD) [111, 112] differ from the REMC/REMD method in that several trajectories are run at a given temperature. Each set of trajectories run at a different temperature constitutes a *layer*. Exchanges are attempted not only within a single layer but also between layers.

In the Hamiltonian Replica Exchange extensions of canonical MC and MD (HREMC and HREMD, respectively) [113–115], M canonical MC/MD simulations are carried out simultaneously at different temperatures and with different potential-energy functions (V), which can differ by the repulsive Lennard-Jones terms, if the purpose is to allow the system to overcome sterical clashes or with different restraint functions, if the purpose is to explore different ranges of order parameters [47, 115]. Usually the Hamiltonian replicas are combined with temperature replicas. Thus, the replicas constitute a two-dimensional (T_i, V_j) grid. For each replica, an exchange is attempted with its neighboring replica in one or two dimensions (up, down or diagonal on the grid, the direction being selected at random). The exchange is accepted based on the probability ω expressed by Eq. 2.15:

$$\omega\left(\mathbf{q}_{ij} \rightarrow \mathbf{q}_{kl}\right) = \min[1, \exp(-\Delta)], \quad (k, l) \in \{(i+1, j), (i, j+1), (i+1, j+1)\} \tag{2.15}$$

with

$$\Delta = \left[\beta_k V_l(\mathbf{q}_{kl}; \beta_k) - \beta_i V_j(\mathbf{q}_{kl}; \beta_i)\right] - \left[\beta_k V_l(\mathbf{q}_{ij}; \beta_k) - \beta_i V_j(\mathbf{q}_{ij}; \beta_i)\right] \tag{2.16}$$

where V_j is the potential-energy function (including the restraining potential) corresponding to the (T_i, V_j) trajectory, and $\mathbf{q}_{ij}$ denotes the variables of the respective conformation of this trajectory at the attempted exchange point.

Multicanonical algorithms [116, 117], also known as entropy sampling [117], are another class of methods, in which the energy is replaced in the Metropolis criterion by the logarithm of the density of states (the microcanonical entropy). A simulation is converged when all energies are sampled with the same frequency; therefore, this method is well suited to overcome energy barriers. Once the density of states is obtained, all ensemble averages can be computed. The multicanonical methods have been applied with the UNRES CG force field [118].

2.3.4 Global Energy Minimization

The methods for finding the global minimum in the potential-energy surface can be divided into deterministic and stochastic ones [95, 119, 120], the deterministic methods being nowadays only of historic significance. An elegant deterministic approach to global optimization are the deformation methods [121–124], in which the original energy surface is deformed to reduce the number of minima, ideally to one minimum, which can easily be found by local minimization. The original energy surface is subsequently restored by means of a reversal procedure, the global minimum in the deformed surface being traced to the global minimum in the original energy surface. The respective methods include the Diffusion Equation Method (DEM) [121, 122], the Distance Scaling Method (DSM) [124], and the imaginary-time Schrödinger equation method [123]. Because of difficulties in determining the appropriate extent of deformation and designing the reversal procedure, these methods have been applied to small systems only, except for DSM which was combined with a stochastic search in the reversal process and was applied to the UNRES CG force field [124].

The stochastic methods combine the Monte-Carlo concept with a systematic search. Monte Carlo Minimization (MCM) is the most straightforward and, at the same time, a very efficient method of this class [125]. In this method, one starts from a local energy minimum, which is perturbed using a large step and the resulting conformation is energy minimized. If a different energy minimum is reached, the new conformation is accepted unconditionally, if the new energy is lower and accepted with the Boltzmann probability, if it is higher. It should be noted that the sampling scheme in MCM violates the detailed-balance condition and, therefore, this method cannot be readily applied to compute ensemble averages. MCM was applied in coarse-grained modeling [126]. More advanced variants of MCM are the basin-hopping [127] and αBB (a branch-and-bound method) [128].

Genetic algorithms are the most efficient in the global-minimum search. The Conformational Space Annealing (CSA) method [129] is one of the most successful algorithms of this class. It combines essential aspects of the build-up procedure and a genetic algorithm. In the early stages of the search, the whole conformational space is explored, and the search is gradually narrowed to smaller regions with low energy. CSA has been adapted to carry out the conformational search with the UNRES [130] and NARES-2P [131] force fields, leading to many outstanding predictions of the protein structures in the CASP exercises [132–134].

2.3.5 Use of Geometrical Restraints in CG Simulations

The present force fields alone, both all-atom and coarse-grained ones, do not yet have a sufficient accuracy to model the structures of biomolecules in the unassisted or *ab initio* mode. Therefore, incorporating information from low-resolution experimental data and bioinformatics is highly advisable. Several experimental techniques can be used [135]. The chemical crosslinking/mass spectrometry (XL-MS) and Fluorescence Resonance Energy Transfer (FRET) experiments provide information about the distance distribution between specific amino-acid residues. Small/Wide-Angle X-ray/Neutron scattering (SAXS, SANS and WAXS) can determine the shape of a molecular system and distance distribution. Nuclear magnetic resonance (NMR) and electron paramagnetic resonance (EPR) allow for the mapping of the interacting regions of biomolecules at the residue/atomic level. Mutagenesis and hydrogen-deuterium exchange experiments provide useful information on the residues that are exposed to the solvent and those which are involved in interactions within complexes. In contrast to the X-ray crystallography, the above-mentioned approaches are not that demanding in terms of time, cost, and sample specificity. The determined observables are included in a penalty function added to the potential energy, which drives the search at the regions of the conformational space consistent with the experimental data.

Different experimental techniques have their strengths and limitations and, therefore, combining them as, e.g., in the Integrative/Hybrid Structure Modeling can give better results than using them individually [136, 137]. In particular, hybrid structural modeling reduces the danger of obtaining wrong structures when only sparse or ambiguous data are available.

2.3.5.1 Chemical Cross-Link Mass-Spectroscopy and Fluorescence Energy Transfer

The data obtained from the XL-MS and FRET experiments provide information on the inter- and intra-molecular contacts or distance distributions between specific atoms (residues) in the system [138, 139]. In the XL experiment, the samples are treated with a reagent that can bridge two amino-acid residues, if they are at a relatively close distance (up to 25 Å). There are specific and nonspecific cross-linking reagents, the specific ones bridging the residues that possess free acidic or basic groups. After an about 30–60 min incubation period, the reaction is quenched by removing the reagent from the solution. Afterwards, the protein is unfolded by urea, the existing disulfide bonds are reduced, the free thiol groups of the cysteine residues are alkalized and the whole molecule is subjected to the process of digestion. At the end, the digested protein samples are purified and analysed by Liquid Chromatography (LC) and Mass Spectroscopy (MS) [140]. The peptide fragments attached to the XL reagent are identified and knowing the size and kind of XL reagent, the distance distribution between specific residues and the surface residues can be estimated.

The XL-MS data can be used to derive distance restraints, if only distance ranges can be estimated or, if the distance distribution can be obtained, they can be converted into pseudopotentials added to the energy function. They can also be used as a filter to select the models compatible with these data.

The distances can be confined within a range determined by XL-MS experiments by using a flat-bottom penalty function, such as that defined by Eq. 2.17 [141–143].

$$V_{cont}(d, d_l, d_u, A) = \begin{cases} A\frac{(d-d_l)^4}{\sigma^4+(d-d_l)^4} & \text{for } d < d_l \\ 0 & \text{for } d_l \le d \le d_u \\ A\frac{(d-d_u)^4}{\sigma^4+(d-d_u)^4} & \text{for } d > d_u \end{cases} \tag{2.17}$$

where d is the distance between the C^{α} atoms of the two crosslinked residues in the computed structure and d_l and d_u are the lower and upper contact-distance boundaries, respectively (usually $d_l = 2.5$ Å, $d_u = 25$ Å), σ is the width of the transition region between zero and the maximum restraint height, and A is the height of the restraint wall, which can also be considered as the confidence of a crosslink. This function quickly approaches the asymptote A, contributing virtually no force when $d \gg d_u$. Thus, the penalty terms do not force incompatible restraints (which usually correspond to false cross-links), preventing a simulation from producing non-protein-like structures.

When the distance distributions can be determined, they are usually approximated by a sum of Gaussians, resulting in the log-Gaussian pseudopotentials such as those expressed by Eq. 2.18 [144].

$$V(d) = -A \log \left\{ \sum_i \frac{a_i}{\sigma_i} \sqrt{\frac{2}{\pi}} \exp \left[-\frac{1}{2} \left(\frac{d - d_{0i}}{\sigma_i} \right)^2 \right] \right\} \tag{2.18}$$

where d is the distance between the C^α atoms of two crosslinked residues, d_{0i} and σ_i, and a_i are the parameters of the ith Gaussian, and A is the confidence of a crosslink. The parameters of the log-Gaussian function vary based on the type of XL experiment (reagent specificity, length of the linker).

FRET is mostly used to study the dynamics of macromolecules (e.g., domain movement or the dynamics of the loops or C- and the N-ends) and in molecular-assembly prediction. Here, it is necessary to know in advance the approximate structure of a molecule, its domain structure or have a reliable theoretical (template-based) model. In the FRET experiments, energy donor and acceptor fluorophores (dyes) are attached to carefully selected sites in the target molecule [145, 146]. The donor dye is excited by a laser pulse, and part of the excitation energy is transferred to the acceptor dyes, which are at a sufficiently close distance. The measured intensities of dye fluorescence provide the information of the distance distribution between them, because the transfer efficiency depends on the distance between the donor and the acceptor fluorophores [147–149].

The information derived from FRET is the average distance or the distance distribution between two residues to which the dyes were attached. Based on several distance distributions from strategically selected labeled pairs of amino-acid residues it is possible to determine the dynamics of the protein.

2.3.5.2 Small Angle X-ray/Neutron Scattering

In a SAXS experiment, a collimated monochromatic X-ray beam is passed through the solution of the target macromolecules and the scattering intensity is measured after passing through the sample. The same procedure is repeated for the solvent alone. The obtained radiation intensities for the reflection angle for the pure solvent are subtracted from those for the sample in the solvent to obtain a distribution of reflected radiation intensities for the sample itself. In the simplest instance, the information obtained from the SAXS experiment enables us to determine a low-resolution shape outline of a molecule [150, 151].

The results of SAXS measurements are used in the form of the radiation scattering intensity [152–156] (which is the Fourier transform of the distance distribution), distance distribution [157, 158] or as the area occupied by a molecule obtained by adjusting the distribution of dummy atoms to obtain same as experimental curve of radiation intensity as a function of scattering vector (shape envelope). Theoretical and practical concepts of using molecular-modeling techniques combined with the scattering data to determine high-resolution structures are described in the literature [151]. Methods of using the scattering data in the determination of biomolecule structure and dynamics are still under development [159, 160]. The scattering-intensity profiles are calculated from atomic or CG-site coordinates by using the Debye formula [152, 153], which has been implemented in the CRYSOL program [161]. A newer version of the CRYSOL program maps the experimental scattering intensities and the associated errors onto a sparser grid [162]. The Pepsi-SAXS program uses cubic-spline interpolation to represent the scattering profiles [154]. FoXS uses a linear approximation to the Debye scattering equation [155, 156].

Because the water molecules of the hydration shell are involved in scattering, they must be taken into account in calculating scattering intensity or distance distribution. Hydration shell can be modelled explicitly [152, 154, 163–165] or implicitly [157, 158, 164]. The explicit approaches include that of Yang et al. [152], in which nonhomogeneous distribution within the hydration layer is obtained by assigning a different scaling factor for dummy water molecules. Zheng et al. [164] model the hydration shell implicitly by combining each residue and its nearby implicit water molecules into a composite representation [164]. In the SASSIM method, the hydration shell is defined in terms of spherical harmonics and is calculated using a Lebedev grid [166]. Watson and Curtis [165] implemented the Euler formula to compute the rotationally averaged scattering intensity. The orientations of the wavevectors ($\mathbf{q}$) are taken from a quasi-uniform spherical grid. The Pepsi-SAXS program [154] computes the scattering contribution of the hydration shell of the molecule by constructing a grid approximation using the linked-cell approach [167]. Several methods use a very simplistic representation of the hydration shell of the sample; e.g., a two-dimensional angular function has been implemented in the CRYSOL program [163]. In the approaches in which the distance distributions are used instead of scattering profiles, the contribution from a given pair of amino-acid residues is approximated by a Gaussian distribution with the standard deviation calculated from the Stokes radii of these residues [157, 158].

There are several ways to combine molecular dynamics simulations with the SAXS data [157, 168–170]. Larsen et al. [169] used the Bayesian/Maximum Entropy method [171] and the pre-calculated SAXS and SANS intensities to reweight the trajectories. Bowerman at al. [170] used Bayesian Analysis of Accelerated Molecular Dynamics Simulations. In the conformational search approach, SAXS restraints can be incorporated in the form of maximum-likelihood function [157] where sampling is guided directly by scattering data. A molecular-docking approach directed by the SAXS-determined shape of a protein complex has also been proposed [172].

2.3.5.3 Mutagenesis and Hydrogen-Deuterium Exchange (HDX)

These two techniques are used to determine the pattern of protein-protein interactions.

In the mutagenesis analysis, the residues potentially responsible for interactions are mutated to interaction-prone residues (usually alanines). If a mutation of a residue results in reduced binding, the residue is likely to be involved in the interactions. This information can be used to map the interfaces or binding sites of interacting macromolecules [173–175].

In the hydrogen-deuterium exchange (HDX) experiments, the sample is treated with D_2O and then NMR or MS measurements are carried out, giving the information about solvent-exposed residues (whose protons are exchanged for deuterons) [176–179]. Consequently, it can also provide information about docking sites.

The information from the mutagenesis and HDX experiments can be used in molecular-docking simulations or as a filter to select the structures of the complexes which are compatible with the experimental data.

2.3.5.4 Nuclear Magnetic Resonance

The nuclear magnetic resonance (NMR) spectroscopy provides atom-atom (usually proton-proton) distance and dihedral-angle restraints. Its advantage is that the measurements are carried out in solution, under nearly-physiological conditions. The first NMR-derived protein structures were reported in the 1980s [180]. NMR restraints are usually quite sparse, therefore the force field used for refinement can have a large impact on the quality of NMR structures [181].

Use of NMR-derived restraints in simulations with coarse-grained force fields is straightforward only with those, which keep some atomistic details such as, e.g., Rosetta in which the all-atom representation of the backbone is used [182]. NMR protein structure refinement based on Monte Carlo minimization has been implemented with the Rosetta force field [183, 184], which incorporates elements of statistical and physical parameter sets. It has been shown to improve the quality and accuracy of the protein structures determined by NMR [185, 186].

For fully coarse-grained simulations, one approach is to reconstruct all-atom structures from the CG representation and then to evaluate the restraints. This approach was used by Latek et al. [187] in connection with the CABS statistical force field [188]. Another approach is to estimate the proton-proton distance based on the site-site distances (e.g., by adding 2 Å to the $C^{\alpha} \cdots C^{\alpha}$ distances to obtain the upper limit). This approach is used with the UNRES force field [189]. In that implementation, the dihedral-angle restraints are also considered, by converting the (ϕ, ψ) backbone angles to the $C^{\alpha} \cdots C^{\alpha} \cdots C^{\alpha} \cdots C^{\alpha}$ backbone virtual-bond-dihedral angles (γ), by using the formulas derived by Nishihawa et al. [190]. The flat-bottom function defined by Eq. 2.17 (ref [141]) is used to handle possible false assignments and another extension has been introduced to treat multiple peak assignments (Eq. 2.19).

$$V_{NMR}(\{d\}; d_l, d_u, A) = -\frac{1}{\alpha} \ln \left\{ \sum_{i=1}^{n_{amb}} \exp[-\alpha V(d_i; d_l, d_u, A)] \right\} \tag{2.19}$$

where $\{d\}$ is the set of distances potentially corresponding to a given ambiguous restraint, d_l and d_u being the lower and the upper distance limits, respectively, and

A is the wall-height of the restraint (cf. Eq. 2.17). With α large enough, this function takes a value of nearly 0 independent of whether one only or all restraints of the ambiguous set are satisfied, thus naturally eliminating the restraints of an ambiguous set, which are incompatible with the structure.

2.3.6 Contact-Distance and Template-Based Restraints

Contact-assisted simulations play an increasing role in modeling protein [191, 192] and RNA structure [193]. The contacts can be obtained from XL-MS, fluorescence-resonance spectroscopy or NMR experiments, as described in Sects. 2.3.5.1 and 2.3.5.4, respectively. However, most of the contact information is acquired through contact prediction [194–196]. One of the most successful methods of contact prediction and implementation of the predicted contacts in protein-structure modeling has been developed by the Zhang group [195, 197, 198]. In this approach, weakly scoring threading templates are reordered by the structural similarity to the *ab initio* folding models, which are then reassembled [197] based on fragment-assembly simulations. The robustness of this approach can stem from the composite fragment-assembly simulations that combine structures from both *ab initio* folding and threading template refinements. Nevertheless, long-range β-strand folding still remains a challenge [195, 198]. This method is used in the I-TASSER [197] and Zhang servers [195].

In the approach used in the CoinFold server [194], a better contact-prediction accuracy was achieved, without using any templates, owing to the integration of a joint multi-family evolutionary coupling analysis and supervised machine learning. This approach is unique because it uses residue coevolution information in the target protein family, and also the related families which may have divergent sequences but similar folds. The supervised learning further improves contact prediction accuracy by making use of sequence profile, contact (distance) potential and other information [194].

In our laboratory, we developed a method of utilizing contact prediction, in which it is assumed that some of the predicted contacts can be false [141]. For this purpose, we implemented the flat-bottom restraint function defined by Eq. 2.17. This approach scored some success in the CASP13 exercise [189].

The deep neural networks are also used in DeepCDpred (deep contact distance prediction) to predict inter-residue contacts and distances. The distances are predicted with a higher accuracy than that of most of contact prediction techniques. The addition of distance restraints improved *de novo* prediction of protein structures, compared to using the best contact prediction methods alone. Moreover, the use of distance prediction allows for the selection of better models without an external model-assessment tool [196].

Recently, another promising approach, termed AlphaFold [199, 200], has been proposed, which is based on estimating residue-residue distance distribution through machine learning. This approach outperformed other methods in the CASP13 experiment. In AlphaFold, a neural network is trained to make accurate predictions of residue-residue distances, which convey more information about the structure than contact predictions. Using this information, a pseudopotential is constructed. The authors claim that local gradient minimization of this potential plus local terms leads to the final structure, without the need of using sophisticated sampling algorithms. The method has achieved high accuracy, even for targets with weak homology to known structures [200].

2.4 Examples of Physics-Based CG Force Fields and Their Applications

In this section, selected physics-based force fields are characterized briefly. For more extensive reviews of physics-based and statistical force fields the reader is referred to the literature [4, 12, 23, 28, 29, 32–34, 36, 201]. The coarse-graining schemes of the models discussed below are shown in Fig. 2.2.

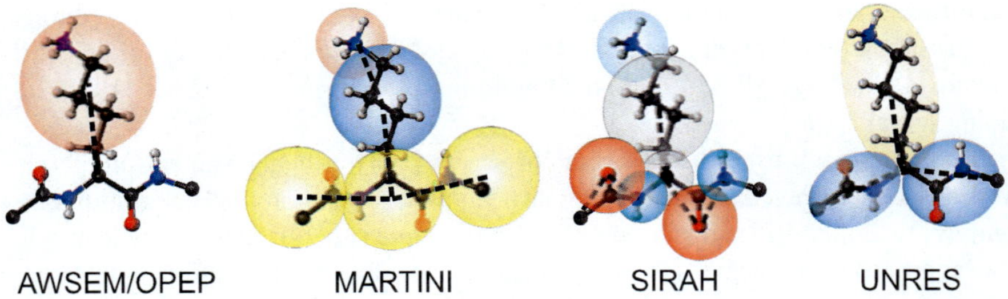

Fig. 2.2 Illustration of the CG schemes implemented in the AWSEM [201], OPEP [202], MARTINI [31], SIRAH [57], and UNRES [48] models of polypeptide chains with the example of an L-lysine residue in the context of the immediate backbone surroundings. Each model is labelled at the bottom of the respective panel. The interaction sites are represented with colored spheres or spheroids centered at the centers of the sites and covering the atoms that are assigned to a site. The virtual bonds linking the site centers are shown as black dashed lines. In AWSEM and OPEP, the backbone is in the all-atom representation, while the side chain is represented as a single spherical particle. In MARTINI, a single spherical particle of polar type encompasses the H, N, C^{α}, H^{α}, C', and O backbone atoms of a given residue, an apolar spherical particle contains the carbon and hydrogen atoms of the side chain, while the protonated amino group is represented by a charged particle. In SIRAH, the division of the side chain is similar as in MARTINI, while the backbone is split into the C^{α}-H, N-H, and C'=O spherical particles. In UNRES, the backbone site encompasses the C', O, N, and H atoms of a peptide group, while the side-chain site encompasses all side-chain atoms and the C^{α}-H^{α} group. The peptide-group and side-chain particles in UNRES have the axial and not the spherical symmetry

2.4.1 AWSEM

The Associative memory, Water mediated, Structure and Energy Model (AWSEM) [201] has been developed in the Wolynes and Papoian labs. The force field applies to polypeptide chains. The geometry of the chain is defined in terms of C^{α}, C^{β} and backbone carboxyl-oxygen atoms, which are interaction sites; the positions of backbone N and H atoms are calculated assuming ideal trans-peptide-group geometry. The energy function is expressed by Eq. 2.20.

$$U_{total} = U_{backbone} + U_{contact} + U_{burial} + U_{HB} + U_{AM} + U_{DSB} \tag{2.20}$$

where $U_{backbone}$ is the backbone potential, which consists of the connectivity, Ramachandran, chirality, and excluded-volume terms, $U_{contact}$ is the contact potential that consists of the pairwise $C^{\beta} \cdots C^{\beta}$ (C^{β} replaced with C^{α} for Gly) and the water-mediated potentials. This potential was optimized to maximize the ratio of the folding-transition to the glass-transition temperature (T_f/T_g) [203]. U_{burial} is the many-body residue-burial potential, U_{HB} is the hydrogen-bonding potential that depends on the oxygen-nitrogen and oxygen-hydrogen distances, peptide-group orientation, and an explicit helical term. U_{AM} is the knowledge-based associative-memory term [204], which encodes alignments to proteins with known structures, and U_{DSB} is the desolvation-barrier term. Most of these terms are sums of Gaussians or Heaviside-like functions in the respective geometric parameters.

The conformational-search engine in AWSEM is coarse-grained molecular dynamics, which is performed using the LAMMPS [205] general MD package, into which AWSEM has been integrated. AWSEM has been used successfully in protein-structure prediction and simulations of protein folding and assembly [206], including protein aggregation [207].

2.4.2 MARTINI

MARTINI [31] is probably the most popular CG force field. Originally designed to simulate lipid systems [50], it has been extended to proteins [27], polysaccharides [208], and nucleic acids [209]. A big advantage of MARTINI is a standardized coarse-graining scheme, in which chain fragments are coarse-grained into sites comprising 4 non-hydrogen atoms on average, while rings are divided into 3-atom fragments. Depending on character, each CG particle is assigned a polar (P), nonpolar (N), apolar (C) or charged (Q) type, with standardized parameters. This scheme enables automatic coarse-graining without user intervention. Water and lipid environments are treated at granular level except for the variant termed dry-MARTINI [62]. Four or eight water molecules are assembled into an extended site.

The MARTINI force field is of neoclassical type, with the same kind of terms as in all-atom force fields, as given by Eq. 2.21.

$$U_{total} = U_{bond} + U_{bend} + U_{tor} + U_{nb} + U_{coul} \tag{2.21}$$

where the consecutive terms stand for the virtual-bond-stretching, virtual-bond-angle-bending, virtual-bond-torsional, nonbonded, and Coulomb-electrostatic terms, respectively.

The conformational search with MARTINI is performed by means of molecular dynamics. MARTINI was originally developed with the GROMACS MD suite [210] but is also used with other standard MD packages.

MARTINI has been used worldwide to simulate a variety of biological systems [31]. It has also been successfully applied in scoring the docked protein structures [211] and in protein-RNA docking [212]. However, the force field is not predictive and, consequently, secondary-structure restraints have to be imposed when simulating systems containing proteins and nucleic acids.

2.4.3 OPEP and HiRe-RNA

Optimized Potential for Efficient protein structure Prediction (OPEP) [30, 202, 213] is a CG force field, which was initially developed to perform simulations of polypeptides. Like in AWSEM, the all-atom representation of polypeptide backbone is used, while each side chain is represented by a single spherical bead. The effective energy function consists of local (bond-stretching, bond-angle, torsional) and long-range term, which include the sidechain-sidechain contact potentials and a sophisticated backbone-hydrogen-bonding term, which reflect the nearly-linear arrangement of the N-H$\cdots$O groups as observed in experimental structures. The H-bond potentials also include four-body terms that promote the formation of regular hydrogen-bond patterns. The force field has been parameterized using a combination of bottom-up and top-down approaches, (cf. Sect. 2.2.4). A variety of conformational-search techniques are used, including Monte Carlo, molecular dynamics and their replica-exchange extensions [30]. Recently [214], the model has been extended to run constant-pH simulations, which is accomplished by Monte Carlo sampling of protonation states.

Owing to careful design and parameterization, unrestrained folding simulations can be performed with OPEP. Based on OPEP, the PEP-FOLD3 server [215] was created, which enables the user to run peptide- and small protein folding simulations.

Based on the concept of OPEP, the HiRe-RNA model of nucleic acids [216] was developed, in which 4 interaction sites (P, O5', C5', C4', and C1') are assigned per backbone unit, a pyrimidine base is represented by 1 bead and a purine base is represented by 2 beads. A special orientation-dependent potential has been designed to keep paired nucleic-acid bases at appropriate geometry.

OPEP/HiRe-RNA have been used in investigating protein dynamics, including the effect of hydrodynamics interactions, small protein and RNA folding, and in investigating protein aggregation, including amyloid formation [30]. The OPEP force field, after training, was able to score structures of protein-protein complexes with higher accuracy than that of ZDOCK [217].

2.4.4 oxDNA and oxRNA

The oxDNA/oxRNA force field developed by Ouldridge, Louis and Doye [218–223] is a low-resolution DNA and RNA model with three beads per nucleotide. These sites are the backbone-repulsion site, the base-repulsion site, and the base hydrogen-bonding/stacking site, respectively, and form a linear fragment perpendicular to the direction of the chain. The interaction potentials have been engineered to reproduce base pairing but depend on the kinds of interacting bases and not on their position in the sequence. Salt effects have been included [222]. The model has been parameterized to reproduce the geometry and thermodynamics of DNA hybridization and has been used with success to model the formation and rearrangements of DNA and RNA nanostructures [223]. The model has been included in the oxDNA package [222]. Both MC and MD simulations can be carried out.

2.4.5 SIRAH

SIRAH [57, 58] is a CG force field to treat proteins, later extended to nucleic acids [224], lipids, and polysaccharides. There are 3 interaction sites per polypeptide backbone, located at the amide-N (GN), C^{α} (GC), and carbonyl-O (GO) atoms, while each side chain is represented by 1 (for Gly) to 7 (for Arg and Trp) beads. Water is explicit in the model. Each "water" particle (WT4) represents 11 tetrahedrally-coordinated water molecules and is comprised of 4 beads. Hydrated ions: Na^{+}, K^{+}, and Cl^{-} are considered explicitly, each extended ion particle comprising the actual ion and the 6 water molecules constituting the first hydration shell.

As in OPEP, the interaction potential consists of local and long-range terms. Coulombic interactions are calculated explicitly and partial charges are present on the beads of the WT4 particles, which results in the appropriate handling of hydration. The parameters have been determined to reproduce the structural and thermodynamic properties of model systems.

SIRAH has been ported to GROMACS [210] and AMBER [225] and, consequently, the conformational-search engine is molecular dynamics. It has been used in the simulations of natively-unfolded proteins and protein aggregation.

2.4.6 UNICORN

The UNIfied COarse gRaiNed model (UNICORN) [38, 48] is the most heavily coarse-grained model of those described in this section. It has been created by merging the models specific for proteins (UNRES) [39, 48, 63], nucleic acids (NARES-2P) [226], and for polysaccharides (SUGRES-1P) [48, 74, 75]. Recently, its extension to treat protein-nucleic-acid complexes has been developed [227]. For each macromolecule type, one bead is placed in the middle of the backbone virtual bond (peptide group, phosphate group or sugar ring, respectively), and one on the side group, if applicable. The respective potentials have been derived by using the recently developed scale-consistent theory [35] (see Sect. 2.2.1), which was evolving together with UNRES [35, 39, 228].

The energy function consists of local (virtual-bond-stretching, virtual-bond-angle-bending, virtual-bond-torsional, side-chain chirality), and long-range terms. The torsional terms depend both on virtual-bond-dihedral angles and on the virtual-bond angles, which is a consequence of the application of the scale-consistent theory [35]. The long-range terms are distance- and orientation-dependent [35, 39, 48, 63] and they also include the terms that couple the long-range and local interactions. Owing to these features, UNICORN can model regular secondary-structure patterns without engineering specific potentials. Solvent is implicit in the interaction potentials. The force field has recently been extended to lipid-bilayer environment, which is treated as a continuous medium [229].

The potentials were parameterized by using a combination of the bottom-up and top-down approaches (see Sect. 2.2.4) [35, 48, 94, 226]. The latest version of UNRES was calibrated with the maximum-likelihood method, by using a set of 9 training proteins with various structural classes [38]. NARES-2P was calibrated with the structural and melting-thermodynamics data of small DNA molecules [84]. UNICORN is probably the only coarse-grained force field in which the effective energy function depends on temperature [47], which is a consequence of the fact that a CG energy function is a simplified and parameterized potential of mean force and not potential energy (Eq. 2.2.1).

Originally UNRES was used with methods of global optimization such as MCM [125] and CSA [129], which enabled us only to find the lowest-energy structures. With introducing coarse-grained molecular dynamics [40, 99, 107] and its REMD and MREMD extensions [112], it became capable of simulating protein folding, association, and functionally-important motions. MD and its extension were implemented from the beginning in NARES-2P [226] and SUGRES-1P [48, 74, 75]. The MD algorithm has been specifically designed for the model, which implies solving the equations of motion with a non-diagonal (albeit constant) inertia matrix (see Sect. 2.3.2) [99, 107].

UNRES has very good prediction capacity in unassisted mode, as proved over the years in CASP exercises [129, 134, 141, 230, 231], which is even more enhanced when calculations are run in data- [232] or bioinformatics-assisted mode [233, 234]. UNRES handles proteins with D-amino-acid residues [71, 94], phosphorylated residues [235] and also takes into account the breaking and formation of disulfide bonds during the course of simulations [236]. An extension to treat the binding of proteins to nanoparticles has also been developed [237]. Recently [238] the UNRES web server was developed with which the users can run the simulations with this force field. Inspired by the success of the CABS-Dock server for coarse-grained protein-peptide docking with the statistical CABS force field [239, 240], we have recently extended UNRES to perform peptide-protein and protein-protein docking simulations with adjustable level of flexibility, achieving satisfactory accuracy with proper sampling [241].

UNRES was used to study protein folding [242], free-energy landscapes [243], and to solve a variety of biological problems [48, 244, 245], including the formation of oligomers and fibrils of amyloidogenic peptides [246–249].

NARES-2P not only reproduces the double-helix structure and folding thermodynamics of DNA and RNA molecules but also the pre-melting transition in DNA [84, 226]. With limited restraints, NARES-2P is capable of modeling complex DNA and RNA folds [131]. NARES-2P was used to investigate telomere stability [250] and the influence of single-strand breaks on the mechanical stability of various DNA chains [251].

SUGRES-1P is still under development [74, 75]. It is being extended [74] to include glycosaminoglycans (GAGs), which are not present in other CG force fields despite their high biological significance. These molecules constitute a special class of linear anionic polysaccharides comprised of disaccharide periodic units that contain an uronic acid and an N-acetylated aminosugar [252]. GAGs are located in the extracellular matrix, where they participate in a number of biologically relevant processes as cell signaling, cell proliferation, cell adhesion, anticoagulation and angiogenesis, by interacting with their protein targets such as collagen [253], growth factors [254], chemokines [255], cathepsins [256], etc. Due to the diversity of these compounds and lack of definite structure, the best way to learn about their interactions with the target proteins and, consequently, the details of their biological functions is through large-scale physics-based coarse-grained simulations.

2.4.7 Structure-Based and Elastic-Network Potentials

The structure-based models, also known as Gō-like models [257, 258], employ a simple interaction model in which the native structure is, by design, the global energy minimum. The long-range interaction potentials between residues that are at contact in the native structure possess a minimum and are modeled by the Lennard-Jones formula. Conversely, the interaction potentials for the pairs of residues, which are not at contact in the native structure do not possess a minimum and, consequently, consist only of the repulsive terms. These models have been used to study protein folding [259–261], the underlying assumption being that the folding pathway is determined by the native-structure topology. Despite their simplicity, the Gō-like models were used to solve important biological problems such as the folding and unfolding of knotted proteins [262] or the mechanostability of virus capsids [263]. In other structure-based models, specific restraints are imposed only on secondary structure [264, 265]. An advantage of the Gō-like models is that they are not expensive computationally and can readily be constructed for a given system.

In the elastic-network models, harmonic or anharmonic (sometimes double-well) potential is imposed on the native structure. Consequently, this method is used to study the fluctuations around the native structures, in particular finding the functionally-important motions [266–268].

2.5 Conclusions and Outlook

Since the original idea of Levitt and Warshel [43], coarse-grained approaches have evolved tremendously and have become a necessary tool with which to model nanoscale systems, including the biological systems [4, 12, 36], enabling the extension of the time- and size-scale of simulations by 3 and more orders of magnitude [12]. Physics-based CG force fields are gradually becoming more popular, which is reflected in the increasing number of workshops and conferences dedicated to their development and applications, e.g., the CECAM workshops in Leiden (The Netherlands), Tel Aviv (Israel), and Clifden (Ireland) in year 2019.

Most of the coarse-grained force fields are still being developed by analogy to the all-atom force fields [12, 31, 36], this resulting in their reduced modeling capacity. However, with carefully implemented restraints on the local (secondary) structure and distances, it is practical to run large-scale simulations with the CG models. Also, despite their simplicity and limitations, the structure-based CG models (the Gō-like and the elastic-network models) are very useful in simulating the structural transformations and dynamics of large systems.

The recently developed scale-consistent theory of the construction of coarse-grained force fields [35, 38], with the resulting analytical expressions embedding the averaged out atomistic details of a system and, thereby, achieving correct dependence of the effective energy expressions on orientation and coupling of the long-range and local interactions necessary to reproduce regular structures, is likely to result in the development of more powerful and accurate CG force fields. Moreover, owing to their rooting in the potential of mean force, the energy expressions derived in such a way will enable us to bridge the coarse-grained representation with the all-atom representation of a system, thus making possible physics-based multiscale simulations.

Apart from the development of the CG force fields and associated methodologies, evaluation of existing approaches to determine their shortcoming, directions of improvement, and needs of the simulation community is necessary. CASP [8] is probably the most extensive initiative undertaken in this direction, which has been followed by the CAPRI [9] and RNA-Puzzles [10] experiments. With the addition of other bio- and nano-system and also mechanical and thermodynamical properties to compare with their simulated counterparts, these initiatives have the potential to result in a standard system of force-field evaluation.

Acknowledgments This work was supported by grants UMO-2017/25/B/ST4/01026, UMO-2017/27/B/ST4/00926, UMO-2017/26/M/ST4/00044, UMO-2018/30/E/ST4/00037, UMO-2015/17/N/ST4/03935, and UMO-2015/17/N/ST4/03937 from the National Science Center of Poland (Narodowe Centrum Nauki). Calculations were carried out using the computational resources provided by (a) the supercomputer resources at the Centre of Informatics—Tricity Academic Supercomputer & networK (CI TASK) in Gdańsk, (b) the supercomputer resources at the Interdisciplinary Center of Mathematical and Computer Modeling (ICM), University of Warsaw (grant GA71-23), (c) the Polish Grid Infrastructure (PL-GRID; grants unres19, unres2021, and gagstr), and (d) our 488-processor Beowulf cluster at the Faculty of Chemistry, University of Gdańsk.

References

1. Frenkel D, Smit B (2000) Understanding molecular simulation: from algorithms to applications. Academic Press, New York
2. Mackerell, Jr AD (2004) J Comput Chem 25:1584–1604
3. Leach AR (2010) Molecular modeling: principles and applications. Pearson Education Limited, Harlow
4. Voth G (2008) Coarse-graining of condensed phase and biomolecular systems, 1st edn. CRC Press, Taylor & Francis Group, Boca Raton
5. Peter C, Kremer K (2009) Soft Matter 5:4357–4366
6. Rapaport DC (2011) The art of molecular dynamics simulations. Cambridge University Press, Cambridge
7. Liwo A (2018) Computational methods to study the structure and dynamics of biomolecules and biomolecular processes from bioinformatics to molecular quantum mechanics. Springer Nature Switzerland AG, Cham
8. Kryshtafovych A, Schwede T, Topf M, Fidelis K, Moult J (2019) Proteins 87:1011–1020
9. Lensink MF, Brysbaert G, Nadzirin N, Velankar S, Chaleil RAG, Gerguri T, Bates PA, Laine E, Carbone A, Grudinin S, Kong R, Liu RR, Xu X-M, Shi H, Chang S, Eisenstein M, Karczynska A, Czaplewski C, Lubecka E, Lipska A, Krupa P, Mozolewska M, Golon Ł, Samsonov S, Liwo A, Crivelli S, Yan Y, Huang S-Y, Rosell M, Rodríguez-Lumbreras LA, Romero-Durana M, Díaz-Bueno L, Fernandez-Recio J, Christoffer C, Terashi G, Shin W-H, Aderinwale T, Raghavendra S, Subraman MV, Kihara D, Kozakov D, Vajda S, Porter K, Padhorny D, Desta I, Beglov D, Ignato M, Kotelnikov S, Moal IH, Ritchie DW, Chauvot de Beauchêne I, Maigret B, Devignes M-D, Ruiz Echartea ME, Barradas-Bautista D, Cao Z, Cavallo L, Oliva R, Cao Y, Shen Y, Baek M, Park T, Woo H, Seok C, Scheidman D, Dapkūnas J, Olechnovič K, Venclovas Č, Kundrotas PJ, S Belkin, Chakravarty D, Badal VD, Vakser IA, Vreven T, Vangaveti S, Borrman T, Weng Z, Guest JD, Gowthaman R, Pierce BG, Xu X, Duan R, Qiu L, Hou J, Merideth BR, Ma Z, Cheng J, Zou X, Wodak SJ (2019) Proteins 87:1200–1221
10. Miao Z, Adamiak RW, Antczak M, Batey RT, Becka AJ, Biesiada M, Boniecki M, Bujnicki JM, Chen SJ, Cheng CY, Chou FC, Ferré-D'Amaré AR, Das R, Dawson WK, Ding F, Dokholyan NV, Dunin-Horkawicz S, Geniesse C, Kappel K, Kladwang W, Krokhotin A, Łach GE, Major F, Mann TH, Magnus M, Pachulska-Wieczorek K, Patel DJ, Piccirilli JA, Popenda M, Purzycka KJ, Ren A, Rice GM, Santalucia J, Sarzynska J, Szachniuk M, Tandon A, Trausch JJ, Tian S, Wang J, Weeks KM, Williams B, Xiao Y, Xu X, Zhang D, Zok T, Westhof E (2017) RNA 23:655–672

11. Bardwell DA, Adjimanxi CS, Arnautova YA, Bartashevich E, Boerrigter SXM, Braun DE, Cruz-Cabeza AJ, Day GM, Valle RGDella, Desiraju GR, van Eijck BP, Facelli JC, Ferraro MB, Grillo D, Habgood M, Hofmann DWM, Hofmann F, Jose KVJ, Karamertzanis PG, Kazantsev AV, Kendrick J, Kuleshova LN, Leusen FJJ, Maleev AV, Misquitta AJ, Mohamed S, Needs RJ, Neumann MA, Nikylov D, Orendt AM, Pal R, Pickard CC Pantelides CJ, Price LS, Price SL, Scheraga HA, van de Streek J, Thakur TS, Tiwari S, Venuti E, Zhitkov IK (2011) Acta Cryst B 67:535–551
12. Kmiecik S, Gront D, Koliński M, Wieteska L, Dawid A, Koliński A (2016) Chem Rev 116:7898–7936
13. Atkins P, Friedman R (2010) Molecular quantum mechanics. Oxford University Press, Oxford
14. Parr RG, Yang W (1989) Density-functional theory of atoms and molecules. Oxford University Press, Oxford
15. Stewart JJ (1990) J Comput Aided Mol Des 4:1–105
16. Friedrichs MS, Eastman P, Vaidyanathan V, Houston M, Legrand S, Beberg AL, Ensign DL, Bruns CM, Pande VS (2009) J Comput Chem 30:864–872
17. Pande VS, Baker I, Chapman J, Elmer S, Kaliq S, Larson SM, Rhee YM, Shirts MR, Snow CD, Sorin EJ, Zagrovic B (2003) Biopolymers 68:91–109
18. Shaw DE, Deneroff MM, Dror RO, Kuskin JS, Larson RH, Salmon JK, Young C, Batson B, Bowers KJ, Chao JC, Eastwood MP, Gagliardo J, Grossman JP, Ho CR, Ierardi DJ, Kolossvary I, Klepeis JL, Layman T, Mcleavey C, Moraes MA, Mueller R, Priest EC, Shan Y, Spengler J, Theobald M, Towles B, Wang SC (2008) Commun ACM 51:91–97
19. Shaw DE, Maragakis P, Lindorff-Larsen K, Piana S, Dror RO, Eastwood MP, Bank JA, Jumper JM, Salmon JK, Shan Y, Wriggers W (2010) Science 330:341–346
20. Kolinski A, Skolnick J (1992) J Chem Phys 97:9412–9426
21. Kolinski A, Skolnick J (2004) Polymer 45:511–524
22. Liwo A, Pincus MR, Wawak RJ, Rackovsky S, Ołdziej S, Scheraga HA (1997) J Comput Chem 18:874–887
23. Tozzini V (2005) Curr Opinion Struct Biol 15:144–150
24. Ayton GS, Noid WG, Voth GA (2007) Curr Opinion Struct Biol 17:192–198
25. Voltz K, Trylska J, Tozzini V, Kurkal-Siebert V, Langowski J, Smith J (2008) J Comput Chem 29:1429–1439
26. Clementi C (2008) Curr Opinion Struct Biol 18:10–15
27. Monticelli L, Kandasamy SK, Periole X, Larson RG, Tieleman DP, Marrink S-J (2008) J Chem Theory Comput 4:819–834
28. Czaplewski C, Liwo A, Makowski M, Ołdziej S, Scheraga HA (2010) Coarse-grained models of proteins: theory and applications. In: Koliński A (ed) Multiscale approaches to protein modeling, Chapter 3. Springer, Berlin, pp 35–83
29. Sauders MG, Voth GA (2013) Annu Rev Biophys 42:73–93
30. Sterpone F, Melchionna S, Tuffery P, Pasquali S, Mousseau N, Cragnolini T, Chebaro Y, St-Pierre J-F, Kalimeri M, Barducci A, Laurin Y, Tek A, Baaden M, Phuong HN, Derreumaux P (2014) Chem Soc Rev 43:4871–4893
31. Marrink SJ, Tieleman DP (2013) Chem Soc Rev 42:6801–6822
32. Ingolfsson HI, Uusitalo JJ, de Jong DH, Gopal S, Periole X, Marrink S-J (2014) WIREs Comput Mol Sci 4:225–248
33. Kar P, Feig M (2014) Adv Protein Chem Struct Biol 96:143–180
34. Papoian GA (2017) Coarse-grained modeling of biomolecules. CRC Press, Boca Raton
35. Sieradzan AK, Makowski M, Augustynowicz A, Liwo A (2017) J Chem Phys 146:124106
36. Singh N, Li W (2019) Int J Mol Sci 20:3774
37. Mirzoev A, Nordenskiöld L, Lyubartsev A (2019) Comput Phys Commun 237:263–273
38. Liwo A, Czaplewski C, Sieradzan AK, Lubecka EA, Lipska AG, Golon Ł, Karczyńka A, Krupa P, Mozolewska MA, Makowski M, Ganzynkowicz R, Giełdoń A, Maciejczyk MD (2020) Scale-consistent approach to the derivation of coarse-grained force fields for simulating structure, dynamics, and thermodynamics of biopolymers. In: Strodel B, Barz B (eds.) Progress in molecular biology and translational science computational approaches

for understanding dynamical systems: protein folding and assembly, vol 170, chapter 2. Academic Press, London, pp 73–122
39. Liwo A, Czaplewski C, Pillardy J, Scheraga HA (2001) J Chem Phys 115:2323–2347
40. Khalili M, Liwo A, Jagielska A, Scheraga HA (2005) J Phys Chem B 109:13798–13810
41. Murarka RK, Liwo A, Scheraga HA (2007) J Chem Phys 127:155103
42. Senn MH, Thiel W (2009) Angew Chem Int Ed 48:1198–1229
43. Levitt M, Warshell A (1975) Nature 253:694–698
44. Levitt M (1976) J Mol Biol 104:59–107
45. Karplus M (2014) Angew Chem Int Ed 53:9992–10005
46. Ben Naim A (1997) J Chem Phys 107:3698–3706
47. Liwo A, Khalili M, Czaplewski C, Kalinowski S, Ołdziej S, Wachucik K, Scheraga HA (2007) J Phys Chem B 111:260–285
48. Liwo A, Baranowski M, Czaplewski C, Gołaś E, He Y, Jagieła D, Krupa P, Maciejczyk M, Makowski M, Mozolewska MA, Niadzvedtski A, Ołdziej S, Scheraga HA, Sieradzan AK, Ślusarz R, Wirecki T, Yin Y, Zaborowski B (2014) J Mol Model 20:2306
49. Izvekov S, Voth GA (2005) J Phys Chem B 109:2469–2473
50. Marrink SJ, Risselada HJ, Yefimov S, Tieleman DP, de Vries AH (2007) J Phys Chem B 111:7812–7824
51. Gay JG, Berne BJ (1981) J Chem Phys 74:3316–3319
52. Kubo R (1962) J Phys Soc Japan 17:1100–1120
53. Ball P (2008) Chem Rev 108:74–108
54. Buckle AM, Henrick K, Fersht AR (1993) J Mol Biol 234:847–860
55. Rhee YM, Sorin EJ, Jayachandran G, Lindahl E, Pande VS (2004) Proc Natl Acad Sci USA 101:6456–6461
56. Ladbury JE (1996) Chem Biol 3:973–980
57. Darré L, Machado MR, Brandner AF, González HC, Ferreira S, Pantano S (2015) J Chem Theory Comput 11:723–739
58. Machado MR, Barrera EE, Klein F, Sóñora M, Silva S, Pantano S (2019) J Chem Theory Comput 15:2719–2733
59. Elezgaray J, Laguerre M (2006) Comp Phys Commun 175:264–268
60. Yesylevskyy SO, Schäfer LV, Sengupta D, Marrink S-J (2010) PLOS Comput Biol 6:e1000810
61. Hadley KR, McCabe C (2012) Mol Simul 38:671–681
62. Arnarez C, Uusitalo JJ, Masman MF, Ingólfsson HI, de Jong DH, Melo MN, Periole X, de Vries AH, Marrink S-J (2015) J Chem Theory Comput 11:260–275
63. Liwo A, Ołdziej S, Pincus MR, Wawak RJ, Rackovsky S, Scheraga HA (1997) J Comput Chem 18:849–873
64. Bashford D, Case DA (2000) Annu Rev Phys Chem 51:129–152
65. Makowski M (2018) Physics-based modeling of side chain-side chain interactions in the UNRES force field. In: Liwo A (ed) Computational methods to study the structure and dynamics of biomolecules and biomolecular processes from bioinformatics to molecular quantum mechanics. Springer Nature Switzerland AG, Cham, pp 89–115
66. Makowski M, Liwo A, Scheraga HA (2007) J Phys Chem B 111:2910–2916
67. Makowski M, Sobolewski E, Czaplewski C, Ołdziej S, Liwo A, No JH, Scheraga HA (2007) J Phys Chem B 111:2925–2931
68. Makowski M, Liwo A, Scheraga HA (2017) J Phys Chem B 121:379–390
69. Pincus MR, Scheraga HA (1977) J Phys Chem 81:1579–1583
70. Sieradzan AK, Scheraga HA, Liwo A (2012) J Chem Theor Comput 8:1334–1343
71. Sieradzan AK, Niadzvedtski A, Scheraga HA, Liwo A (2014) J Chem Theory Comput 10:2194–2203
72. Kozłowska U, Maisuradze GG, Liwo A, Scheraga HA (2010) J Comput Chem 31:1154–1167
73. Kumar S, Bouzida D, Swendsen RH, Kollman PA, Rosenberg JM (1992) J Comput Chem 13:1011–1021

74. Samsonov SA, Lubecka EA, Bojarski KK, Ganzynkowicz R, Liwo A (2019) Biopolymers 110:e23269
75. Lubecka EA, Liwo A (2017) J Chem Phys 147:115101
76. Izvekov S, Voth GA (2005) J Chem Phys 123:134105
77. Thorpe IF, Goldenberg DP, Voth GA (2011) J Phys Chem B 115:11911–11926
78. Liwo A, Czaplewski C (2020) J Chem Phys 152:054902
79. Soper AK (1996) Chem Phys 202:295–306
80. Reith D, Püt M, Müller-Plathe F (2003) J Comput Chem 24:1624–1636
81. Lyubartsev AP, Laaksonen A (1995) Phys Rev E 52:3730–3737
82. Lyubartsev AP, Naômé A, Vercauteren DP, Laaksonen A (2015) J Chem Phys 143:243120
83. Mirzoev A, Lyubartsev AP (2013) J Chem Theory Comput 9:1512–1520
84. He Y, Liwo A, Scheraga HA (2015) J Chem Phys 143:243111
85. Brungelson JD, Wolynes PG (1987) Proc Natl Acad Sci USA 84:7524–7528
86. Eastwood MP, Hardin C, Luthey-Schulten Z, Wolynes PG (2002) J Chem Phys 117:4602–4615
87. Eastwood MP, Hardin C, Luthey-Schulten Z, Wolynes PG (2003) J Chem Phys 118:8500–8512
88. Fujitsuka Y, Takada S, Luthey-Schulten ZA, Wolynes PG (2004) Proteins Struct Funct Genet 54:88–103
89. Seetharamulu P, Crippen GM (1991) J Math Chem 6:91–110
90. Ołdziej S, Liwo A, Czaplewski C, Pillardy J, Scheraga HA (2004) J Phys Chem B 108:16934–16949
91. Ołdziej S, Łągiewka J, Liwo A, Czaplewski C, Chinchio M, Nanias M, Scheraga HA (2004) J Phys Chem B 108:16950–16959
92. Zaborowski B, Jagieła D, Czaplewski C, Hałabis A, Lewandowska A, Żmudzińska W, Ołdziej S, Karczyńska A, Omieczynski C, Wirecki T, Liwo A (2015) J Chem Inf Model 55:2050–2070
93. Krupa P, Hałabis A, Żmudzińska W, Ołdziej S, Scheraga HA, Liwo A (2017) J Chem Inf Model 57:2364–2377
94. Liwo A, Sieradzan AK, Lipska AG, Czaplewski C, Joung I, Żmudzińska W, Hałabis A, Ołdziej S (2019) J Chem Phys 150:155104
95. Wales DJ, Scheraga HA (1999) Science 285:1368–1372
96. Liwo A, Czaplewski C, Ołdziej S, Rojas AV, Kaźmierkiewicz R, Makowski M, Murarka RK, Scheraga HA (2008) Simulation of protein structure and dynamics with the coarse-grained UNRES force field. In: Voth G (ed) Coarse-graining of condensed phase and biomolecular systems, Chapter 8. CRC Press, Boca Raton, pp 1391–1411
97. Metropolis N, Rosenbluth AW, Rosenbluth MN, Teller AH, Teller E (1953) J Chem Phys 21:1087–1092
98. Kolinski A, Godzik A, Skolnick J (1993) J Chem Phys 98:7420–7433
99. Khalili M, Liwo A, Rakowski F, Grochowski P, Scheraga HA (2005) J Phys Chem B 109:13785–13797
100. Paterlini MG, Ferguson DM (1998) Chem Phys 236:243–252
101. Swope WC, Anderson HC, Berens PH, Wilson KR (1982) J Chem Phys 76:637–649
102. Hoover WG (1985) Phys Rev A 31:1695–1697
103. Nosé S (2001) J Phys Soc Jpn 70:75–77
104. Tuckerman M, Berne BJ, Martyna GJ (1992) J Chem Phys 97:1990–2001
105. Martyna GJ, Tuckerman ME, Tobias DJ, Klein ML (1996) Molec Phys 87:1117–1157
106. Ciccotti G, Kalibaeva G (2004) Phil Trans R Soc Lond A 362:1583–1594
107. Rakowski F, Grochowski P, Lesyng B, Liwo A, Scheraga HA (2006) J Chem Phys 125:204107
108. Hansmann UHE, Okamoto Y (1994) Physica A 212:415–437
109. Sugita Y, Okamoto Y (2000) Phys Rev Lett 329:261–270
110. Shen H, Liwo A, Scheraga HA (2009) J Phys Chem B 113:8738–8744
111. Rhee YM, Pande VS (2003) Biophys J 84:775–786
112. Czaplewski C, Kalinowski S, Liwo A, Scheraga HA (2009) J Chem Theor Comput 5:627–640

113. Fukunishi H, Watanabe O, Takada S (2002) J Chem Phys 116:9058–9067
114. Lee KH, Chen J (2015) J Comput Chem 37:550–557
115. Karczyńska AS, Czaplewski C, Krupa P, Mozolewska MA, Joo K, Lee J, Liwo A (2017) J Comput Chem 38:2730–2746
116. Berg BA, Neuhaus T (1992) Phys Rev Lett 68:9–12
117. Lee J (1993) Phys Rev Lett 71:211–214
118. Nanias M, Czaplewski C, Scheraga HA (2006) J Chem Theory Comput 2:513–528
119. Scheraga HA, Lee J, Pillardy J, Ye Y-J, Liwo A, Ripoll DR (1999) J Glob Optim 15:235–260
120. Scheraga HA, Pillardy J, Liwo A, Lee J, Czaplewski C, Ripoll DR, Wedemeyer WJ, Arnautova YA (2002) J Comput Chem 23:28–34
121. Piela L, Kostrowicki J, Scheraga HA (1989) J Phys Chem 93:3339–3346
122. Kostrowicki J, Scheraga HA (1992) J Phys Chem 96:7442–7449
123. Amara P, Hsu D, Straub JE (1993) J Phys Chem 97:6715–6721
124. Pillardy J, Liwo A, Groth M, Scheraga HA (1999) J Phys Chem 103:7353–7366
125. Li Z, Scheraga HA (1987) Proc Natl Acad Sci USA 84:6611–6615
126. Liwo A, Pincus MR, Wawak RJ, Rackovsky S, Scheraga HA (1993) Protein Sci 2:1715–1731
127. Wales DJ (1992) J Chem Soc Faraday Trans 88:653–657
128. Androulakis IP, Maranas CD, Floudas CA (1995) J Glob Optimiz 11:337–363
129. Lee J, Scheraga HA (1999) Int J Quant Chem 75:255–265
130. Lee J, Liwo A, Scheraga HA (1999) Proc Natl Acad Sci USA 96:2025–2030
131. Sieradzan AK, Golon Ł, Liwo A (2018) Phys Chem Chem Phys 20:19656–19663
132. Lee J, Liwo A, Ripoll DR, Pillardy J, Scheraga HA (1999) Proteins: Struct Funct Genet Suppl 3:204–408
133. Pillardy J, Czaplewski C, Liwo A, Lee J, Ripoll DR, Kaźmierkiewicz R, Oldziej S, Wedemeyer WJ, Gibson KD, Arnautova YA, Saunders J, Ye Y-J, Scheraga HA (2001) Proc Natl Acad Sci USA 98:2329–2333
134. Ołdziej S, Czaplewski C, Liwo A, Chinchio M, Nanias M, Vila JA, Khalili M, Arnautova YA, Jagielska A, Makowski M, Schafroth HD, Kaźmierkiewicz R, Ripoll DR, Pillardy J, Saunders JA, Kang YK, Gibson KD, Scheraga HA (2005) Proc Natl Acad Sci USA 102:7547–7552
135. Koukos PI, Bonvin AMJJ (2020) J Mol Biol 432:2861–2881
136. Różycki B, Boura E (2014) J Phys Condens Matter 26:463103
137. Sali A, Berman HM, Schwede T, Trewhella J, Kleywegt G, Burley SK, Markley J, Nakamura H, Adams P, Bonvin AM, Chiu W, Peraro MD, Di Maio F, Ferrin TE, Grünewald K, Gutmanas A, Henderson R, Hummer G, Iwasaki K, Lawson CL, Johnson G, Meiler J, Marti-Renom MA, Montelione GT, M23 MNilges, Nussinov R, Patwardhan A, Rappsilber J, Read RJ, Saibil H, Schröder GF, Schwieters CD, Seidel CA, Svergun D, Topf M, Ulrich EL, Velankar S, Westbrook JD (2015) Structure 23:1156–1167
138. Sinz A, Arlt C, Chorev D, Sharon M (2015) Protein Sci 24:1193–1209
139. Leitner A, Fainiand M, Stengel F, Aebersold R (2016) Trends Biochem Sci 41:20–32
140. Heck AJR (2008) Nat Meth 5:927–933
141. Lubecka EA, Liwo A (2019) J Comput Chem 40:2164–2178
142. Joo K, Lee J, Sim S, Lee SY, Lee K, Heo S, Lee IH, Lee SJ, Lee J (2014) Proteins: Struct Funct Bioinf 82:188
143. Sieradzan AK, Jakubowski R (2017) J Comput Chem 38:553
144. Fajardo JE, Shrestha R, Gil N, Belsom A, Crivelli SN, Czaplewski C, Fidelis K, Grudinin S, Karasikov M, Karczyńska AS, Kryshtafovych A, Leitner A, Liwo A, Lubecka EA, Monastyrskyy B, Pages G, Rappsilber J, Sieradzan AK, Sikorska C, Trabjerg E, Fiser A (2019) Proteins 87:1283–1297
145. Dimura M, Peulen TO, Hanke CA, Prakash A, Gohlke H, Seidel CA (2016) Curr Opin Struct Biol 163–185
146. Hoefling M, Lima N, Haenni D, Seidel CAM, Schuler B, Grubmüller H (2011) PLOS One 6:e19791
147. Kalinin S, Felekyan S, Valeri A, MSeidel CAM (2008) J Phys Chem B 112:8361–8374

148. Wozniak A, Nottrott S, Kuhn-Holsken E, G GSchröder, Grubmüller H, Lührmann R, Seidel CAM, Oesterhelt F (2005) RNA 11:1545–1554
149. Sindbert S, Kalinin S, Nguyen H, Kienzler A, Clima L, Bannwarth W, Appel B, Müller S, Seidel CA (2011) J Am Chem Soc 133:2463–2480
150. Graewert MA, Svergun DI (2013) Curr Opin Struct Biol 23:748–754
151. Brosey CA, Tainer JA (2019) Curr Opin Struct Biol 58:197–213
152. Yang S, Park S, Makowski L, Roux B (2009) Biophys J 96:4449–4463
153. Stovgaard K, Andreetta C, Ferkinghoff-Borg J, Hamelryck T (2010) BMC Bioinf 11:429
154. Grudinin S, Garkavenko M, Kazennov A (2017) Acta Cryst D D73:449–464
155. Schneidman-Duhovny D, Hammel M, Sali A (2010) Nucleic Acids Res 38:W540–W544
156. Schneidman-Duhovny D, Hammel M, Tainer JA, Sali A (2013) Biophys J 105:962–974
157. Karczyńska AS, Mozolewska MA, Krupa P, Giełdoń A, Liwo A, Czaplewski C (2018) Proteins 86:228–239
158. Gorba C, Miyashita O, Tama F (2008) Biophys J 94:1589–1599
159. Hura GL, Hodge CD, Rosenberg D, Guzenko D, Duarte JM, Monastyrskyy B, Grudinin S, Kryshtafovych A, Tainer JA, Fidelis K (2019) Proteins 87:1298–1314
160. Rambo RP, Tainer JA (2013) Annu Rev Biophys 42:415–441
161. Svergun D, Barberato C, Koch MHJ (1995) J Appl Cryst 28:768–773
162. Petoukhov MV, Franke D, Shkumatov AV, Tria G, Kikhney AG, Gajda AG, Gorba C, Mertens HDT, Konarev PV, Svergun DI (2012) J Appl Cryst 45:342–350
163. Stuhrmann HB (1970) Acta Cryst A26:297–306
164. Zheng W, Tekpinar M (2011) Biophys J 101:2981–2991
165. Watson MC, Curtis JE (2013) J Appl Cryst 46:1171–1177
166. Merzel F, Smith JC (2002) Acta Cryst D58:242–249
167. Artemova S, Grudinin S, Redon S (2011) J Comput Chem 32:2865–2877
168. Chen P, Shevchuk R, Strnad FM, Lorenz C, Karge L, Gilles R, Stadler A, Hennig J, Hub JS (2019) J Chem Theory Comput 15:4687–4698
169. Larsen AH, Wang Y, Bottaro S, Grudinin S, Arleth L, Lindorff-Larsen K (2020) PLOS Comput Biol 16:e1007870
170. Bowerman S, Rana AS JB, Rice A, Pham GH, Strieter ER, Wereszczynski J (2017) J Chem 13:2418–2429
171. Bottaro S, Bengtsen T, Lindorff-Larsen K (2020) Integrating molecular simulation and experimental data: a bayesian/maximum entropy reweighting approach. In: Gáspari Z (ed) Structural bioinformatics methods in molecular biology, New York, , pp 219–240
172. Karaca E, Rodrigues JP G LM, Graziadei A, Bonvin AMJJ, Carlomagno T (2017) Nat Meth 14:897–902
173. Vajdos FF, Adams CW, Breece TN, Presta LG, de Vos AM, Sidhu SS (2002) J Mol Biol 320:415–428
174. Yu EW, Aires JR, McDermott G, Nikaido H (2005) J Bacteriol 187:6804–6815
175. Ashkenazi A, Presta LG, Marsters SA, Camerato TR, Rosenthal KA, Fendly BM, Capon DJ (1990) Proc Natl Acad Sci USA 87:7150–7154
176. Englander SW, Kallenbach NR (1983) Q Rev Biophys 16:521–655
177. Konermann L, Pan J, Liu Y-H (2011) Chem Soc Rev 40:1224–1234
178. Masson GR, Burke JE, Ahn NG, Anand GS, Borchers C, Brier S, Bou-Assaf GM, Engen JR, Englander SW, Faber J, Garlish R, Griffin PR, Gross ML, Guttman M, Hamuro Y, Heck AJR, Houde D, Iacob RE, Jørgensen TJD, Kaltashov IA, Klinman JP, Konermann L, Man P, Mayne L, Pascal BD, Reichmann D, Skehel M, Snijder J, Strutzenberg TS, Underbakke ES, Wagner C, Wales TE, Walters BT, Weis DD, Wilson DJ, Wintrode PL, Zhang Z, Zhengand J, Schriemer DC, Rand KD (2019) Nat Methods 16:595–602
179. Rey M, Sarpe V, Burns KM, Buse J, Baker CAH, van Dijk M, Wordeman L, Bonvin AMJJ, Schriemer DC (2014) Structure 22:1538–1548
180. Williamson MP, Havel TF, Wüthrich K (1985) J Mol Biol 182:295–315
181. Linge J, Nilges M (1999) J Biomol NMR 13:51–59
182. Rohl CA, Strauss CEM, Misura KMS, Baker D (2004) Meth Enzymol 383:66

183. Das R, Baker D (2008) Annual Rev Biochem 77:363–382
184. Kuhlman B, Baker D (2000) Proc Natl Acad Sci USA 97:10383–10388
185. Mao B, Tejero R, Baker D, Montelione G (2014) J Am Chem Soc 136:1893–1906
186. Ramelot TA, Raman S, Kuzin AP, Xiao R, Ma L-C, Acton TB, Hunt JF, Montelione GT, Baker D, Kennedy MA (2009) Proteins: Struct Funct Bioinf 75:147–167
187. Latek D, Koliński A (2011) J Comput Chem 32:536–544
188. Kolinski A (2004) Acta Biochim Pol 51:349–371
189. Lubecka EA, Karczyńska AS, Lipska AG, Sieradzan AK, Zięba K, Sikorska C, Uciechowska U, Samsonov SA, Krupa P, Mozolewska MA, Golon Ł, Giełdoń A, Czaplewski C, Ślusarz R, Ślusarz M, Crivelli SN, Liwo A (2019) J Molec Graphics Modell 92:154–166
190. Nishikawa K, Momany FA, Scheraga HA (1974) Macromolecules 7:797–806
191. Modi V, Xu Q, Adhikari S, Dunbrack Jr RL (2016) Proteins: Struct Func Bioinf 84:200–220
192. Lafita A, Bliven Sp, Kryshtafovych A, Bertoni M, Monastyrskyy B, Duarte JM, Schwede T, Capitani G (2018) Proteins 86:247–256
193. Dawson WK, Maciejczyk M, Jankowska EJ, Bujnicki J (2016) Methods 103:138–156
194. Wang S, Li W, Zhang R, Liu S, Xu J (2016) Nucl Acids Res 44:W361–W366
195. Zhang W, Yang J, He B, Walker SE, Zhang H, Govindarajoo B, Virtanen J, Xue Z, Shen H-B, Zhang Y (2016) Proteins: Struct Funct Bioinf 84:76–86
196. Ji S, Oruç T, Mead L, Rehman MF, Thomas CM, Butterworth S, Winn P (2019) PLOS One 14:e0205214
197. Jianyi Y, Zhang Y (2015) Nucl Acids Res 43:W174–W181
198. Zhang C, Mortuza SM, He B, Wang Y, Zhang Y (2018) Proteins 86:136–151
199. Senior AW, Evans R, Jumper J, Kirkpatrick J, Sifre L, Green T, Qin C, Židek A, Nelson AWR, Bridgland A, Penedones H, Petersen S, Simonyan K, Crossan S, Kohli P, Jones DT, Silver D, Kavukcuoglu K, Hassabis D (2019) Proteins 87:1141–1148
200. Senior AW, Evans R, Jumper J, Kirkpatrick J, Sifre L, Green T, Qin C, Žídek A, Nelson AWR, Bridgland A, Penedones H, Petersen S, Simonyan K, Crossan S, Kohli P, Jones DT, Silver D, Kavukcuoglu K, Hassabis D (2020) Nature 577:706–710
201. Davtyan A, Schafer NP, Zheng W, Clementi C, Wolynes PG, Papoian GA (2012) J Phys Chem B 116:8494–8503
202. Maupetit J, Tuffery P, Derreumaux P (2007) Proteins 69:394–408
203. Papoian G, Ulander J, Eastwood M, Luthey-Schulten Z, Wolynes P (2004) Proc Natl Acad Sci USA 101:3352–3357
204. Sasai M, Wolynes PG (1990) Phys Rev Lett 65:2740–2743
205. Plimpton SJ (1995) J Comput Phys 117:1–19
206. Chen M, Lin X, Lu W, Onuchic JN, Wolynes PG (2017) J Phys Chem B 121:3473–3482
207. Chen X, Chen M, Schafer NP, Wolynes PG (2020) Proc Natl Acad Sci USA 117:4125–4130
208. Lopez CA, Rzepiela A, de Vries AH, Dijkhuizen L, Hunenberger PH, Marrink SJ (2009) J Chem Theor Comput 5:3195–3210
209. Uusitalo JJ, Ingólfsson HI, Akhshi P, Tieleman DP, Marrink SJ (2015) J Chem Theory Comput 11:3932–3945
210. van der Spoel D, Lindahl E, Hess B, Groenhof G, Mark AE, Berendsen HJC (2005) J Comput Chem 26:1701–1718
211. Hou Q, Lensink MF, Heringa J, Feenstra KA (2016) PLOS One 11:e0155251
212. Honorato RV, Roel-Touris J, Bonvin AMJJ (2016) Front Mol Biosci 6:102
213. Shen Y, Maupetit J, Derreumaux P, Tufféry P (2014) J Chem Theor Comput 10:4745–4758
214. Silva FLBarroso, Sterpone F, Derreumaux P (2019) J Chem Theory Comput 15:3875–3888
215. Lamiable A, Thevenet P, Rey J, Vavrusa M, Derreumaux P, Tuffery P (2016) Nucl Acids Res 44:W449–W454
216. Cragnolini T, Laurin Y, Derreumaux P, Pasquali S (2015) J Chem Theory Comput 11:3510–3522
217. Kynast P, Derreumaux P, Strodel B (2016) BMC Biophys 9:4
218. Ouldridge TE, Louis AA, Doye JPK (2010) Phys Rev Lett 104:178101
219. Ouldridge TE, Louis AA, Doye JPK (2011) J Chem Phys 134:085101

220. Šulc P, Romano F, Ouldridge TE, Rovigatti L, Doye JPK, Louis AA (2012) J Chem Phys 137
221. Šulc P, Romano F, Ouldridge TE, Doye JPK, Louis AA (2014) J Chem Phys 140:235102
222. Snodin BEK, Randisi F, Mosayebi M, Šulc P, Schreck JS, Romano F, Ouldridge TE, Tsukanov R, Nir E, Louis AA, Doye JPK (2015) J Chem Phys 142:234901
223. Snodin BEK, Romano F, Rovigatti L, Ouldridge TE, Louis AA, Doye JPK (2016) ACS Nano 10:1724–1737
224. Brandner A, Schüller A, Melo F, Pantano S (2018) Biochem Biophys Res Commun 498:319–326
225. Pearlman DA, Case DA, Caldwell JW, Ross WS, Cheatham TE, III, DeBolt S, Ferguson D, Seibel G, Kollman P (1995) Comp Phys Commun 91:1–41
226. He Y, Maciejczyk M, Ołdziej S, Scheraga HA, Liwo A (2013) Phys Rev Lett 110:098101
227. Sieradzan AK, Giełdoń A, Yin Y, He Y, Scheraga HA, Liwo A (2018) J Comput Chem 39:2360–2370
228. Liwo A, Kaźmierkiewicz R, Czaplewski C, Groth M, Ołdziej S, Wawak RJ, Rackovsky S, Pincus MR, Scheraga HA (1998) J Comput Chem 19:259–276
229. Zięba K, Ślusarz M, Ślusarz R, Liwo A, Czaplewski C, Sieradzan AK (2019) J Phys Chem B 22:4758
230. He Y, Mozolewska MA, Krupa P, Sieradzan AK, Wirecki TK, Liwo A, Kachlishvili K, Rackovsky S, Jagieła D, Ślusarz R, Czaplewski CR, Ołdziej S, Scheraga HA (2013) Proc Nat Acad Sci USA 110:14936–14941
231. Krupa P, Mozolewska MA, Wiśniewska M, Yin Y, He Y, Sieradzan AK, Ganzynkowicz R, Lipska AG, Karczyńska A, Ślusarz M, Ślusarz R, Giełdoń A, Czaplewski C, Jagieła D, Zaborowski B, Scheraga HA, Liwo A (2016) Bioinformatics 32:3270–3278
232. Karczyńska AS, Mozolewska MA, Krupa P, Giełdoń A, Liwo A, Czaplewski C (2018) Proteins 86:228–239
233. Karczyńska A, Mozolewska MA, Krupa P, Giełdoń A, Bojarski KK, Zaborowski B, Liwo A, Ślusarz R, Ślusarz M, Lee J, Joo K, Czaplewski C (2018) J Mol Graphics Modell 83:92–99
234. Karczyńska A, Zięba K, Uciechowska U, Mozolewska MA, Krupa P, Lubecka EA, Lipska AG, Sikorska C, Samsonov SA, Sieradzan AK, Giełdoń A, Liwo A, Ślusarz R, Ślusarz M, Lee J, Joo K, Czaplewski C (2020) J Chem Inf Model 60:1844–1864
235. Sieradzan AK, Bogunia M, Mech P, Ganzynkowicz R, Giełdoń A, Liwo A Makowski M (2019) J Phys Chem B 119:8526-8534
236. Chinchio M, Czaplewski C, Liwo A, Ołdziej S, Scheraga HA (2007) J Chem Theory and Comput 3:1236–1248
237. Sieradzan AK, Mozolewska MA (2018) J Mol Model 24:121
238. Czaplewski C, Karczyńska AS, Sieradzan AK, Liwo A (2018) Nucl Acids Res 46:W304–W309
239. Kurcinski M, Jamroz M, Blaszczyk M, Kolinski A, Kmiecik S (2015) Nucl Acids Res 43:W419–W424
240. Kurcinski M, Badaczewska-Dawid A, Kolinski M, Kolinski A, Kmiecik S (2020) Prot Sci 29:211–222
241. Krupa P, Karczyńska AS, Mozolewska MA, Liwo A, Czaplewski C (2021) Bioinformatics 37:1613–1615
242. Zhou R, Maisuradze GG, Sunol D, Todorovski T, Macias MJ, Xiao Y, Scheraga HA, Czaplewski C, Liwo A (2014) Proc Natl Acad Sci USA 111:18243–18248
243. Maisuradze GG, Senet P, Czaplewski C, Liwo A, Scheraga HA (2010) J Phys Chem A 114:4471–4485
244. Golas EI, Maisuradze GG, Senet P, Ołdziej S, Czaplewski C, Scheraga HA, Liwo A (2012) J Chem Theor Comput 8:1334–1343
245. Mozolewska M, Krupa P, Scheraga HA, Liwo A (2015) Proteins: Struct, Funct, Bioinf 83:1414–1426
246. Rojas A, Liwo A, Browne D, Scheraga HA (2010) J Mol Biol 404:537–552
247. Rojas A, Liwo A, Scheraga HA (2011) J Phys Chem B 115:12978–12983
248. Rojas AV, Maisuradze GG, Scheraga HA (2018) J Phys Chem B 122:7049–7056

249. Nguyen HL, Krupa P, Hai NM, Linh HQ, Li MS (2019) J Phys Chem B 123:7253–7269
250. Sieradzan AK, Krupa P, Wales DJ (2017) J Phys Chem B 121:2207–2219
251. Krupa P, Wales DJ, Sieradzan AK (2018) J Phys Chem B 122:8166–8173
252. Esko JD, Kimata K, Lindahl U (2009) Proteoglycans and sulfated glycosaminoglycans. In: Varki A, Cummings RD, Esko JD, Freeze HH, Stanley P, Bertozzi CR, Hart GW, Etzler ME (eds) Essentials of Glycobiology, 2nd ed. Cold Spring Harbor Laboratory Press, Cold Spring Harbor
253. Peng Y, Yu Y, Lin L, Liu X, Zhang X, Wang P, Hoffman P, Kim SY, Zhang F, Linhardt RJ (2018) Glycoconj J 35:119–128
254. Shute J (2012) Handb Exp Pharmacol 207:307–324
255. Li Z, Yasuda Y, Li W, Bogyo M, Katz N, Gordon RE, Fields GB, Brömme D (2004) J Biol Chem 279:5470–5479
256. Sankaranarayanan NV, Nagarajan B, Desai UR (2018) Curr Opin Struct Biol 50:91–100
257. Taketomi H, Ueda Y, Gō N (1975) Int J Peptide Protein Res 7:445–459
258. Hills RD, Brooks CL (2009) Int J Mol Sci 10:889–905
259. Yang SC, Onuchic JN, Levine H (2004) J Chem Phys 125:054910
260. Schug A, Hyeon C, Onuchic JN (2008) Coarse-grained structure-based simulations of proteins and RNA. In: Voth G (ed) Coarse-graining of condensed phase and biomolecular systems, Chapter 9. CRC Press, Boca Raton, pp 123–140
261. Hoang TX, Cieplak M (2000) J Chem Phys 112:6851–6862
262. Sułkowska JI, Sułkowski P, Szymczak P, Cieplak M (2010) J Am Chem Soc 132:13954–13956
263. Cieplak M (2018) Mechanostability of virus capsids and their proteins in structure-based models. In: Liwo A (ed) Computational methods to study the structure and dynamics of biomolecules and biomolecular processes—from bioinformatics to molecular quantum mechanics. Springer, Heidelberg, pp 295–315
264. Brown S, Fawzi NJ, Head-Gordon T (2003) Proc Natl Acad Sci USA 100:10712–10717
265. Brown S, Head-Gordon T (2004) Prot Sci 13:958–970
266. Sinitskiy AV, Voth GA (2008) Chem Phys 422:165–174
267. Trylska J (2010) J Phys Cond Mat 22:453101
268. Kmiecik S, Badaczewska-Dawid A, Kouza M, Kloczkowski A, Kolinski A (2018) Int J Mol Sci 19:3496

Chapter 3
First-Principles Modeling of Non-covalent Interactions in Molecular Systems and Extended Materials

Pabitra Narayan Samanta, Devashis Majumdar, Szczepan Roszak, and Jerzy Leszczynski

Abstract The intermolecular non-covalent interactions through van der Waals or dispersion forces are pervasive in nature and play a fundamental role in regulating the structure and function of molecular systems ranging from solid state materials to biological systems. The atomistic modeling of non-covalent interactions is incredibly difficult, as they often require exact treatment of long-range electron correlation which in turn demand to go beyond second-order perturbation theory. As for example, the prediction of induction that stems from the response of a molecular system to a permanent multipole necessitate the precise evaluation of molecular polarizabilities. The computation of dispersion interaction also appears to be a formidable task as they involve Coulomb interaction between the instantaneous correlated fluctuations of electrons. Therefore, a systematic and unified theoretical framework for isolating non-covalent interactions is essentially required to reliably model the structure, energetics, and reactivities of realistic molecular systems. In this review, the fundamental theoretical principles and computational aspects for the estimation of strong and weak non-covalent interactions are discussed by emphasizing studies of classic examples such as hydrogen bonding and related properties of small water clusters, halide-water clusters, fatty acid dimers and their amides; several gas-phase and dihydrated cation-π complexes comprising benzene, p-methylphenol, and 3-methylindole as the π-donor systems and Mg^{2+}, Ca^{2+}, and NH_4^+ cations as the acceptor units; the π-π interactions between benzene and monosubstituted benzenes in parallel face-to-face stacking configuration, as well as the supramolecular complexes. A comprehensive picture of the accuracy of the most widely used first-principles approaches including dispersion-corrected

P. N. Samanta · D. Majumdar (✉) · J. Leszczynski (✉)
Interdisciplinary Center for Nanotoxicity, Department of Chemistry, Physics and Atmospheric Sciences, Jackson State University, Jackson, MS, USA
e-mail: pabitra.samanta@icnanotox.org; devashis@icnanotox.org; jerzy@icnanotox.org

S. Roszak
Department of Physical and Quantum Chemistry, Faculty of Chemistry, Wroclaw University of Science and Technology, Wroclaw, Poland
e-mail: szczepan.roszak@pwr.wroc.pl

J. Leszczynski, M. K. Shukla (eds.), *Practical Aspects of Computational Chemistry V*, https://doi.org/10.1007/978-3-030-83244-5_3

density functional approximations, second order Møller-Plesset and symmetry-adapted perturbation theory, as well as non-canonical coupled cluster theory in predicting van der Waals and dispersion interactions has also been presented. The discussion culminates through the conceptual and mathematical ingredients required to establish structure-property relationships e.g., the correlation between hydrogen-boning and the vibrational modes, impact of electrostatic interactions on charge transfer to solvents, and the relation between Hammett substituent constants and the dispersion interactions in extended π-systems.

3.1 Introduction

Various kind of forces are associated with the formation of molecular structures and crystals. They can be generally classified in terms of strong and weak interactions. The strong interactions are responsible for the formation of molecular frame and are generally classified as covalent and electrostatic forces. There are, of course, two sub-classes of these two forces and are characterized in terms of co-ordinate and metallic bonding. Pure electrostatic bond is generally considered as stronger than the co-ordinate and metallic bonds, while a covalent bond can be much stronger than these three forces. Such a comparison is, of course, purely qualitative and the magnitude of such binding forces can vary depending on the bonding condition. The weak interactions, on the other hand, exert much weaker forces (than these four strong forces) and are responsible for the shape and properties of various molecular systems (as well as crystals) and, like strong interactions, are important forces of nature.

The natures of strong forces are quite well-defined using bonding theories of quantum chemistry. The covalent bonding, for example, was explained from Pauling's hybridization theories [1], and through molecular orbital (MO) theories [2]. The MO theories were later modified in present day quantum chemistry for very accurate description of molecular systems in relation to their shape and bonding characteristics. These are mostly based on Hartree-Fock (HF) theories [3] with the inclusion of basis set concepts of Roothan and Hall [3, 4]. More accurate HF based theories, for example Møller-Plesset perturbation theories [5–7], coupled-cluster theories [6, 8], and configuration interaction (CI) techniques [3, 6, 8] include electronic correlation effect and are widely used for present-day structure and bonding analysis of molecular systems. The alternative density functional theory (DFT) based approaches, which are based on Kohn-Sham variational principle [9, 10], are also very successful and widely used techniques to explain molecular structures in gas-phase, solvated conditions, and their excited state properties [6, 11, 12]. The DFT techniques find further importance in calculating the structures and electronic properties of solid systems and in present days are widely used in the field of materials science research. All the techniques described above are well developed and could be used directly to explain the nature of the four major strong forces occurring in different forms of the molecular systems. These are, of course, not in the purview of the present article. In the present review, we will discuss the nature of weak interactions, and the use of the present-day theories (MP, CC, DFT etc. theories) to assess the various structural and binding characteristics of the weakly-bonded species.

Weak interactions, also called non-covalent interactions in chemistry, usually occur between two molecular species. Some molecular systems, for example inert gas dimers (He_2, Ar_2 etc.) are also weakly bonded as such. These interactions are of medium to long range in nature and arise from the charge distribution patterns around a molecule (or an atom). Such a charge distribution is not constant and fluctuates due to movement of electrons. In the case of weakly interacting systems, the transient asymmetry of the charge distribution around one of the component species can induce a complementary asymmetry in the electronic distribution around the neighboring partner through electrostatic interactions and produce a weak attractive force to hold them together. This force of attraction is operative up to a certain distance r between these two systems and at smaller distance (with respect to r), the forces become repulsive because of the overlap of the outer electron clouds. Thus, this attractive force is of long range in nature, but at sufficiently large separation between the two species (much larger than r), it disappears due to simple electrostatic law. The important point related to the weakly bound system is that the individual components retain their characteristics more or less unchanged. This is a very qualitative oversimplified way to explain the origin of weak interactions. The actual situation is more complex when someone would try to identify the nature of such forces through proper quantitative analysis of the operating forces.

The origin of weak interactions, as introduced here, are special forms of electron-correlation effects and fall under the category of van der Waals interactions. Dipole-dipole, dipole-induced dipole, and London dispersion forces (r^{-6} dependent forces) are mostly responsible for such van der Waals interactions. Although these terms have their common classical mechanics definitions, they could be explicitly treated in quantum chemical methods through introduction of electron correlation effects. Formation of hydrogen boded systems, e.g., water, methanol, ammonia etc. are mostly due to the van der Waals interactions arising from dipole-dipole interactions effects, since the individual molecular components in these systems have permanent dipole moments. The interactions inside the H_2-He or noble gas dimers, on the other hand, has binding forces arising purely through London dispersion types. The forces are generally weak in nature (0.1–0.2 kcal/mol), although hydrogen bonding interactions are much stronger (e.g., 2–5 kcal/mol for hydrogen bonding involving N-H and O-H units).

Long-range π-π and cation-π are two different type of non-covalent interactions, which are usually stronger than regular hydrogen bonds. In both the cases dispersion interactions are generally operative, but the interaction-nature cannot be explained so simply. In the case of cation-π interactions, dipole-induced dipole and higher order quadrupole-dipole, quadrupole-quadrupole interactions etc. interactions are also operative. These types of interactions were predicted initially by Kier and coworkers [13, 14] and, almost 20 years later, were fully established by Dougherty and coworkers [15, 16]. These interactions have emerged as a very common binding force to interpret binding nature between cation and aromatics in various biomolecular systems including proteins, receptor-ligand complexes, molecular recognition, drug actions and protein folding [15–21]. The π-π dispersion interactions are long known, and this type of weak interactions are important in many naturally occurring materials. The most common example is graphite, and multilayered graphene sheets.

The present review is primarily oriented around the quantification of these non-covalent weak interactions. These would be based on various case specific analyses to understand the energetics of such interactions. There are, of course, several other characteristics of such systems which are available through experimental and theoretical analyses. Low-frequency vibrations along with the high frequency vibrational modes were found to modulate the strength of hydrogen bonds in several hydrogen-bonded systems [22]. Furthermore, the weakly bound systems demonstrate several molecular properties e.g., additive properties of interaction energies and charge transfer to solvents (CTTS, in halide-water clusters) [23]. In the case of π-π interactions, Hammett equation criteria was found to be important to understand the nature of interactions in specific cases, when various π-derivative systems were allowed to interact with a specific π-scaffold [24]. These properties have been experimentally studied in weakly bound systems using finite sized weak clusters [25]. The non-covalent interactions are also the building blocks for the artificial designing of several bioinspired materials [26]. The present review will also address these properties of the weakly bound systems with specific examples. A brief review related to the theoretical background of the energy quantification, both classical and quantum mechanical, would be presented prior to the discussion of the individual type of non-covalently bound systems.

3.2 Theoretical Models of Non-covalent Interactions

Non-covalent interactions between atoms or molecules could be treated through regular molecular interaction calculations. The only difference is that the interaction energy values are low because of the larger separation of the interacting systems. Electrostatic interactions are central to all these attractive forces, but it cannot account for the whole interaction energies. Interactions between the noble gas atoms are typical examples, since they have no dipole moment or higher moments. On the other hand, there must be some clear interactions between these atoms. The molecular beam experiments [27, 28] were used to measure interactions between the noble gas atoms and Fig. 3.1 schematically shows the nature of such interactions in the case of argon dimer. The computed interaction energy surface reaches a minimum at 3.8 Å (i.e., it shows an attractive nature) and this energy tends to vanish at a very large separation [29]. At shorter distances, this interaction energy curve shows repulsive character. This trend of non-covalent interactions has already been discussed in Sect. 3.1 and such an energy curve is general for all the noble gas atoms dimers also. The force between the atoms, which could be defined as the negative of

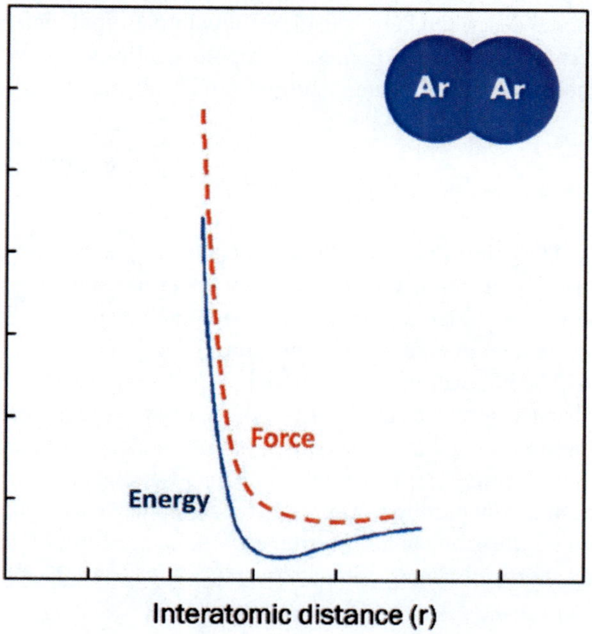

Fig. 3.1 Schematic diagram of the intermolecular energy and the force between two argon (Ar) atoms

the potential energy derivative with respect to the interatomic distances (r) (shown schematically in Fig. 3.1), also shows similar trend with respect to r.

The curves in Fig. 3.1 are considered as a balance between two forces viz., attractive dispersive force and the repulsive force of exchange type. Both forces have their quantum mechanical origin. The dispersive force arises from the generation of instantaneous dipole from the fluctuating electron clouds. This instantaneous dipole can induce a dipole in the neighboring molecule producing an attractive inductive effect. The dispersive force could be developed from Drude model based on traditional Schrödinger equation. For two interacting molecules, the Drude model generates interaction energy of the form,

$$E(r) = -\frac{1}{2}\frac{\alpha^4 \hbar\omega}{(4\pi\varepsilon_0)^2 r^6} \tag{3.1}$$

Where α is the polarizability, $\omega/2\pi$ is the frequency of an isolated Drude molecule, r is the separation between the two molecules, and $\hbar(=h/2\pi)$ is the Planck's constant. In three dimensions, this energy is given by,

$$E(r) = -\frac{3}{4}\frac{\alpha^4 \hbar\omega}{(4\pi\varepsilon_0)^2 r^6} \tag{3.2}$$

The complete derivation of the interaction energies is not needed in this review. Interested readers can see the ref. [30] for the complete derivation. The Drude model described here considers only dipole-dipole interaction term. If higher order terms arising from dipole-quadrupole, quadrupole-quadrupole etc. interactions are included, the interaction energy from Drude model could be represented as [30, 31]

$$E(r) = \sum_n \frac{A_n}{r^n} \quad (n = 6, 8, 10, \ldots\ldots) \tag{3.3}$$

The determination of the coefficients (A_n) is discussed in detail in ref. [31]. These coefficients are negative due to attractive nature of the interaction terms. If just A_6 term is included for Ar case, the computed dispersion is ~25% smaller with respect to the total interaction energies from experiment.

The interaction energy curve in Fig. 3.1 shows that slight decrease in interatomic distance between Ar-atoms (~3 Å) causes a large increase in energy to the repulsive region. This increase has quantum mechanical origin and could be explained from Pauli's principle. It formally prevents two electrons in a system having same set of quantum numbers. The short-range repulsive force here, is arising from electrons with same spin, and is referred to as exchange forces (also known as overlap forces). At short r, the interaction energy varies as $1/r$ due to nuclear repulsion and at larger r, the energy decays exponentially as $\exp\left(-\frac{2r}{a_0}\right)$ (a_0 : *Bohr Radius*).

3.2.1 Modeling van der Waals Interactions

The dispersive and repulsive (exchange-repulsion) interactions between atoms and molecules could be calculated through quantum mechanics. These calculations are far from trivial and require electron correlation with large basis sets. We will discuss such methods in connection with the non-covalent interactions in DFT calculations. The simpler form of the van der Waals interaction as would be outlined here are the basis for the interpretation of non-covalent interactions computed through more complicated quantum-mechanical calculations.

The basic criteria of the dispersive and repulsive forces in a non-covalent interaction between two molecules or atoms should fit to the general functional form,

$$E_{VW}\left(r^{IJ}\right) = E_{repulsive}\left(r^{IJ}\right) - \frac{C^{IJ}}{\left(r^{IJ}\right)^6} \tag{3.4}$$

It is not possible to classically derive the functional form of the repulsive interactions. The interaction energy should go to zero as $(r^{IJ}) \rightarrow \infty$ and should approach zero faster than $(r^{IJ})^{-6}$ term.

The most popular potential, which satisfy this requirement, is the Lennard-Jones (LJ) potential [32]. Here, the repulsive part has $(r^{IJ})^{-12}$ dependence,

$$E_{LJ}(r) = \frac{C_1}{\left(r^{IJ}\right)^{12}} - \frac{C_2}{\left(r^{IJ}\right)^{6}} \tag{3.5}$$

where, C_1 and C_2 are suitable constants. The alternative form of LJ potential has the following form,

$$E_{LJ}(r) = \varepsilon\left[\left(\frac{r_0}{r^{IJ}}\right)^{12} - 2\left(\frac{r_0}{r^{IJ}}\right)^{6}\right] \tag{3.6}$$

Where, r_0 is the minimum distance, and ε is the depth of the minimum. There is no theoretical basis for the choice of the repulsive part, this is purely for computational convenience. Sometimes exponent of 9 or 10 can generate better results.

Considering the exponential decay of the repulsive term, a E_{VW} potential, known as Buckingham or Hill type potential [33] was developed in the following form,

$$E_{VW}(r) = C_1 \exp\left(-C_2 r^{IJ}\right) - \frac{C_3}{\left(r^{IJ}\right)^{6}} \tag{3.7}$$

Here, C_1, C_2 and C_3 are suitable constants. Equation (3.7) is sometimes written in the following convoluted form also.

$$E_{VW}(r) = \zeta\left[\frac{6}{\alpha-6}\exp\left\{\alpha\left(1-\frac{r^{IJ}}{r_0}\right)\right\} - \frac{\alpha}{\alpha-6}\left(\frac{r_0}{r^{IJ}}\right)^{6}\right] \tag{3.8}$$

Where, α is a force parameter and choosing $\alpha = 12$, Eq. (3.8) generates LJ potential. Figure 3.2 schematically shows the attractive part of the LJ(12-6), LJ(9-6) and Buckingham potentials for the H_2–He interactions. The Buckingham potential shows a better description of the attractive part with respect to the other two potentials, which may be due to the presence of three parameters (Eq. 3.7) in the potential. The LJ-potentials have only two adjustable parameters in this respect.

The hydrogen bond interaction potential can also be represented in terms of LJ potentials, but this potential needs some adjustments in the attractive part, as the interaction energies are stronger in such cases. The function form of $E_{VW}(r)$ is commonly represented in such cases through the following 12-10 potential (Eq. 3.9).

$$E_{HB} = \varepsilon\left[\left(\frac{r_0}{r^{IJ}}\right)^{12} - 6\left(\frac{r_0}{r^{IJ}}\right)^{10}\right] \tag{3.9}$$

E_{HB} represents the hydrogen bond potential. In some cases of hydrogen bond potential, directional terms like $(1 - \cos\theta)$ or simply $\cos\theta$ [34] (θ : *a HB angle*) are multiplied with the distance dependent part of Eq. (3.9). The various E_{VW} terms

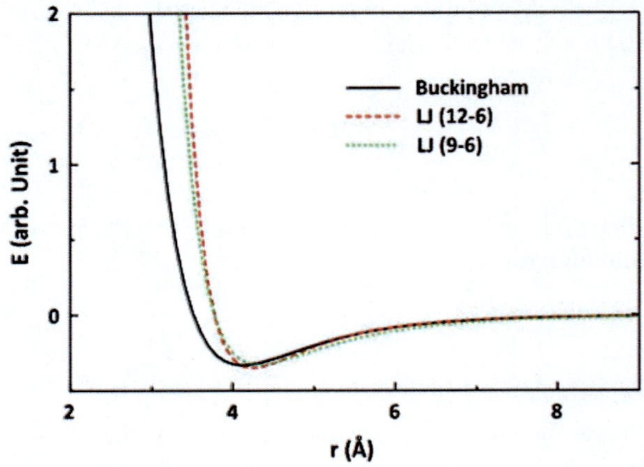

Fig. 3.2 Schematic representations of Buckingham, Lennard-Jones (12-6) and Lennard-Jones (9-6) potentials

discussed here comprise only the basic form of the potentials to be used in molecular mechanics (MM) or more commonly known *force field* calculations. These force fields are central to the classical molecular dynamics (MD) simulations and there are many different forms of such force fields in dealing with diverse molecular systems including simple molecules to more complex systems like proteins, enzymes, nucleic acids, membranes, polysaccharides etc. This is not a place to discuss these large collection of force fields applicable to various diverse systems, since they are not the objective of the present review article. We will discuss only several forms of dispersion potentials related to their use in combination with quantum chemical methods to show the diversity of the force field parameters.

Let us start with the rare gas interaction cases. The interactions between the rare gas atoms are usually very weak. The potential energy for interactions varies from 0.08 kcal/mol ($1.29 \times 10^{-4} a.u.$) for He_2 to 0.78 kcal/mol ($12.5 \times 10^{-4} a.u.$) for the Rn_2 dimers. Such interaction energy calculations need specially designed van der Waals potentials to account for the binding distance and energies of such dimers. In recent times, Tang-Toennies (TT) potentials were developed [35] to account such interactions. The potential has usual attractive and repulsive parts. The repulsive potential part is short-ranged Born-Mayer type, while the long-range attractive potential is added as damped dispersion series [36]. Mathematically, it is written as,

$$E_{VW}(r) = E_{rep} + E_{att} = Ae^{-br^{IJ}} - \sum_{n=3}^{N} f_{2n}\left(br^{IJ}\right)\frac{C_{2n}}{\left(r^{IJ}\right)^{2n}} \tag{3.10}$$

where C_{2n} terms are dispersion coefficients. The term b is the Born-Mayer range parameter and the only parameter in the damping function $f_{2n}(br^{IJ})$. This function

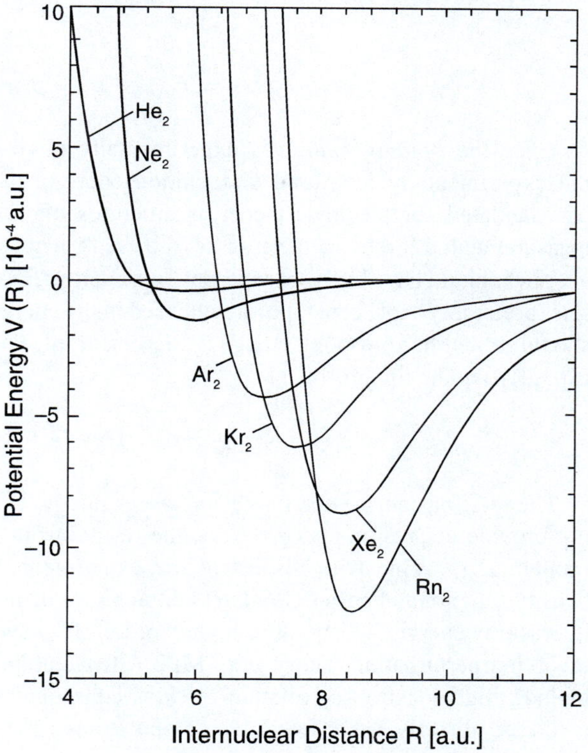

Fig. 3.3 Potential energy curves of the homogeneous rare gas dimers calculated with the Tang–Toennies (TT) potential model (reprinted from ref. [35], with the permission of AIP Publishing)

can be expressed as

$$f_{2n}\left(br^{IJ}\right) = 1 - e^{-br^{IJ}} \sum_{i=0}^{2n} \frac{\left(br^{IJ}\right)^i}{i!} \tag{3.11}$$

This function can be computed from incomplete gamma function also [35]. It is to be noted that the function has some similarity with the Eq. (3.3) originating from Drude model. With proper knowledge of the parameters in Eq. (3.10), potential energy curves for the rare atom dimers could be very accurately computed. The example related to the dimers He_2 through Rn_2 are shown in Fig. 3.3. The details related to the other different rare gas atom interactions are also available in ref. [35].

3.2.2 Quantum Chemical Approaches for Non-covalent Interactions

The calculations of non-covalent interaction energies between two interacting systems A and B to form the weakly bound complex is straightforward and given by,

$$\Delta E_{AB} = E_{AB} - E_A - E_B \tag{3.12}$$

Since the *binding energy* is experimentally observable quantity (negative of the experimentally measured dissociation energy), the E_{AB}, E_A and E_B should be calculated for the lowest energy structures of the individual species for its measurement. Moreover, because of the weak nature of the binding force, the energies should have basis set superposition energy (BSSE) correction, if large basis sets (preferably triple zeta type) are not used in the energy computations. The energy should further be zero-point energy (ZPE) corrected. Thus, Eq. (3.12) would be like Eq. (3.13) in the final form,

$$\Delta E_{AB} = E_{AB} - E_A - E_B - \Delta BSSE - \Delta ZPE \tag{3.13}$$

There is, of course, one caveat in such strategy. If the techniques involved do not provide dispersion energy effect, the results from Eq. (3.13) will not provide proper ΔE_{AB} value with respect to the experiments. Pure HF and Kohn-Sham density functional theories (KS-DFT) do not have dispersion correction to the total interaction energies. Generally, higher order HF-wavefunction based techniques involving perturbation theory, e.g., MP2, CC-techniques (CCSD, CCSD(T)) etc., include high electron-correlation effects (consequently dispersion effects), and could be directly used for such computations. The DFT-calculations become successful when a proper density functional is augmented with empirical dispersion term. The quantum chemical methods also provide the techniques to breakdown the total interaction energies in terms of dispersion, exchange, electrostatic etc. to understand the role of such interaction energy components to shape up a non-covalently bound system. In the following subsections we will discuss these features in more detail.

3.2.3 Dispersion Computations in DFT

The KS-DFT functionals mostly include Grimme's D2 [37], D3 [38], or D3BJ [39] dispersion terms to compute dispersion energy part in DFT calculations. The total energy through any dispersion modified density functional could be expressed as,

$$E(DFT - D) = E(KS - DFT) + E_D \tag{3.14}$$

Here, $E(KS - DFT)$ is usual Kohn-Sham energy as obtained from a chosen density functional, and E_D is the empirical dispersion correction.

The D2, D3 or D3BJ empirical dispersion terms are similar to TT-potentials, as discussed earlier, but generate the dispersion part of the non-covalent molecular interactions in a different way. The D2 dispersion correction is written as [37],

$$E_D = -S_6 \sum_{I=1}^{N_{atom}-1} \sum_{J=I+1}^{N_{atom}} \frac{C_6^{IJ}}{(r_{IJ})^6} f(r_{IJ}) \tag{3.15}$$

Here, N_{atom} is the number of atoms in the system; C_6^{IJ} is the dispersion coefficient for the atom pair IJ; S_6 is the global scaling factor (that depends on the density functional used) and r_{IJ} is the distance between the atom pair I and J. The $f(r_{IJ})$ term is a damping function, and it is important to avoid near-singularities for small r_{IJ}. The f term is expressed as,

$$f(r_{IJ}) = \frac{1}{1 + e^{-d\left(\frac{r_{IJ}}{R} - 1\right)}} \tag{3.16}$$

Where, R is the sum of atomic van der Waals radii. The term d is a preset parameter and $d = 20$ is usually set here to provide larger corrections at intermediate distances and generating negligible dispersion energy at covalent binding region.

Further developments of these dispersion potentials in DFT calculations were introduced later in the D3 and D3BJ techniques [38, 39]. The basic equation is the same as Eq. (3.14); only the dispersion calculations were modified. The D3 calculation [37] uses the following form of $E(D)$,

$$E(D) = E^{(2)} + E^{(3)} \tag{3.17}$$

The $E^{(2)}$ and $E^{(3)}$ are the two-body and three-body dispersion interaction terms, respectively. The philosophy for using such equation comes from the definition of interaction energies in terms of many body interactions [40].

The term $E^{(2)}$ and $E^{(3)}$ are usually expressed as follows,

$$E^{(2)} = \sum_{IJ} \sum_{n=6,8,10\ldots} S_n \frac{C_n^{IJ}}{r_{IJ}^n} f_{d,n}(r_{AB}) \tag{3.18}$$

$$f_{d,n}(r_{AB}) = \frac{1}{1 + 6\left(\frac{r_{IJ}}{S_{r,n} r_0^{IJ}}\right)^{-\alpha_n}} \tag{3.19}$$

$$E^{(3)} = \sum_{IJK} f_{d,(3)}(r_{IJK}) E^{IJK} \tag{3.20}$$

f terms in Eqs. (3.18) and (3.20) are damping terms and the terms given in Eq. (3.19). The nonadditive dispersion term E^{IJK} in Eq. (3.20) is computed from the third-order perturbation theory for the three atoms I, J, K and is discussed in detail in the ref. [38]. The three-body term is insignificant for small molecular systems, and usually has some contribution when the system is quite large. The D3BJ potential is similar to D3 potential. It differs from D3 only in the contribution of damping factor [39]. The detailed discussions related to these methods are available in the respective papers, and they include the determination of such factor (S_n) and assignment of the proper values of C_n^{IJ} and r_0^{IJ} terms. We will discuss below only the basic features of the D3BJ potential to show its difference with D3.

Becke and Johnson (BJ) [41] proposed a rational damping procedure for the dispersion potential in the following form,

$$E(D) = -\frac{1}{2}\sum_{I \neq J} \frac{C_n^{IJ}}{r_{IJ}^n + const.} \tag{3.21}$$

Based on such a damping approach, the modified form of DFT-D3 method could be represented in the following form of Eqs. (3.22) and (3.23).

$$E(D) = -\frac{1}{2}\sum_{I \neq J} S_6 \frac{C_6^{IJ}}{r_{IJ}^6 + f\left[r_0^{IJ}\right]^6} + S_8 \frac{C_8^{IJ}}{r_{IJ}^8 + f\left[r_0^{IJ}\right]^8} \tag{3.22}$$

with

$$f\left(r_0^{IJ}\right) = x_1 r_0^{IJ} + x_2 \tag{3.23}$$

Where, x_1 and x_2 are the fit parameters introduced by BJ [40]. The BJ-damping leads to a constant contribution of E(D) to the total correlation energy for each bonded atom-pairs. This damping procedure seems theoretically more justified over a normal zero-damping [42], although it was found to change the thermochemical description of the underlying density functional (DF). Adjustment of standard correlation functionals are required to overcome such a problem. The related computer codes are usually equipped with these parameters for the computation of the dispersion terms for the proper thermochemical description of DF. It has been shown that, although the damping procedures in D3BJ differs from D3, both the procedures produce almost equivalent results [39].

3.2.4 Dispersion Computation Through MP2 and Higher Correlation Methods

The binding energy of non-covalently bound complexes could be computed through MP2, CCSD, CCSD(T) etc. type of quantum chemical techniques using Eq. (3.13). In such techniques, it is not needed to include the dispersion term separately. This energy is already included in these methods through higher electron-correlation effects. Let us consider, the case of MP2 method. We begin with the consideration of the effect of perturbation $\lambda \hat{V}$ on the generalized Hamiltonian $\hat{H}_\lambda$.

$$\hat{H}_\lambda = \hat{H}_0 + \lambda \hat{V} \tag{3.24}$$

Here, $\hat{H}_0$ is the unperturbed HF Hamiltonian. The Rayleigh-Schrödinger perturbation theory tells us that the effect of perturbation renders the following expansion effect on the wave function (ψ_λ) and energy (E_λ) corresponding to $\hat{H}_\lambda$.

$$\psi_\lambda = \psi^{(0)} + \lambda \psi^{(1)} + \lambda^2 \psi^{(2)} + \ldots \tag{3.25}$$

$$E_\lambda = E^{(0)} + \lambda E^{(1)} + \lambda^2 E^{(2)} + \ldots \tag{3.26}$$

The Møller-Plesset (MP) energy up to the first order perturbation term is the HF-energy. The second order contribution ($E^{(2)}$) to the MP-energy could be written as [5–7],

$$E^{(2)} = \sum_{i<j}^{occ}\sum\sum_{a<b}^{virt}\sum \frac{\langle ij||ab\rangle^2}{(\varepsilon_a + \varepsilon_b - \varepsilon_i - \varepsilon_j)} \tag{3.27}$$

Here, ε terms are orbital energies of the occupied (i,j) and virtual (a,b) orbitals and $\langle ij||ab\rangle$ is a two-electron integral over spin-orbitals (and obviously involves double substitution). Now, if we consider non-covalent interactions between two molecular systems or atoms, $E^{(2)}$ is the correlation correction to the HF-energy and represents the effect of dispersion energy (with higher correlation terms). The dispersion interaction energy at the MP2 level could then be written as [43, 44],

$$\varepsilon^{(2)}_{MP} = E^{(2)}_{AB} - E^{(2)}_{A} - E^{(2)}_{B} \tag{3.28}$$

Where, AB is the overall system generated from the species A and B. The Pure HF-interaction energy in such a case could be written as

$$\Delta E_{HF} = E^{AB}_{HF} - E^{A}_{HF} - E^{B}_{HF} \tag{3.29}$$

This ΔE_{HF} could be partitioned further into Heitler-London $\left(\Delta E^{ex}_{HL}\right)$ energy components as

$$\Delta E^{ex}_{HL} = \varepsilon^{(10)}_{EL} + \varepsilon^{(10)}_{EX} \tag{3.30}$$

where $\varepsilon^{(10)}_{EL}$ and $\varepsilon^{(10)}_{EX}$ are the electrostatic and exchange contribution to the ΔE^{ex}_{HL}. The difference between ΔE_{HF} and ΔE_{HL} generates the delocalization component of the total HF interaction energies. Thus,

$$\Delta E_{HF}^{del} = \Delta E_{HF} - \Delta E_{HL}^{ex} \tag{3.31}$$

Such partitioning implies that total MP2 interaction energy ΔE_{MP2} could be partitioned into four components in the following way [43, 44].

$$\Delta E_{MP2} = \varepsilon_{EL}^{(10)} + \varepsilon_{EX}^{(10)} + \Delta E_{HF}^{del} + \varepsilon_{MP}^{(2)} \tag{3.32}$$

The $\varepsilon_{EL}^{(10)}$ can further be divided into short-range penetration and long-range multipolar components [44]. Thus, in quantum chemical analysis, the interaction energies can not only be computed with accuracy but also the various energy components affecting the interactions can be deduced from energy-component analysis as outlined here. The details of such analyses are available in the refs. [43, 44]. The coupled cluster level of computations can be used for binding energy calculations through Eq. (3.13) for additional accuracy, and energy components can also be measured for the total interaction energies through symmetry adopted perturbation analysis (SAPT) [45]. In recent times, SAPT analysis has also been developed for energy component measurements for DFT techniques [46]. The DFT energy component analysis computes total interaction energy as the sum of $\Delta E_D + \Delta E_{ex} + \Delta E_{ind}$ (ΔE_{ex}: exchange component; ΔE_{ind}: induction component). The ΔE_{ex} in DFT and ΔE_{HL}^{ex} in the MP2 energy component analysis [as in Eq. (3.32)] are similar. The ΔE_{HF}^{del} in Eq. (3.31) and ΔE_{ind} in DFT-SAPT calculations are closely related. The difference is, while ΔE_{HF}^{del} is associated with the relaxation of electron densities of monomers upon interactions restrained by Pauli principle [47] (charge delocalization with charge transfer interactions), ΔE_{ind} represents interactions arising from the charges due to deformation of the monomer units.

We have so far narrated a brief description of the various quantum chemical techniques used to analyze the non-covalent interactions between two molecular units. The rest part of the review article would be devoted to various non-covalent weak interaction analysis based on techniques described in this section.

3.3 Non-covalent Interactions in Hydrogen-Bonded (HB) Systems

A hydrogen bond is formed when hydrogen atom covalently bound to an electronegative atom or group, generally termed as donor (D) group that interacts with another electronegative atom with lone pair of electrons (the so-called acceptor (A) group) forming a non-covalent interaction of the type D-H . . . A. The hydrogen bond formed is often described as electrostatic (dipole-dipole type), although it has some features of covalent bonding. The interaction distance is usually shorter than the sum of the van der Waals radii of the interacting components. The hydrogen bond strength depends on the electronegativities of D and A, and such bond could be as strong as 38.6 kcal/mol [48]. Examples of very weak hydrogen bond are also found in cases of S, Cl and C (~1.0 kcal/mol) [48, 49]. They are also ubiquitous and have importance in control of receptor-ligand interactions in medicinal chemistry, and intra-/intermolecular interactions in materials science [50, 51]. Most common examples of hydrogen bonding are found in water, ammonia, hydrogen fluoride, organic fatty acids, alcohols etc. Hydrogen bonding in such solvent systems is important in manifestation of several physical properties viz., melting point, boiling point, solubility, viscosity, and azeotropic properties of solvent mixtures. Such interactions are also one of the most important building blocks in various polymeric materials including DNA, proteins, cellulose, synthetic polymers, and the bonding features in these systems are explored mostly through crystallography, NMR, and IR experiments [52, 53].

Quantum chemical (QC) techniques and molecular dynamics (MD) simulations are important in silico methods to understand the nature of hydrogen bonding. The theoretical findings are important in explaining various experimental observations in this context. Small clusters are usually important to understand such interaction phenomenon, since these clusters are generated in recent times in gas phase and their structural and bonding features could be observed through IR spectra. For example, gas phase IR spectroscopic techniques viz., extensive terahertz laser vibration-rotation-tunneling (VRT) spectra and mid-IR laser spectra [54–57] were used to understand the origin of hydrogen bond formation in small water clusters. QC-techniques are very effective in elucidating structural behavior and in the present section we will discuss two such examples. The discussions would be oriented around the structural, binding, and spectroscopic properties of several small water clusters and fatty acid dimers. The discussions on water clusters would also include the effect of inclusion of halide ions in several of such water clusters and their charge transfer properties to solvent (CTTS) [23]. The fatty acid dimers were generated in gas phase and apart from their binding properties, they have unique low and high frequency vibrational modes which are involved in the relative stability of such clusters [22]. Furthermore, these fatty acid clusters have unique fragment energy additive properties, which could be used to predict the binding energies of higher fatty acids. This characteristic is unique to such fatty acids and were not explored before.

3.3.1 Hydrogen Bonding and Related Properties of Small Water Clusters

The change of structural and spectroscopic properties of small water clusters with increasing cluster size have been investigated through theory and experiments for a long time [23, 54–63]. The primary objectives of such instigations were to understand the nature of hydrogen bonding in such clusters, how such changes are related to their vibrational characteristics, effect of adding halide ions to such clusters [23, 64] and related changes in hydrogen bonding and spectroscopic properties. Experimental structure analysis of the small water clusters, $(H_2O)_n$, n = 2–6, have been reported from the VRT spectroscopy [54–57], while the vibrational spectra of n = 1–8 water clusters bound the benzene and water clusters of n = 7–10 are available for their O-H vibrational modes [59, 60, 63]. The water dimer is linearly hydrogen bonded [65], while water trimer to pentamer are cyclic rings [61, 62]. The higher clusters starting from n = 8 and above are multiring types [59, 60]. These structures were ascertained from the theoretical and experimental results. The low-energy structures of water hexamer and heptamer are also mostly multiring type, although they can have two-dimension (2D) and three-dimensional (3D) structures. A 3D-cage structure of $(H_2O)_6$ is believed to be the lowest energy structure from both theory and experiments [57, 58, 61]. On the other hand, extensive QC calculations on water hexamer has also suggested the presence of open-book like structure [65].

With this brief resumé on the structural aspect of the small water clusters, we turn our attention to the more specific properties related to the hydrogen bonding, i.e., relative stabilities and IR characteristics arising from hydrogen bonds. We are taking the specific example of water heptamer, as the low energy water clusters formed in this case show various structural possibilities including directional hydrogen bond properties due to the formation of 3D structures. The small water clusters fall under two broad structural types viz., 2D ring and 3D cage (e.g., prism, or cube). Furthermore, the water monomers inside a water cluster could be of single proton donor–single acceptor (*da*), single donor–double acceptor (*daa*), double donor–single acceptor (*dda*), and double donor–double acceptor (*ddaa*) types. In the water hexamer, for example, these types of water monomers were found to be related to the O–H spectra [66]. Water clusters presented in Fig. 3.4 tries to explain these features through specific examples of $(H_2O)_7$-clusters [67]. It contains twelve optimized water heptamer structures (C to R7, all in C_1-symmetry) together with four stable hexamer (cage, prism, book, and ring in C_1-symmetry) [66] and two stable octamer clusters (D2d, S4) [59, 60]. These hexamer and octamer structures are shown here, since they are geometrically related to the heptamer structures for their formation. The cage and prism structures of $(H_2O)_6$-clusters can generate the C, D, E, and F structures through addition of a water molecule. The octamer structure could also affect these structure formations in an alternative way. The global minima D2d and S4 octamers are degenerate and contain *dda* and *aad* types of water monomers. Exclusion of these water molecules from S4 would generate the C and D structures, whereas removal of similar water molecules from the D2d cluster would be responsible for the E and F cluster formation. The formation of the G and H structures could similarly be explained from the S4 structure. The ring structures R5, R4 and R7 could be generated from the ring and book structures of water hexamers through addition of one water molecule [67].

The number of hydrogen bonds (HBs) play a major role in relative stability and strength of hydrogen bond in water clusters. As a result, such properties together with the polarity of hydrogen bonds can influence the observed –OH frequencies of

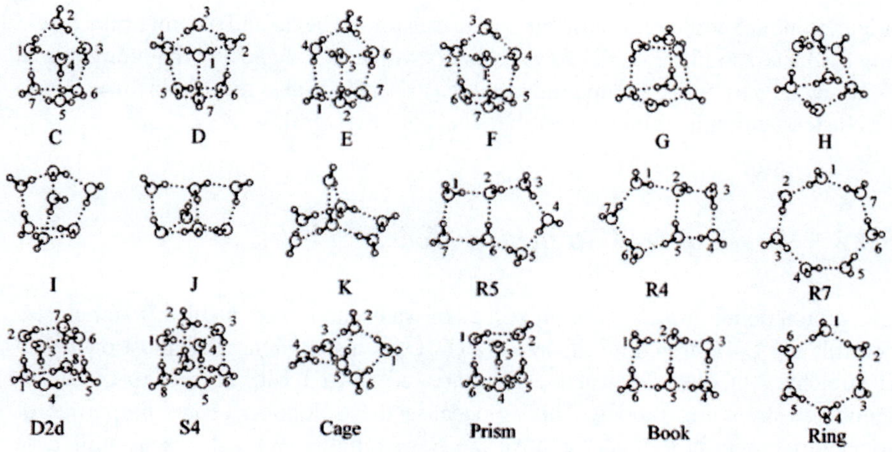

Fig. 3.4 The optimized geometries of water heptamer (C–R7), hexamer (ring, book, cage, and prism), and octamer (D2d and S4) clusters. All the structures are in C_1 symmetry except D2d, S4, and ring (S6). The sequential numbers in each figure represent the direction of the unidirectional H orientations (reprinted from ref. [67], with the permission of AIP Publishing)

these clusters. The number HBs in C, D, E, F, G, and J are ten, while they are nine in H, I, and K, eight in R5, R4, and seven in R7. The average HB distances vary between 2.84 Å (cluster F) to 2.76 Å (cluster R4). These results were available from the fully optimized structures of these clusters (MP2/TZ2P++ calculations) and are slightly shorter than the normal O... O distance (2.98 Å) in water. These average bond distances of the individual clusters together with the number of hydrogen bonds determine the following relative stability order of the clusters.

$$C_{10}^{2.82} > D_{10}^{2.82} > G_{10}^{2.83} > H_{9}^{2.83} > E_{10}^{2.83} > F_{10}^{2.84} > J_{10}^{2.82} > I_{9}^{2.80} > R5_{8}^{2.76} > R4_{9}^{2.76}$$

The subscripts in the cluster notation indicate the number of HBs, and the superscript represents average HB-length in Å (in terms of O-H-O distance). There are few deviations in the above stability order in relation to the aforementioned hypothesis. This could be attributed to the strain in the structure (Fig. 3.4), which alters the stability order in terms of average HB-number and length considerations. The relative stability order was verified through Gibbs free energy change measurements in molecular beam experiments [54–58]. The energy difference between C and D is only 0.5 kcal/mol and the other clusters except R7 and K are within 2.5 kcal/mol. The average HB energies for the lowest energy clusters C and D were found to be 3.8 and 3.7 kcal/mol respectively (MP2/TZ2P++) and these values are weaker than normal hydrogen bond energy in water (~5.0 kcal/mol). The average HB-energy in other clusters is all within 4.2 kcal/mol (MP2//B3LYP/6-311++G** level) [67].

Unidirectionality of HBs also determine the relative stability of the water clusters. The low-energy clusters C, D, and R5 in Fig. 3.4 show that their HB orientations are unidirectional. This is also true for the hexa- and octamer (n = 6, and n = 8) clusters in Fig. 3.4. The lower clusters with n = 3, 4 and 5 with unidirectional HBs (cyclic structures) are the most stable [61]. This is also true for the n = 9, and 10 clusters with nine- and ten-membered cyclic rings [60].

3.3.2 Nature of O-H Stretching Modes

The proton donor and acceptor properties of water monomers inside a water cluster controls the O-H stretching modes of a $(H_2O)_n$ cluster. Generally, most of the O-H stretching modes in a water cluster are red shifted with respect to the water monomer stretching modes. This is because these donor-acceptor properties of water monomers in a cluster control the HB-strengths. We will discuss this case for the low-energy water clusters with n = 6, 7, and 8. Figure 3.5 shows the calculated vibrational spectra of these clusters (B3LYP/6-311++G**) [67]. The spectra are presented with respect to the water monomer frequencies computed at the same level of theory (3921 cm^{-1}, 3816 cm^{-1}, and 1603 cm^{-1}). As it could be seen from the spectra (Fig. 3.5), the $H_2O)_n$ clusters have n number of asymmetric (n_3) and symmetric (n_1) O-H stretching modes. Most of these peaks are red shifted with respect to the monomer frequencies, except one (slightly blue shifted). The classification of water monomers based on their donor-acceptor properties in n = 7 cluster, as discussed earlier, are as follows: C, D, E, F, G (1 *da*, 3 *aad*, and 3 *daa*); H, I (3 *da*, 2 *aad*, 2 *daa*); J (2 *da*, 2 *aad*, 2 *dda*, 1 *ddaa*); K (5 *da*, 2 *ddaa*); R5, R4 (5 *da*, 1 *aad*, 1 *dda*); R7 (7 *da*). The general principle is that the n_3-band of the *dda* or *ddaa* types are lower (in energy) than those of the *da* types. On the other hand, in the case of n_1-band, the *dda* and *ddaa* type O-H stretches are higher than those of the *da* and *aad* types. The spectral shifts in the whole spectral data in Fig. 3.5 can be explained from this principle.

The experimental determination of the IR spectra of $(H_2O)_7$-cluster [63] indicated two structural isomers. These are based on the number of experimental peaks These peaks in terms of frequency shift with respect to the experimental average O-H stretching frequencies of the water monomers, are –757 (2), –627 (2), 397, 287 (2), –147 (4), –57, and 13 cm^{-1}) (the numbers within parentheses are splitting pattern of the peaks). These spectral shifts are similar to the calculated shifts of the mixture of C and D isomers (Fig. 3.4), indicating the presence to two competitive lowest energy structures of the water heptamer. It is to be mentioned in this connection that the computed absolute values of frequencies were not in very good agreement with experiment, since anharmonicity effect was not explicitly considered in such calculations. The anharmonicity effect, of course, do not alter the binding energy (also the HB-energy) of these clusters, as was found in a later study for n = 2–10 clusters [68].

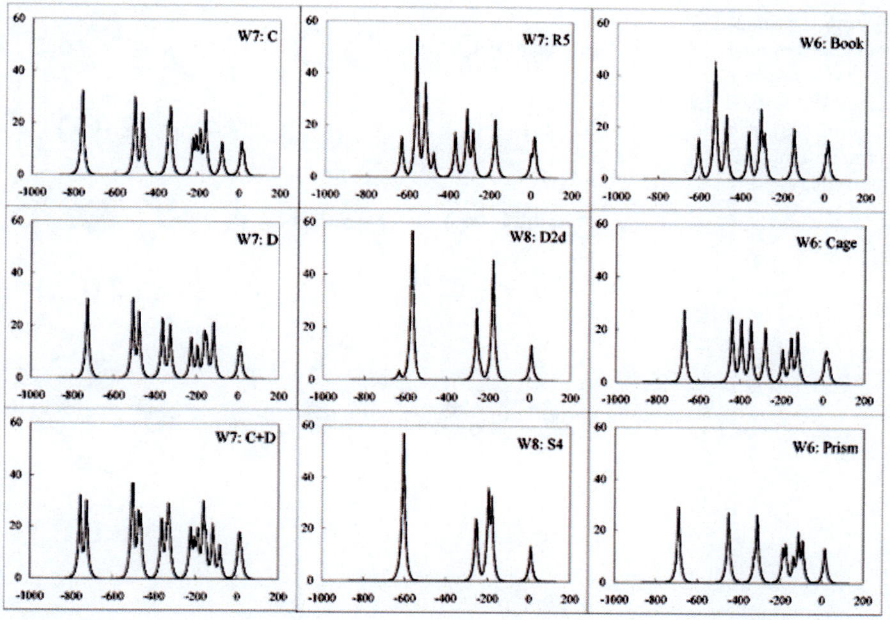

Fig. 3.5 Computed IR spectra for various low energy clusters of the water heptamer, hexamer, and octamer at the B3LYP/6-311++G** level of theory (reprinted from ref. [67], with the permission of AIP Publishing)

3.3.3 Effect of Halide Ion Interactions with Small Water Clusters

The structural characteristics of the halide ion inclusions in water clusters were monitored effectively through photoelectron spectroscopy (PES) [69, 70]. The PES analysis showed that $X(H_2O)_n$ ($X = F^-$, Cl^-, Br^-, and I^-), surface structures are favored for X= Cl^-, and Br^- form small n. On the other hand, for $X = I^-$, the $n = 6$ cluster has an internal state with first solvation layer of 6. The QC calculations on $X(H_2O)_n$ ($X = Cl^-$, Br^-, and I^-) [64, 71] indicated that for n = 2–6, the Cl^-, and Br^- ions resides on the surface of the cluster, while the I^- resides on the surface for n = 2–5, but for the n = 6 cluster, the I^- tends to move from the surface to the interior site. The $F^-(H_2O)_n$ clusters are different from the other halide-water clusters, since the F^- ion interactions are much stronger than the rest of the halide ions. Because of the exceptionality of HB-interactions, we have chosen to discuss the structural features and binding characteristic of such clusters in more details. Moreover, all these halide-water clusters show CTTS properties [23]. These special features of $X(H_2O)_n$ (n = 1–4) would also be reviewed here.

The $F^-(H_2O)_n$ (n = 1–6) clusters were analyzed through QC calculations using various approaches, and F^--binding energies of these clusters are experimentally available. Figures 3.6 and 3.7 contain different structural possibilities of these

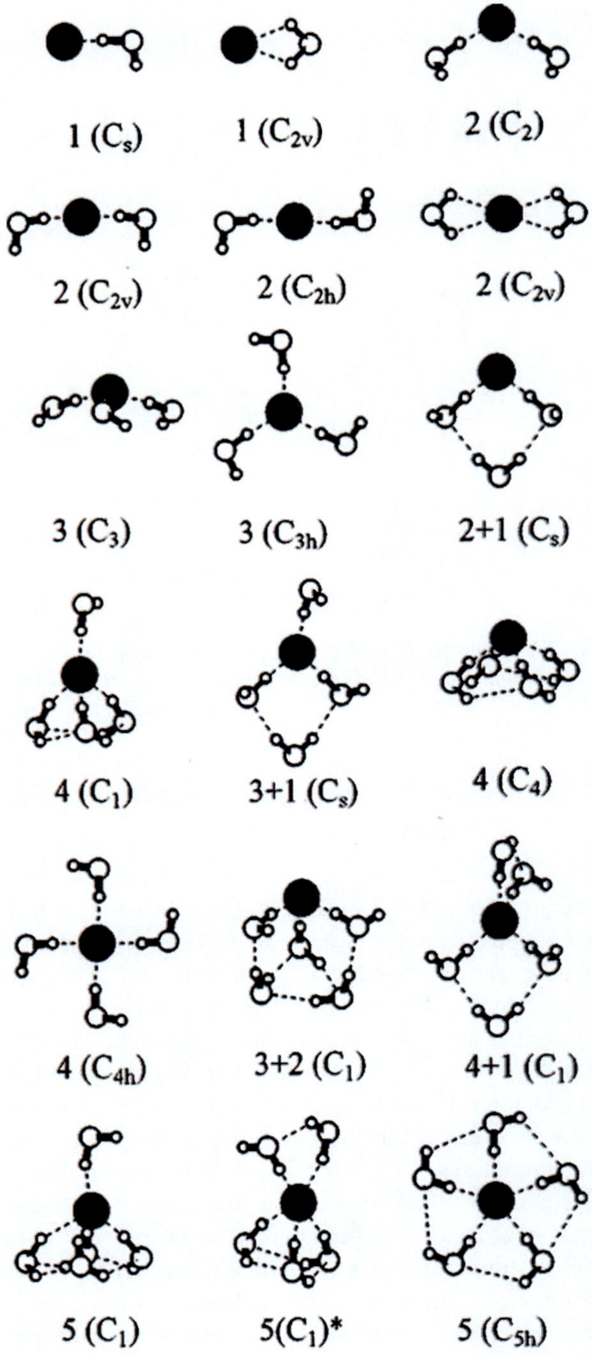

Fig. 3.6 Optimized geometries of the $F^{-}(H_2O)n$, n = 1–5, clusters obtained by using MP2 method (reprinted from ref. [72], with the permission of AIP Publishing)

Fig. 3.7 Optimized geometries of the $F^-(H_2O)_6$ clusters predicted by employing MP2 method (reprinted from ref. [72], with the permission of AIP Publishing)

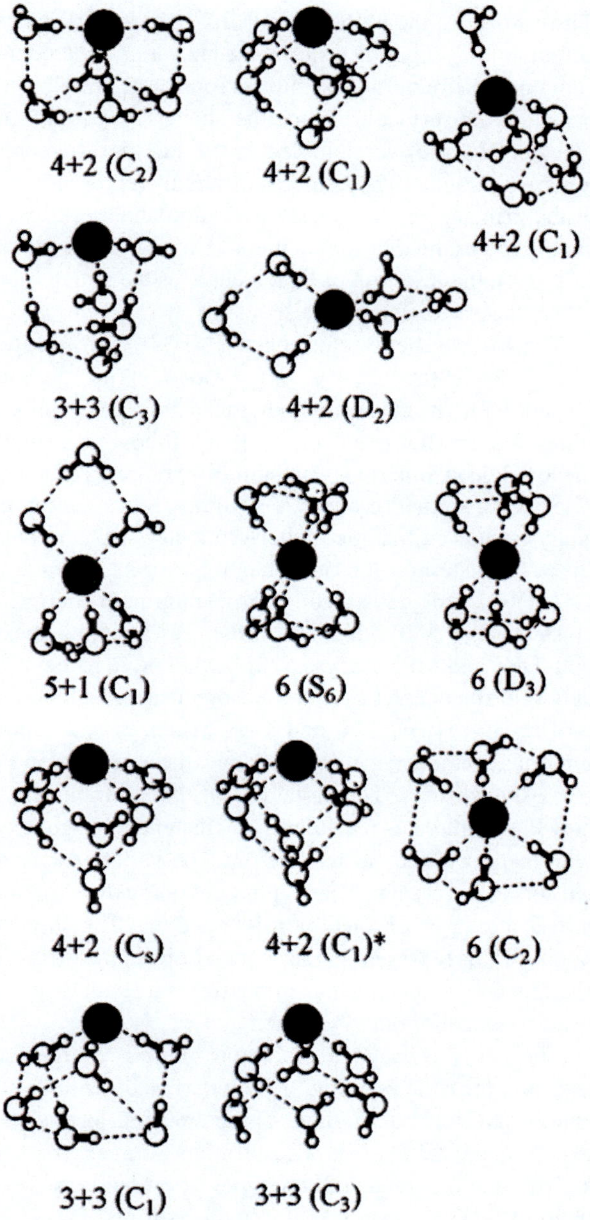

clusters in different QC calculations [72]. There were not many disagreements related to the cluster properties, and most of the calculations agreed on the minimum energy structures, and these are important to compute the parameters related to experiments. We have chosen to discuss the results based on DFT/B3LYP and MP2 calculations using large basis sets (6-311++G**). The results do not differ much with respect to the higher level MP2 (and larger basis set) and CCSD(T) [72, 73] computations. Figure 3.6 shows the cluster arrangements for n = 1–5, while Fig. 3.7 contains the structural possibilities for n = 6. The F^- ion in the higher clusters could orient on the surface or inside the cluster. A specific ($n_1 + n_2$) notation is chosen, together with the overall molecular symmetries, to represent such possibilities of F^- ion arrangements. The numbers n_1 and n_2 represent the number of water molecules in the primary and secondary hydration shells, and when $n_2 = 0$, the structure is simple represented as n_1. As it could be seen from the structures of $(H_2O)_6$ clusters (Fig. 3.4), the insertion of F^- ion changes the structural patterns. This is true for the other clusters (different n) also.

The lowest energy clusters of $F^-(H_2O)_n$ (n = 1–6) are $1(C_s)$ (n = 1), $2(C_2)$ (n = 2), $3(C_3)$ (n = 3), $4(C_1)$ (n = 4), $5(C_1)$ (n = 5), and $(4 + 2)(C_2)$ (n = 6) (Figs. 3.6 and 3.7). The clusters with higher symmetry are mostly high-energy or transition states. The smaller sized clusters ($n \leq 3$) show very small energy difference between the low-lying isomers (~0.2 kcal/mol). The cases of n = 3 and 4 are quite interesting. The $3(C_{3h})$ isomer of n = 3 has slightly lower energy than $3(C_3)$, although careful analysis showed that the higher symmetry structure is a transition state. The $4(C_1)$ (n = 4) cluster, on the other hand, is not an unambiguous global minimum. The $(3 + 1)(C_s)$ is actually a competitive minimum energy isomer. In the case of higher clusters (N = 5, 6), global minimum (as assigned above) is unambiguous through both DFT and MP2 analysis [72], and in more recent calculations also [73]. Further details of the energetics of these isomers are available in ref. [72].

The F. . . H, F. . . O, and O-H distances and related H-O-H bond angle are important parameters to understand the strength of F^- ion interactions in these $F^-(H_2O)_n$ clusters. The variations of these parameters with increasing n represent how F^- ion interacts with the water cluster part. Figure 3.8 represents such variations with respect to the cluster size (n). The variations are related to the lowest energy clusters of different n. The graphs containing the variations of different bond-lengths reflect the variation of F^- ion interactions. The shortest F. . . H, F. . . O and O-H bond lengths represent higher interactions in this respect. The variation of the H-O-H angle with respect to the cluster size is a result of such interactions, and this angle is always smaller than free water.

The actual variations of F^--ion binding energies with respect to the cluster size are presented in Fig. 3.9 in terms of enthalpy (ΔH^{298K}) and Gibbs free energy (ΔG^{298K}) of binding. These are MP2-level data and are compared with the experimental results from two different sources (marked as Expt. a [74] and Expt. b [75]), and the trends of both types of estimations are consistent. The MP2 results (6-3111++G**) are presented with both BSSE-corrected and uncorrected ways, and it could be observed that BSSE-corrected values somewhat underestimate the binding energies for $n \geq 4$. The interactions could also be analyzed through many-body interaction theories [72], since they are quite important for bigger systems (especially n = 5 and 6 clusters) to estimate interactions as well as binding energies. Both 2-body and 3-body interactions contribute to the total binding energies in these clusters. The higher body interactions (4 and above) are not important. These are in nut-shell the nature of F^--ion *non-covalent* interactions in $F^-(H_2O)_n$ clusters. Such

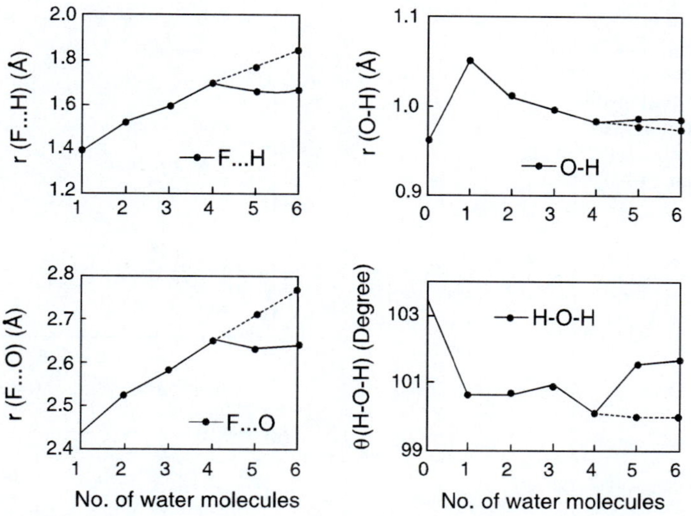

Fig. 3.8 Variation of the lengths r(F... H), r(F... O), and r(O–H) and H–O–H angle [θ(H–O–H)] with increasing water molecules in clusters $F^{-}(H_2O)_n$. The parameters are chosen for the minimum energy (MP2) geometries of each cluster. The dotted lines indicate the values for the minimum energy n_1 cluster ($n_2 = 0$) of $n = 5$ and 6 clusters (reprinted from ref. [72], with the permission of AIP Publishing)

interactions have significant effect on the ionization potential and O-H vibrational characteristics of these clusters. A specific effect is the higher red-shift of the O-H stretching frequencies more than the normal water clusters. The detailed discussion is available in ref. [72]. We will conclude this section after a general discussion of the CTTS properties of halide-water cluster as HB-properties influence such CT-spectra.

3.3.4 CTTS Properties of Halide-Water Clusters

The UV-spectra of halide ions in water demonstrates a unique type of charge-transfer spectra. These specific spectral characteristics occur due to the electron injection from the halide to the solvent in the UV region and generated excited state is known as CTTS state. Thus, it is not the property of halide ion itself. The bound CTTS state is created due to the stabilizing potential of the surrounding solvent molecules. The non-covalent interactions of halide ions with the surrounding solvent are an operative factor for such a CTTS phenomenon. For example, the aqueous solution of iodine exhibits broad charge transfer band due to electron ejection from iodide to the solvent. The spectrum consists of two bands at ~ 2100 Å separated by characteristic spin-orbit splitting (0.94 eV) of neutral iodine [76].

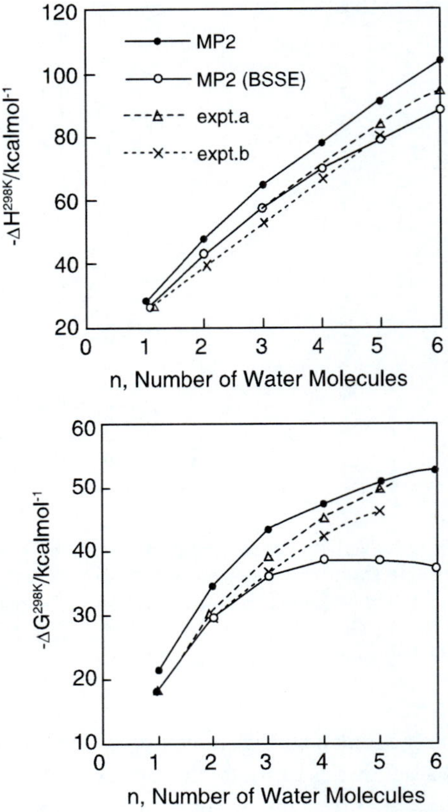

Fig. 3.9 Plots of experimental and calculated (MP2) ΔH and ΔG values of $F^{-}(H_2O)n$ clusters with increasing n (n = 1–6) (reprinted from ref. [72], with the permission of AIP Publishing)

The measurements of the CTTS bands of $I^{-}(H_2O)_n$ (n = 1–4) through photo-detachment spectra [77], and studies on the dynamics of electron solvation in the photo excited states of the $I^{-}(D_2O)_n$ (n = 4–6) and $I^{-}(H_2O)_n$ (n = 2–4) [78] unraveled the importance of water-cluster…halide ion interactions. The experimental CTTS spectra of Cl^{-}, and Br^{-} are only available in bulk water [79]. The theoretical calculations determined the CTTS spectra of $X^{-}(H_2O)_n$ (X = F, Cl, Br, I; n = 1–4) using their lowest energy clusters [80, 81]. The excited state calculations at the TD-DFT level produced sufficiently accurate results. The first excited singlet state (S_1) of these clusters were found to represent the CTTS states through computed charge transfer (Δq) data from $S_0 \rightarrow S_1$ states (Table 3.1). The computed CTTS states were verified with respect to the experiment for the case of $I^{-}(H_2O)_n$ (n = 1–4) clusters (Table 3.1). The minimum energy clusters used in such calculations have halide ions on the surface of the water cluster networks. All these structures are similar to the those of $F^{-}(H_2O)_n$ (n = 1–4) clusters in Fig. 3.4. The only difference is that the $F^{-}(H_2O)_2$ cluster is a $2(C_2)$ isomer, while the rest of the minimum energy n = 2 halide-water clusters are $2(C_1)$ isomers (Table 3.1). The further details of these structures are available in ref. [81].

Table 3.1 Computed vertical transition energies $[\Delta E^{VT}(S_1), eV]$, experimental ΔE_{CTTS}(eV), enthalpy of binding (ΔH, *kcal/mol*), dipole moment (μ, *Debye*) and charge transfer from S_0 to S_1 (Δq, *a. u.*) for the clusters $X^-(H_2O)_n$ ($X = F, Cl, Br, I; n = 1-4$)

Ion	n	Cluster	$\Delta E^{VT}(S_1)$	ΔE_{CTTS}	$-\Delta H$	$\mu(H_2O)$	Δq
F^-	1	$1(C_s)$	4.61 (4.69)	–	26.3	2.35	0.518
	2	$2(C_2)$	5.34 (5.34)	–	44.8	0.14	0.048
	3	$3(C_3)$	5.77 (6.07)	–	60.1	1.35	0.002
	4	$4(C_4)$	6.02 (6.02)	–	39.1	1.24	0.001
Cl^-	1	$1(C_s)$	4.28 (4.28)	–	13.9	2.33	0.857
	2	$2(C_1)$	4.52 (4.52)	–	24.7	3.85	0.828
	3	$3(C_3)$	5.21 (5.21)	–	36.2	3.33	0.715
	4	$4(C_4)$	5.45 (5.50)	–	46.8	3.91	0.654
Br^-	1	$1(C_s)$	4.00 (4.04)	–	12.2	2.32	0.496
	2	$2(C_1)$	4.19 (4.19)	–	23.5	3.93	0.561
	3	$3(C_3)$	4.78 (4.78)	–	35.3	3.98	0.493
	4	$4(C_4)$	5.01 (5.01)	–	46.3	4.19	0.495
I^-	1	$1(C_s)$	3.74 (3.78)	3.60	10.1	2.32	0.395
	2	$2(C_1)$	4.08 (3.82)	3.95	20.0	4.13	0.487
	3	$3(C_3)$	4.29 (4.29)	4.25	30.9	3.98	0.439
	4	$4(C_4)$	4.44 (4.44)	4.50	41.6	4.84	0.466

All the values are reproduced from refs. [80, 81], with the permission of AIP Publishing

The CTTS bands of $I^-(H_2O)_n$ (n = 1–4) show blue-shift with respect to the stepwise increment of n (Table 3.1) in both experiment and theoretical calculations. The computed CTTS bands of other halide-water clusters also show similar trends, indicating the role of water... halide interactions in such spectra. It has been argued that CTTS bands of iodide-water cluster correspond to the excitation of electron from an orbital localized in I^- to a delocalized state with support from the water network. Generally, a neutral molecule with around 2.5 D dipole moment can bind an electron in a dipole-state through exchange repulsion between excess electron and electrons in molecules [82, 83]. Considering the dipole moments of water molecules induced by halide (Table 3.1), it can be safely assumed that the initial upper state of $I^-(H_2O)_n$ (n = 1–4) cluster could be a short-lived $I(^3P_2).[(H_2O)_n]^-$ (n = 1–4) state in pump-pulse experiment. Here electron is transferred from iodide to the dipole-bound state of the water network. This argument could also be generally applied to the CTTS spectra other halide-water clusters. The formation of $X.[(H_2O)_n]^-$ actually might take place through several steps involving dissociation of $X^-(H_2O)_n$ to X^- and $(H_2O)_n$, followed by recapture of electron from X^- to the water network. A thermodynamic cycle was proposed [23] in this respect and was found to work well to interpret CTTS spectral positions [80]. This is a way to interpret the CTTS bands (vertical S_1-state energy) through its dissociation into several thermodynamic components, indicating the importance of non-covalent interactions of halide-water clusters in such phenomena.

3.3.5 *Effect of Low-Frequency Vibrations of HBs in Fatty Acid Dimers and Their Amides*

HB is associated with interesting vibrational properties in molecular systems. They are related to the relative stabilities of various hydrogen bonded isomers generated through intermolecular interactions between the monomers of the same species. Generally, high-frequency vibrations of the O-H/N-H bond, associated with HB formation, are considered to influence the HB-strengths. Specific low-frequency vibrations of several molecular systems, forming polymeric hydrogen bonded systems through OH/NH bonds, were also found to correlate with the HB-strengths/binding energies (ΔE^B) of such systems through coupling with the with the associated high-frequency modes. These correlations could be verified in systems where several hydrogen bonded isomers could be identified. Small organic fatty acids like, formic acid, acetic acid and their amides form hydrogen bonded dimers. Involvement of low-frequency mode/s in the stabilities of such dimers were identified through experiments and QC calculations [22, 84–86]. These correlations further facilitated the development of a local fragment energy-based (related to the HB) additive property to predict binding energies of such dimers. The concept is extendable to other homologs of these acids, e.g., propionic, and n-butyric acids [86].

Formic acid monomer predominantly exists in *trans*-form [87–89], although the less common rotamer *cis*-form has also been characterized [89]. This conformational behavior of formic acid has opened up the possibilities of this molecule to form several dimeric forms through hydrogen bonding. These include *trans-trans*, *trans-cis*, and *cis-cis* combination of monomers, and experiments based on the vibrational excitations of ground state *trans*-formic acid has detected most of these isomers [90–92]. Acetic acid, like formic acid, exhibits rotational isomerism through C-OH bond. *Trans*-form is the predominant rotamer, while the less probable *cis*-variety also exists in the gas-phase [93, 94]. The *cis*-form is ~5.3 kcal/mol above the *trans*-form with *trans to cis* conversion barrier of 13.2 kcal/mol. Thus, this *cis*-form is unlikely to take part in dimer formation. Six dimers in trans-trans combination could be constructed, three of them were found to exist through experiments [84, 95].

QC calculations are available on the various *trans-trans* and *trans-cis* isomers of formic acid dimer and the global minimum was ascertained to be *trans-trans* (TT-1, Fig. 3.10) [85]. This structure is similar to the other theoretical results [96] and experiment [22, 90]. As it could be seen from Fig. 3.10, most of the isomers have dihydrogen bond with a few exceptions (TC-3 and TC-4). High level QC calculations are usually needed for accurate estimation of the ΔE^B of such hydrogen bonded dimers. A ΔE^B value of 13.4 kcal/mol (MP2/aug-cc-pVTZ) [85] was found to be comparable with the experiment (14.2 ± 0.2 kcal/mol) [97]. The theoretical result was estimated using anharmonicity corrections and CCSD(T) (aug-cc-pVTZ) calculations in this respect also generated satisfactory answer (13.23 kcal/mol). The *trans-trans* dimer of acetic acid (AA-1, Fig. 3.11), out of the six probable

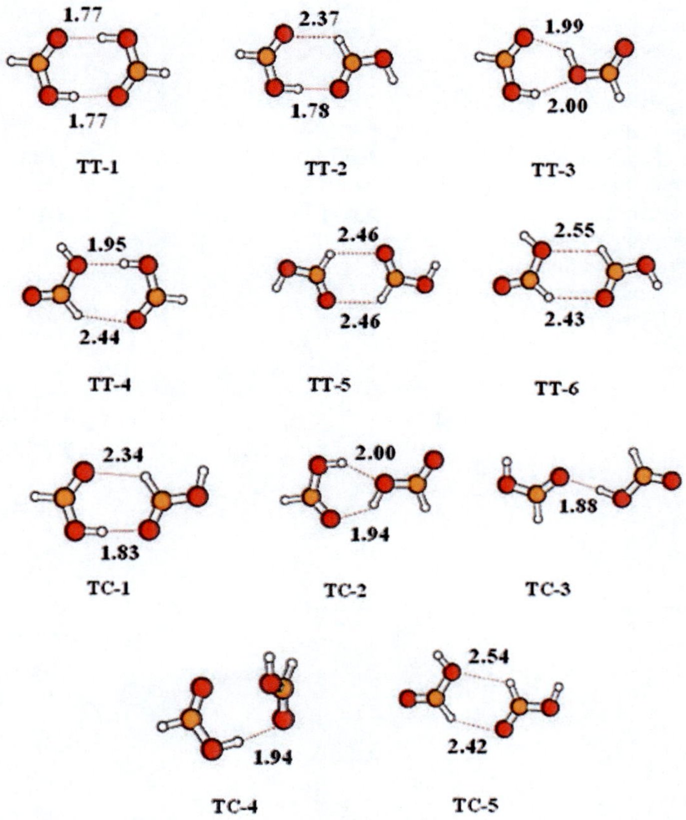

Fig. 3.10 Optimized structures of formic acid dimers (at the CCSD level) with the computed hydrogen-bond distances (Å). These dimers are formed through trans–trans (TT-1 to TT-6) and trans–cis (TC-1 to TC-5) combinations of the monomers (reprinted with permission from ref. [85] Copyright (2013) American Chemical Society)

isomers, is of lowest energy [86]. This isomer was also found to be the most stable isomer through experiment [84, 98]. All these probable acetic acid dimers have dihydrogen bonds, i.e., they form a closed ring system. The computed ΔE^B at different theoretical levels using aug-cc-pVTZ basis sets generate similar results (DFT/B3LYP: −14.7 kcal/mol, MP2: −14.4 kcal/mol, CCSD(T): −14.8 kcal/mol). The observed dissociation energy in this context is a combined theoretical and experimental data (16.2 kcal/mol) [98]. It has been argued that if experimental thermal contribution (~0.7 kcal/mol) is considered, the computed values would be compatible with experiment.

The amides of these fatty acids do not possess the rotamerism properties of the corresponding acids. The situation limits the formation of the hydrogen-bonded dimers to five [85, 86]. In the case of formamide, two of the constructed structures have been characterized through IR spectra (FMAD-A and FMAD-C,

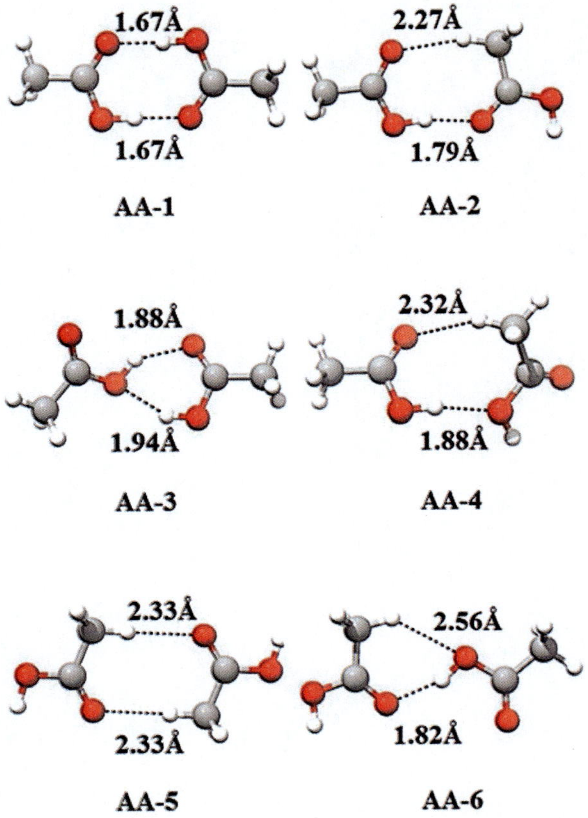

Fig. 3.11 Optimized structures of acetic acid dimers (at the MP2/aug-cc-pVDZ level) with the computed hydrogen-bond distances (Å). These dimers are formed through trans–trans (AA-1 to AA-6) combinations of the monomers (reproduced from ref. [86] with permission from the PCCP Owner Societies)

Fig. 3.12) [99], the rest of them are predicted structures. These isomers, like formic acid dimers, are mostly formed through dihydrogen bond (except FMAD-D). The FMAD-A is the minimum energy isomer and the computed ΔE^B (MP2/aug-cc-pVTZ: –12.86 kcal/mol, CCSD(T)/aug-cc-pVTZ: –12.92 kcal/mol) is close to that of AA-1 (Fig. 3.11). The weaker NH . . . O HB-strength (with respect to O-H . . . O) is reflected in these ΔE^B values. The experimental binding energy of formamide dimer is not known. The presence and abundance of FMAD-A isomer was predicted from strong red-shift of the $n_s(NH_2)$ (~387 cm^{-1}) and $n_{as}(NH_2)$ (~171 cm^{-1}) modes with respect to the monomer frequencies [99]. Acetamide dimer also does not have experimental ΔE^B values. Five possible isomers (Fig. 3.13) were assigned as the probable acetamide dimers, and AMD-1 was found to be the most abundant isomer from the strong red-shift data of $n_s(NH_2)$ (~299 cm^{-1}) and $n_{as}(NH_2)$ (~37 cm^{-1}) modes with respect to the corresponding monomer frequencies [100]. The computed red-shift data of the AMD-1 ($n_s(NH_2)$: ~299 cm^{-1}) and $n_{as}(NH_2)$: ~37 cm^{-1}) compares [86] well with the experiment, and this isomer was found to be the global minimum also. The ΔE^B values (MP2/aug-cc-pVTZ: –12.5 kcal/mol, CCSD(T): –14.3 kcal/mol) are close to the formamide dimer. The structural details related to the other isomers of formic acid, acetic acid, formamide and acetamide dimers are

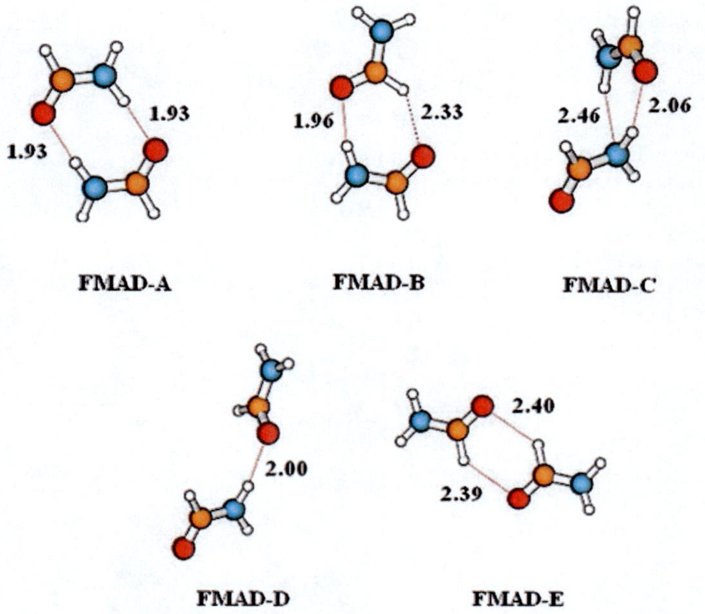

Fig. 3.12 Optimized structures of formamide dimers (at the CCSD level) with the computed hydrogen-bond distances (Å). These dimers are formed through different orientations of the monomers (FMAD-A to FMAD-E) (reprinted with permission from ref. [85] Copyright (2013) American Chemical Society)

available in refs. [85, 86]. We will discuss only the effect of vibrational frequencies on the relative stabilities of these dimers through the ΔE^B values.

The primary characteristics of formic acid, acetic acid and their amide dimers is that they form dihydrogen bonded ring structures in their lowest energy geometry and except a few cases all the isomeric forms of these dimers maintain this HB geometry. These specific non-covalent interaction patterns among these isomers generate unique vibrational characteristics related to their relative stabilities and provide information about the local properties of the hydrogen bonded groups. The computed stretching vibrational data of –OH (n(OH)) group of the acid dimers (TT-1 and AA-1) and $-NH_2$ ($n_s(NH_2)$) group of their amide dimers (FMAD-A and AMD-1) are presented in Table 3.2. These data show that these frequencies are red shifted with respect to the acid (trans-formic and acetic acids) and amide (formamide and acetamide) monomers. The intensities (I) of such modes are substantially enhanced in these respects (Table 3.2), and the computed values are compatible with respect to the experimental vibrational data [85, 86]. The vibrational data for the rest of the isomers of these dimers (not discussed here) are available in refs. [85, 86], and they have the same characteristics of the minimum energy isomers in Table 3.2. These modes are usually called marker bands for such dimers as their intensities decrease regularly with respect to their binding

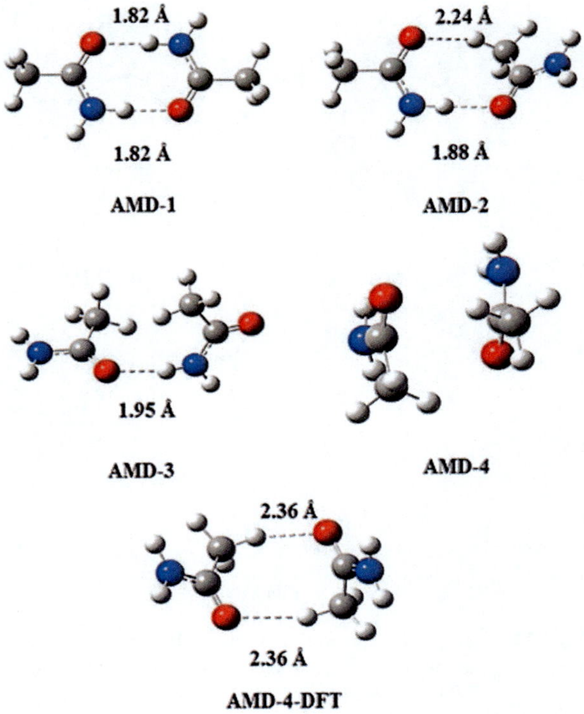

Fig. 3.13 Optimized structures of acetamide dimers (at the MP2/aug-cc-pVDZ level) with the computed hydrogen-bond distances (Å). These dimers are formed through different orientations of the monomers (AMD-1 to AMD-4) (reproduced from ref. [86] with permission from the PCCP Owner Societies)

characteristics (ΔE^B) and main linear correlations (Eqs. 3.33–3.36).

$$\text{Formic acid dimers}: I = -164.0\Delta E^B - 359.0 \quad R = 0.97 \tag{3.33}$$

$$\text{Acetic acid dimers}: I = -248.0\Delta E^B - 729.0 \quad R = 0.98 \tag{3.34}$$

$$\text{Formamide dimers}: I = -107.0\Delta E^B - 333.0 \quad R = 0.96 \tag{3.35}$$

$$\text{Acetamide dimers}: I = -233.0\Delta E^B - 1150.0 \quad R = 1.00 \tag{3.36}$$

These correlations are at the MP2 level. It could be seen from the regression coefficients (R) that these correlations are quite convincing to predict linear correlations.

Experimentally six low-frequency vibrational modes were observed for formic acid lowest energy dimer TT-1 [22]. The modes with u-symmetry (two A_u and one B_g) are IR-active and three g-symmetry modes (two A_g and one B_g) are Raman active. These low-frequency modes were determined theoretically for TT-1 isomer through anharmonicity corrections, since these vibrations were experimentally assigned as the lowest fundamentals of the overtone band. For example, n_I

Table 3.2 –OH and NH_2 stretching modes ($\nu(OH)$ and $\nu_S(NH_2)$ cm^{-1}) of the lowest energy isomers of TT-1, FMAD-A, AA-1, and AMD-1. The table also includes intensities (*I*, KM/MOL), red-shift ($\Delta\nu$, cm^{-1}) and change of intensity (ΔI) of each mode with respect to monomers of formic acid, acetic acid, formamide and acetamide

Isomer	Stretching mode	ν	I	$\Delta\nu$	ΔI
TT-1	$\nu(OH)$	3119	1998	–473	1923
FMAD-A	$\nu_S(NH_2)$	3276	1078	–251	1019
AA-1	$\nu(OH)$	2950	3437	–593	2880
AMD-1	$\nu_S(NH_2)$	3104	1789	–335	1743

All the values are reproduced with permission from ref. [85] Copyright (2013) American Chemical Society and ref. [86] with permission from the PCCP Owner Societies

Table 3.3 Low frequency vibrations of TT-1 and FMAD-A ($\nu_1 - \nu_6$), AA-1 and AMD-1 ($\nu_1 - \nu_8$) dimers. The intensities (*I*, KM/MOL) of the vibrational modes responsible for hydrogen bonding are also included. The results are presented at the MP2/6-311++G** level and the values within parentheses are experimental data

	TT-1	FMAD-A	AA-1	AMD-1
ν_1	65 (A_u)(69)	64 (A_u)	45 (A_u)(–)	–
ν_2	162 (A_u)(169)	144 (A_u)	65 (A_u)(56)	–
ν_3	237 (B_u)(248)	195 (B_u)	66 (A_u)(~50)	24 (A_u)
ν_4	182 (A_g)(174)	134 (A_g)	173 (B_u)(170)	78 (A_u)
ν_5	150 (A_g)(194)	154 (A_g)	60 (B_g)(73)	145 (B_u)
ν_6	234 (B_g)(242)	194 (B_g)	114 (B_g)(99)	84 (B_g)
ν_7	–	–	150 (A_g)(152)	125 (A_g)
ν_8	–	–	169 (A_g)(163)	144 (A_g)
I	57 (ν_3)	84 (ν_3)	30 (ν_4)	45 (ν_5)

All the values are reproduced with permission from ref. [85] Copyright (2013) American Chemical Society and ref. [86] with permission from the PCCP Owner Societies

band was determined as the lowest fundamental (n_1, A_u) of the experimentally assigned overtone band ([$2n_1$], A_g). Table 3.3 contains all these six low-frequency fundamentals along with their theoretically determined values (MP2 level). The computed values for the rest of the isomers are available in ref. [85]. It is only important to note here that the intensities of the in-plane-bending mode (n_3), related to the low-frequency OH-bending of various formic acid dimers, show regular change with respect to their ΔE^B through the following linear correlation (Eq. 3.37).

$$I = -4.62\Delta E^B - 8.49 \quad R = 0.96 \tag{3.37}$$

Formamide dimer (FMAD-A) also have six low-frequency modes and only the intensities of the stretch-bend mode (n_3, B_u) (related to the $n_s(NH_2)$ mode) (Table 3.3) together with the intensities of the similar modes of the other isomers [85] show linear correlation with their ΔE^B,

$$I = -9.54\Delta E^B - 34.8 \quad R = 0.99 \tag{3.38}$$

These correlations were validated using other high-level QC techniques including DFT/B3LYP, G4MP2, CBS-QB3, and G2MP2 methods [85].

Acetic acid dimers have eight such low-frequency vibrational modes [84, 86], and these modes with their symmetries are shown in Table 3.3 for the lowest energy AA-1 isomer. These bands are either IR-active (u-symmetry) or Raman-active (g-symmetry). The acetamide dimer do not have any experimental data and six such modes (instead of eight) could be computationally assigned through theoretical computations (Table 3.3). They are also IR and Raman active and shown in Table 3.3 for the lowest energy AMD-1 isomer. The data for the other isomers of acetic acid and formic acid dimers are available in ref. [86]. Analysis of such data revealed that the in-plane-bending or stretch-bend (in plane) modes of AA-1 (Raman active $n_7(A_g)$ and $n_8(A_g)$ and the IR-active $n_4(B_u)$) modes showed importance in hydrogen bonding. The n_7 and n_8 bands have very low intensities, while the n_4 band have quite large intensity (Table 3.3). The n_4 band also showed regular change of intensities for various acetic acid dimers and maintain a linear correlation with corresponding ΔE^B values (Eq. 3.39, MP2 results).

$$I = -2.52\Delta E^B - 6.41 \quad R = 0.99 \tag{3.39}$$

Similar analysis for the acetamide dimers generated the following correlation (Eq. 3.40, MP2) using intensities of the IR-active n_5 band.

$$I = -5.79\Delta E^B - 22.5 \quad R = 0.91 \tag{3.40}$$

These correlations, like formic acid and formamide dimers, were validated using similar high-level QC techniques [86].

3.3.6 Empirical Additive Relations of ΔE^B for Fatty Acid and Amide Dimers

The important aspect of the individual linear correlations between the ΔE^B of the fatty acid (and amides) dimers and intensities (I) of the high frequency n(OH)-modes, as discussed above, could be extended for their combined cases. For example, the I-values of both trans-trans formic acid (TT-1–TT-5, Fig. 3.10) and acetic acid (AA-1–AA-6, Fig. 3.11) dimers, when combinedly plotted against their respective ΔE^B values, a linear correlation is again prevailed (R= 0.97, MP2 results). Similar linear correlation was also observed for the formamide and acetamide dimers (R = 0.96; for the I [$n_s(NH_2)$] and ΔE^B plot: MP2 results). The low-frequency hydrogen bonding mode also showed similar features [85, 86]. These linear dependencies led to an *empirical* additivity relation of ΔE^B among these dimers due to transferable local character of the individual hydrogen bonding fragments. Such relations were established through analysis of the local hydrogen-bonding/binding energies of the fragments ($E_{X\dots Y}$) (X and Y are the atoms or groups involved in hydrogen bonding). It is assumed that the ΔE^B values are originating solely due to the contribution of $E_{X...Y}$ terms and they are also transferable. In the case of formic acid and acetic acid dimers these fragments are OH . . . O, CH . . . O, OH . . . O(H), and CH . . . O(H), while for formamide and acetamide dimers these fragments are NH . . . O and CH . . . O (see Figs. 3.10, 3.11, 3.12, and 3.13 for the definition of these fragments). The hydrogen within parentheses belong to the O-H group not involved in hydrogen bonding. The $E_{X...Y}$ values of the fragments could be evaluated by inspecting the nature of HB interactions and ΔE^B values of an individual dimer and setting a simple additive relation from these data. The procedure could be explained using the following examples of the different dimers.

In the case of formic acid dimers, the hydrogen bonds in TT-1 and TT-5 isomers are solely due to OH . . . O and CH . . . O fragment interactions (Fig. 3.10). Thus $E_{OH...O}$ and $E_{CH...O}$ fragment energies in these cases are simply half of their ΔE^B values. The evaluations of the $E_{OH...O(H)}$ and $E_{CH...O(H)}$ fragment energies could be computed directly from the following relations for TT-4 and TT-6 isomers (Eqs. 3.41 and 3.42)

$$E_{CH...O} + E_{OH...O(H)} = \Delta E^B\,(TT-4) \tag{3.41}$$

$$E_{CH...O} + E_{CH...O(H)} = \Delta E^B\,(TT-6) \tag{3.42}$$

Once the fragment energies are known, the ΔE^B values of the other dimers could be easily evaluated using these key $E_{X...Y}$ values. In the case of *trans-cis* dimers, the *trans-* to *cis*-formic acid conversion energy (E_{TC}) would be needed to evaluate the binding energies. The following Eq. (3.43) could be used as an example for the use of E_{TC} data to compute ΔE^B of TC-1 isomer (Fig. 3.10).

$$E_{CH...O} + E_{O...HO} + E_{TC} = \Delta E^B\,(TC-1) \tag{3.43}$$

The predicted ΔE^B values (MP2 level) of the isomers TT-2 (–8.28 kcal/mol), TC-1 (–4.20 kcal/mol), TC-3 (–2.60 kcal/mol = TC-4), and TC-5 (–1.66 kcal/mol) were within 0.5 kcal/mol of the computed results at the MP2/aug-cc-pVTZ level. The computed E_{TC} value of 4.08 kcal/mol (MP2 level) was used in such calculations for the *trans-cis* isomers [85].

The acetic acid dimers (Fig. 3.10), like formic acid dimers, have OH . . . O (AA-1), CH . . . O (AA-5), and OH . . . O(H) (AA-4) fragments involved in hydrogen bonding. The evaluations of the related fragment energies, like formic acid dimers, are quite straight forward. These values are not the same with respect to the formic acid dimers due to the difference of ΔE^B values. The AA-3 dimer in this respect presents a special hydrogen bonding situation. Here, O-H center on fragment is bound to O (O-H . . . O) and O-H (HO . . . HO) centers of the second fragment (Fig. 3.11). This HO . . . HO fragment is not like OH...O(H) fragment of AA-4 and is marked as OH...O_1. The ΔE^B of AA-3 isomer is used to evaluate the OH...O_1 fragment energy using the value of $E_{OH...O(H)}$ in the following relation (3.44).

$$E_{OH...O_1} + E_{OH...O(H)} = \Delta E^B (AA - 3) \quad (3.44)$$

These known values of fragments energies predict the ΔE^B AA-2 and AA-6 isomers within 1.0 kcal/mol of the computed values using MP2/aug-cc-pVTZ technique [86]. These additive relations were used in cases of formamide and acetamide dimers also. They needed knowledge of the fragment energies of NH...O and CH...O fragment energies from the structural patterns of formamide (FMAD-A and FMAD-E) and acetamide (AMD-1, and AMD-4) dimers. These fragment energy values differ slightly because of the differences of structures, and ΔE^B values of these two different amides. The predicted values for the other isomers using these fragment values were quite satisfactory [85, 86].

The results discussed so far are not restricted to the dimers of specific fatty acids and their amides. The idea was found to be useful to other homologs also, provided they have similar hydrogen bonding features. The propionic acid and n-butyric acid dimers have similar dihydrogen bonded structures of AA-1. They differed only in the size of the alkyl group (-C_2H_5 for propionic acid and –C_3H_7 for n-butyric acid). Considering the difference of binding energies of TT-1 (–13.4 kcal/mol) and AA-1 (–14.4 kcal/mol) isomers at the MP2/aug-cc-pVTZ level [85], the fragment energy of two –CH_3 $\left(E_{CH_3}\right)$ is estimated to be 1.0 kcal/mol (since rest parts of the two dimers are similar). Assuming the fragment energies –C_2H_5 and –C_3H_7 to be equivalent to two and three –CH_3 groups, the ΔE^B of propionic acid and n-butyric acid dimers could be predicted to be –15.4 and –16.4 kcal/mol [86]. These results are very impressive against the respective experimental values of 15.2 ± 0.2 and 17.2 ± 0.8 kcal/mol. Thus, the additive nature of binding energies, as discussed here, seems quite natural for such dimers. It shows some predictive nature on binding energies in a homologous series as well.

3.4 Molecular Modeling of Strong and Weak Cation-π Interactions

The cation-π interactions, which were fundamentally coined by Kier and coworkers [13, 14] as a non-covalent interaction mediated by ion-induced dipole foreseeable via the molecular modeling of acetylcholinesterase inhibition reactions and subsequently rationalized by Dougherty and coworkers [15, 16] in diverse chemical and biological systems, have appeared as a very prevalent restraining force to explicate the crucial factor responsible for non-covalent binding in small gas-phase ion–molecule complexes as well as macromolecular protein-ligand systems. There are mainly two types of cation-π interactions that falls into the category of weak and strong interactions for the sake of essence of electrostatics in molecular fragments.

The quantitative estimation of such cation-π interactions in biological systems is the paramount concern to comprehend the underlying factors for the molecular recognition processes. To interpret the molecular recognition pattern contributed by the cation-π interactions stemming from the side chain of phenylalanine, tyrosine, and tryptophan with arginine and lysine in 1718 typical protein structures, *Minoux and Chipot* [17] performed quantum mechanical calculations by accounting the interactions of ammonium and guanidinium ions with the toluene, *p*-cresol, and methyl-indole as prime models of large molecular assemblies. The inclusion of polarization effects in predicting cation-π interactions seems to be indispensable, as evident by the computed binding energies with the basis-set superposition error (BSSE)-corrected MP2/6-311++G(d,p)//MP2/6-31G(d,p) level of approximation. Furthermore, the commercial force field such as Amber is demonstrated to be a reliable and efficient approach for evaluating the cation-π interactions in macromolecular assemblies of biological concern. The trends in calculated binding energies for the non-covalent interactions of toluene, *p*-cresol, and methyl-indole with the ammonium and guanidinium ions using the Amber force field compare well with the MP2 results even though the molecular mechanics method based on Amber force field predicts shorter interaction distances between the cations and the centroid of the aromatic ring with respect to the quantum mechanical calculations.

The cation-π interactions also play a critical role in stabilizing the coordination complexes of alkali-metal cations together with the aromatic systems. Nicholas et al. [101] have estimated the strength of cation-π interaction between the alkali-metal cations (Li^+ to Cs^+) and the benzene ring by accounting the consequences of incomplete basis sets within the framework of restricted Hartree-Fock (RHF) and second-order Møller-Plesset perturbation theory (MP2) levels. The predicted binding energies for the three heavier cations at the SVWN/TZ94p level are found to be 15–20% higher compared to the MP2 results, while the computed binding energies using BP96/TZ94p level are shown to be ~20% reduced when compared with the MP2 data. For such cation-π interactions, the binding enthalpies are usually underestimated compared to the experimental measurement, albeit the calculated binding enthalpies using BP96/TZ94p level are in accordance with the MP2 results. In another study by Sunner et al. [102], the ion-quadrupole and ion-induced dipole attractions due to the interaction between the potassium ion and benzene were evaluated by performing ab initio calculations (at the STO-3G level) as well as the classical electrostatic calculations, and the consideration of quadrupole moment of the aromatic ring is suggested to be the pivotal parameter to describe the electrostatics of cation-π interaction. In practice, the total electrostatic interaction could be judged as the sum of charge–charge, charge-dipole, charge–quadrupole, charge–octupole, and higher order terms. Kim et al. [103] have estimated charge-dipole, charge–quadrupole and charge-polarizability interactions using MP2 level in conjunction with 6-311+G(d,p) basis sets, to comprehend the nature of cation-π interactions for the binding of ammonium and tetramethylammonium cations with benzene and water. The obtained results clearly demonstrate that the interaction between tetramethylammonium cation and benzene ring is crucially contributed by the charge-quadrupole and charge-polarizability interactions. However, the contribution of other electrostatic terms has not been properly accounted in such cation-π interactions involving charged amino group or metal ions and aromatic systems.

The pertinency of computing various multipolar electrostatic energy terms in describing both the strong and weak cation-π interactions has been critically analyzed by Kadlubanski et al. [104]. The nature of cation-π interactions is assessed via the comprehensive survey of two-body interaction energy decomposition components of several gas-phase and dihydrated cation-π complexes comprising benzene, p-methylphenol, and 3-methylindole as the π-donor systems and Mg^{2+}, Ca^{2+}, and $NH_4{}^+$ cations as the acceptor units by implementing a hybrid variational–perturbational interaction energy decomposition scheme. The first-order electrostatic and higher order delocalization energy components of the interaction energy are indicated to be the critical parameters in elucidating the strong and weak binding of cation-π complexes. To ascertain the reliability of the computational approach for evaluating the energetics of cation-π interactions, the computed interacting distances between the donor and acceptor units, binding energies, and thermochemical properties such as enthalpy and Gibbs free energy of binding for each complex using local DFT method at the B3LYP level after the counterpoise (CP) and zero-point energy (ZPE) corrections are further compared with those obtained from the G4MP2 and CCSD(T) level of theory in combination with the aug-cc-pVDZ basis sets. The calculated binding energies and thermochemical properties for the gas-phase cation-π complexes using DFT/B3LYP level are in the immediate vicinity of the G4MP2 results. However, in case of dihydrated cation-π complexes, the predicted binding energies using G4MP2 method are usually lowered by 3–7 kcal/mol compared to those obtained by the DFT/B3LYP level, which is in accordance with the calculated shorter interaction distance using G4MP2 method. The binding energies are further improved by the CCSD(T) level of approximation especially for the binding of dihydrated Ca^{2+} ion with benzene and 3-methylindole. The Mulliken population analysis manifests that a significant amount of electronic charge is transferred from the aromatic systems to the Mg^{2+} and Ca^{2+} ions. Furthermore, a linear correlation between the calculated binding energy and the charge transport is obtained, thereby, indicating identical provenance of electrostatics of the cation-π interactions and demanding meticulous assessment of diverse electrostatic interaction components to untangle the dominating factors for the strong and weak cation-π interactions. Within the framework of Onsager reaction field model, the total energy of such a cation-π complex in the presence of a homogeneous electric field (V) induced by the acceptor atoms of the complex could be expressed as,

$$
\begin{aligned}
E = E_0 - \mu_i^0 V_i - \tfrac{1}{2}\alpha_{ij} V_i V_j - \tfrac{1}{3!}\beta_{ijk} V_i V_j V_k - \tfrac{1}{4!}\gamma_{ijkl} V_i V_j V_k V_l \ldots \\
\frac{1}{3} A_{i,jk} V_i \nabla V_{jk} \cdots - \frac{1}{6} B_{ij,kl} V_i V_j \nabla V_{kl} \cdots - \frac{1}{15} C_{i,jkl} V_i \nabla^2 V_{jkl}
\end{aligned}
\tag{3.45}
$$

where E_0 refers to the energy of the molecular system in the absence of external perturbation; μ^0 and α are the dipole moment and polarizability, and β and γ

represent the higher-order polarizabilities, respectively. $A_{i,jk}$ defines the dipole–dipole-quadrupole hyperpolarizability. The dipole–quadrupole and dipole–octupole polarizabilities are denoted by $B_{ij,kl}$ and $C_{i,jkl}$, respectively.

The essence of total electrostatic interaction energies (T-El) and the multipolar components of electrostatic terms including charge–charge (C–C), dipole-charge (D–C), quadrupole-charge (Q–C), and octupole–charge (O–C) as a function of interacting distance (r) in the cation-π complexes of Mg^{2+}, Ca^{2+}, and NH_4^+ are delineated in Figs. 3.14 and 3.15. The T-El curves corresponding to the cation-π complexes of Mg^{2+} exhibit a minimum, while those curves are dispersive in nature for the binding of Ca^{2+} ion with benzene and 3-methylindole (Fig. 3.14).

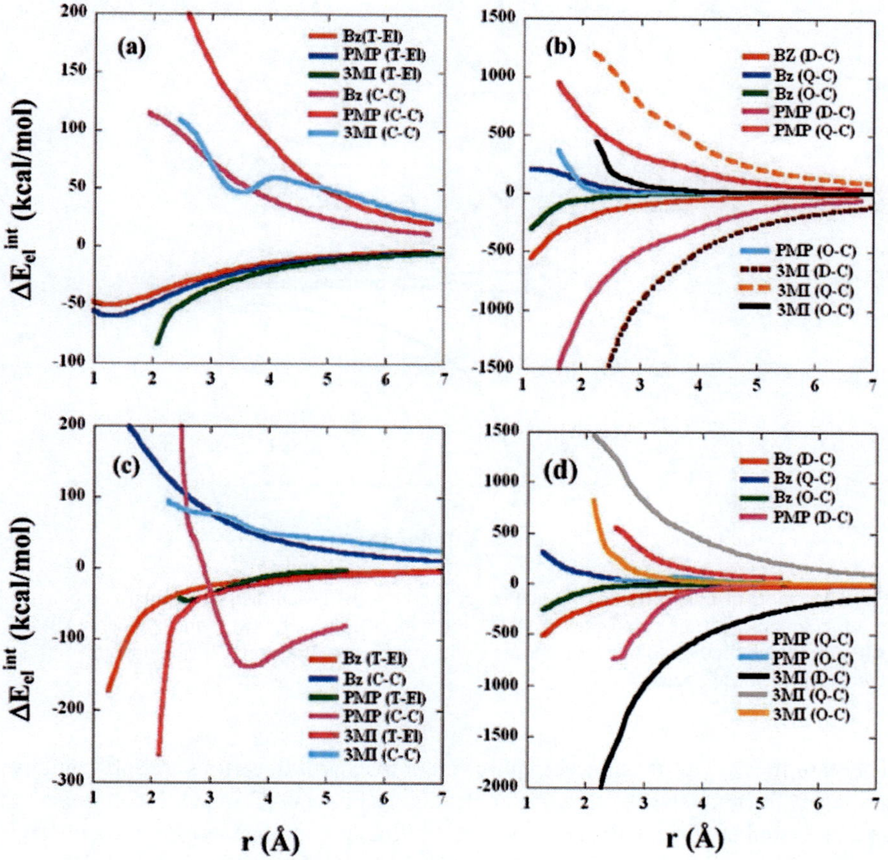

Fig. 3.14 Plots of total electrostatic interaction energies (T-El) and its other multipolar components as a function of r for the complexes of Mg^{2+} and Ca^{2+} ions with benzene (Bz), p-methylphenol (PMP), and 3-methylindole (3MI). Panels **a** and **b** represent the curves for the Mg^{2+} complexes, while panels **c** and **d** are for the Ca^{2+} ion complexes. In the figures, C–C, D–C, Q–C, and O–C represent the multipolar electrostatic interaction energy components (C–C, charge–charge; D–C, dipole–charge; Q–C, quadrupole–charge; O–C, octopole–charge) (reprinted with permission from ref. [104] Copyright (2013) American Chemical Society)

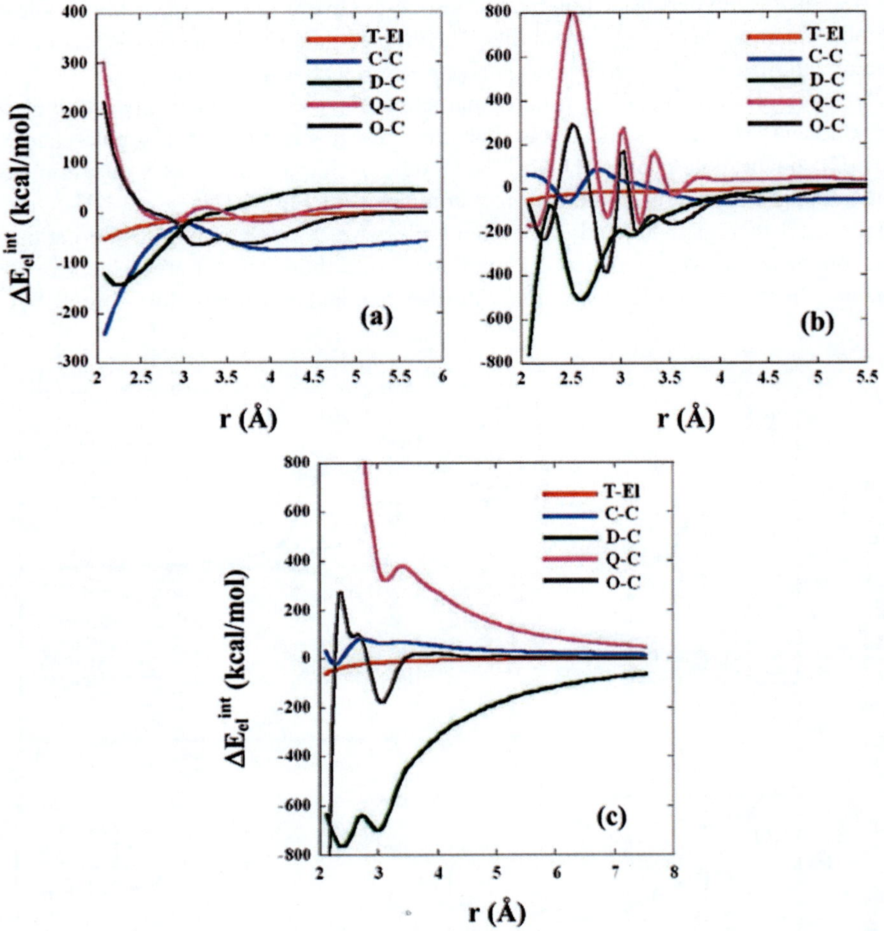

Fig. 3.15 Plots of total electrostatic interaction energies (T-El) and their multipolar components as a function of r for the complexes of NH_4^+ ion with benzene (Bz) (**a**), p-methylphenol (PMP) (**b**), and 3-methylindole (3MI) (**c**) (reprinted with permission from ref. [104] Copyright (2013) American Chemical Society)

The stabilities of strong and weak cation-π complexes of these ions are substantially guided by the repulsive or attractive contributions of the Q–C and O–C components, as manifested by the calculated curves of multipolar components of the Mg^{2+}- and Ca^{2+}-complexes shown in Fig. 3.14. The distinct nature of cation-π interactions for the NH_4^+-complexes is certainly evident by the T-El curves as well as anisotropic potential expansion of the multipolar components around the equilibrium distance r as displayed in Fig. 3.15. The contribution of multipolar components of the interaction energy is further demonstrated to be intrinsically linked to the cation-π vibrational modes specifically the intramolecular stretching frequency (S_z) assigned to the back-and-forth motion of the cation coupled with out-of-plane twisting mode of the aromatic moiety, and the out-of-plane C-H bending mode of the π-system (ν_{opCHb}). The augmentation of IR intensity of a particular mode in a cation-π complex could be further estimated from the alteration in dipole derivative expressed as

$$\partial\left(\frac{\partial\mu}{\partial Q}\right)=\left(\frac{\partial\mu}{\partial Q}\right)_{complex}-\left(\frac{\partial\mu}{\partial Q}\right)_{isolated} \tag{3.46}$$

where Q corresponds to the normal coordinates of the cation–π mode; and the dipole derivative for a given vibrational mode of the molecular system could be computed by neglecting higher order induced moment terms as follows,

$$\left(\frac{\partial\mu_i}{\partial Q}\right)=\left(\frac{\partial\mu_i^0}{\partial Q}+\frac{\partial\alpha_{ij}}{\partial Q}V_j+\frac{1}{2}\frac{\partial\beta_{ijk}}{\partial Q}V_jV_k\ldots\right)+\left(\alpha_{ij}\frac{\partial V_i}{\partial Q}+\beta_{ijk}\frac{\partial V_i}{\partial Q}V_k+\ldots\right) \tag{3.47}$$

The predicted intramolecular stretching mode of the respective cation-π complex corroborates well with the binding strength of the studied systems (Fig. 3.16a). The increase in IR intensities of these vibrational modes is found to correlate well with the multipolar electrostatic comportment of cation-π interactions as anticipated from the computed higher order quadrupolar and octupolar terms (Fig. 3.16c, d), albeit the linear correlation between the calculated blue shift of the C-H bending mode with respect to the isolated π-system ($\Delta\nu_{opCHb}$) and the binding energy (ΔE_B) shows strong dependence on the nature of π-system (Fig. 3.16b).

3.5 Molecular Modeling of π-π Interactions

The estimation of non-covalent interaction between π-systems constitute a basis for understanding binding mechanism of protein-ligand systems [105, 106]. The arene-arene interactions resulting from the edge-to-face or the parallel-displaced stacking orientations are found to play a leading role in stabilizing organic heterostructures and bio-macromolecules [107–109]. The high-level quantum mechanical calculations [110–115] anticipate that these arene-arene interactions are primarily assisted by the dispersion forces, although an earlier investigation led by Hunter and Sanders demonstrated that the essential contributions to the π-π interactions emanate from the electrostatic interactions [116]. A detailed theoretical investigation of the interactions between benzene and monosubstituted benzenes in parallel face-to-face stacking configuration revealed that the computed binding energies using hybrid DFT method correlate well with the Hammett σ_m values for the substituents. The parity of stacking interaction energies with the σ_m parameters could be further

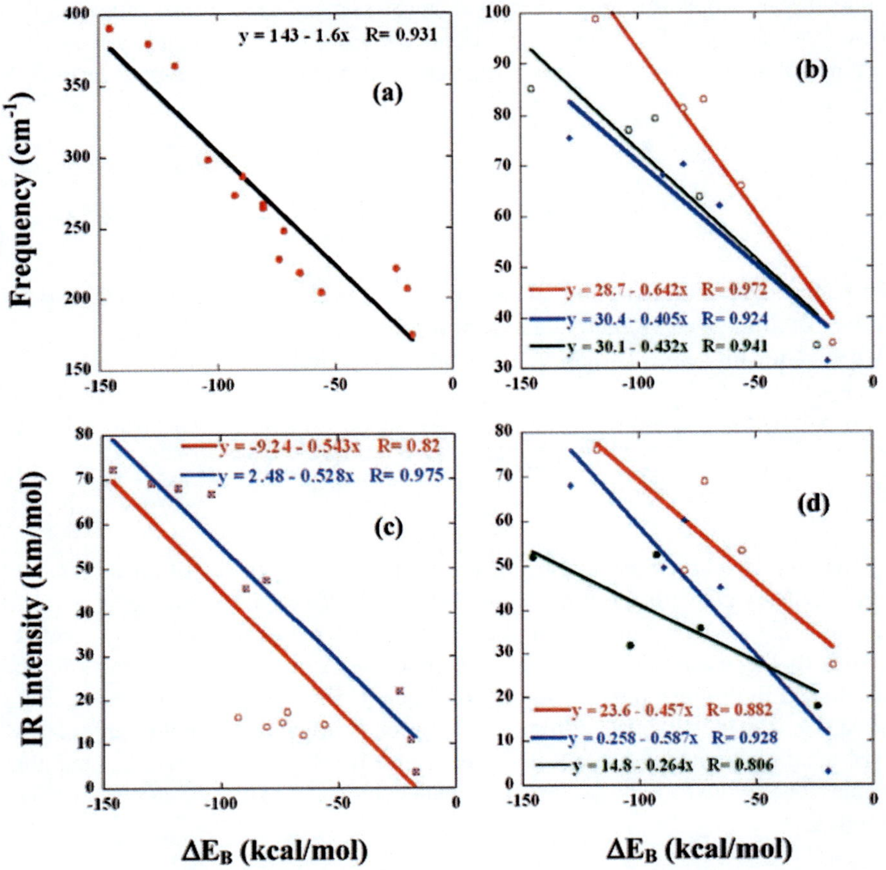

Fig. 3.16 (**a**, **b**) Correlation of ΔE_B with the intermolecular stretching (S_z, panel a) and blue-shifted out-of-plane CH bending mode of aromatic moiety ($\Delta\nu_{opCHb}$, panel **b**) modes of various cation–π complexes. (**b**) Red, blue, and green lines respectively, represent the complexes of the cations with benzene, p-methylphenol, and 3-methylindole. (**c**, **d**) Correlation of ΔE_B with IR_I for the respective Sz (panel **c**) and the intensity enhancement for the respective $\Delta\nu_{opCHb}$ (panel **d**) modes of the various cation–π complexes. The red line in panel **c** represents the correlation with all the complexes, while the blue line represents correlation excluding the hydrated metal ions. The red, blue, and green lines in panel **d** represent correlations for the similar cases in panel **b** (reprinted with permission from ref. [104] Copyright (2013) American Chemical Society)

explicated from the direct electrostatic and dispersive interactions of the benzene and substituted benzene systems as evident by the symmetry-adapted perturbation theory (SAPT) results. Recently, the influences of electrostatic and dispersive interactions in establishing the correlation between binding energies and Hammett σ-parameters for the offset face-to-face (OSFF) stacking interactions of diverse nitrobenzene derivatives with the model graphene systems have been assessed by Khan et al. [117]. An extensive comparative study of the arene-arene interactions is conducted for the interactions of substituted nitrobenzene dimers, $X–C_6H_4–NO_2$

(X= H, CH_3, OCH_3, OC_2H_5, Cl, Br, I, OH, CN, and NH_2 substituents at m- and p-positions) as well as di- and tri-substituted nitrobenzene derivatives [3-NO_2-4-OH, 3,5-di-NO_2-4-OH, and 3,5-di-NO_2-4-CH_3] with the model 5,5-graphene (GR) and its B- and N-doped scaffolds (3BGR and 3NGR) within the framework of DFT using M06 exchange-correlation functional and accounting the impact of large basis sets namely cc-pVDZ, cc-pVTZ and sp-aug-cc-pVTZ. The calculated BSSE- and ZPE-corrected binding energies (ΔE_B) and related thermodynamic parameters (ΔH_B^0 and ΔG_B^0) using DFT methods are further compared with those obtained from the MP2 level. Albeit the estimated binding energies at higher basis sets and the perturbative MP2 technique are found to be somewhat lower compared to the results derived from the M06/cc-pVDZ level, the predicted trends in ΔE_B values at diverse levels closely resemble to each other. The computed values of ΔE_B for the interaction of m- and p-substituted nitroaromatics with GR/BGR/NGR are revealed to be remarkably higher than that of benzene as well as pristine nitrobenzene. The changes in ΔH_B^0 and ΔG_B^0 are shown to be enhanced with the rise in the number of substituents in the nitroaromatic ring. It is noteworthy to mention that the predicted values of ΔG_B^0 for the adsorption of nitrobenzene, m-nitrobenzene, and p-nitrotoluene on the graphene surface compare well with the experimental results obtained from the Freundlich and Langmuir isotherms. The small disagreement in magnitude with respect to the experimental adsorption free energies is mainly attributable to the overestimated CP correction via double-ς basis set. The robust interaction of these extended π-systems stems from the electrostatic and dispersion interactions, as manifested by the molecular electrostatic potential (MEP) maps. The correlation between computed binding energy and Hammett parameter ($\sum|\sigma_m|$ or $\sum|\sigma_p|$) is further illustrated in Fig. 3.17a, b. The values of calculated ΔE_B using M06/cc-pVDZ level of theory are in conformity with the $\sum|\sigma_m|$ for all the nitroaromatics involved in π-π interactions with the GR/3BGR/3NGR scaffold, and the estimated correlation coefficient is found to be > 0.92 for each scaffold. The correspondence between ΔE_B and $\sum|\sigma_m|$ is retained to a greater or lesser extent for using diverse DFT methods and MP2 approach. The linear correlation between ΔE_B and $\sum|\sigma_p|$ parameters is also conserved but to a slightly lesser extent compared to the $\sum|\sigma_m|$ parameters especially for the interaction of nitrobenzene derivatives with the 3NGR. Similar correlations are maintained for the computed thermodynamic parameters ΔH_B^0 and ΔG_B^0 with the Hammett constants $\sum|\sigma_m|$ and $\sum|\sigma_p|$, as portrayed in Fig. 3.18.

Moreover, the dominant contribution of dispersion and electrostatic components to the ΔE_B is justified from the multiple regression analysis by accounting the molar refractivity ($\sum M_r$) or the changes in polarizability ($\Delta\alpha$) (with respect to benzene) in conjunction with the $\sum|\sigma_m|$ and $\sum|\sigma_p|$ parameters. The contributions of electrostatic $\left(E_{el}^{(10)}\right)$, and dispersive interactions (ΔE_{disp}: obtained from MP2) together with empirical dispersion stemming from the M06/cc-pVDZ level are further evaluated to explicate the nature of π-π interaction due to the attachment of nitroaromatics on the surface of GR, 3BGR and 3NGR, as represented in Fig. 3.19. The involvement of $E_{el}^{(10)}$ and ΔE_{disp} terms appear to be substantial in stabilizing

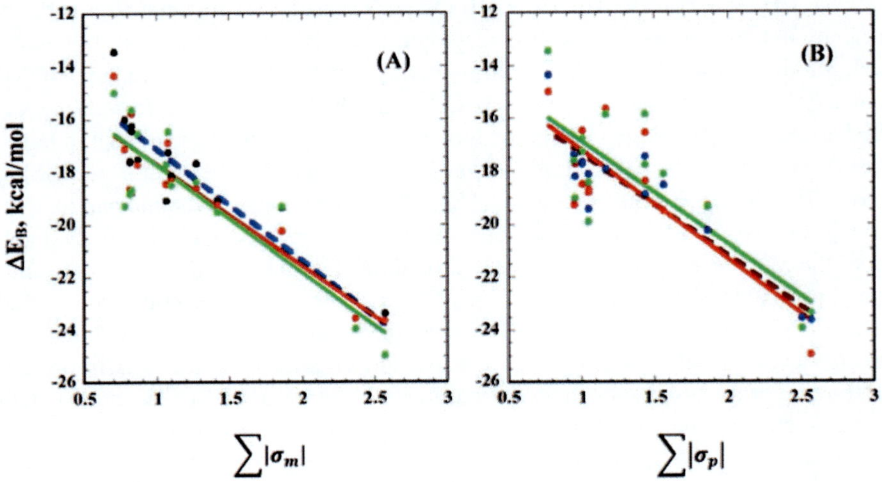

Fig. 3.17 Correlations of ΔE_B with $\Sigma|\sigma_m|$ (panel **a**) and $\Sigma|\sigma_p|$ (panel **b**) for the interactions between GR, 3BGR, and 3NGR with various m-, p- and several di- and tri-substituted nitrobenzene derivatives at the DFT/M06/cc-pVDZ level. Panel **a**: correlations of GR (blue line and black dots; r = 0.9), 3BGR (red line; r = 0.91) and 3NGR (green line; r = 0.87) (r: correlation coefficient) interacting with m-, and other higher substituted nitrobenzene derivatives. Panel **b**: correlations of GR (blue line; r = 0.91), BGR (red line; r = 0.91) and NGR (green line; r = 0.80) interacting with p-, and other higher substituted nitrobenzene derivatives (reprinted with permission from ref. [117] (https://pubs.acs.org/doi/10.1021/acsomega.7b01912). Copyright (2018) American Chemical Society; "Further permissions related to the material excerpted should be directed to the ACS")

offset face-to-face stacking interactions in such extended π-systems. The $E_{el}^{(10)}$ term is found to increase with the stronger electron-withdrawing and electron-donating substituents and furnish approximately equal contribution when compared with the sum of energy components emanating from the ΔE_{disp}, the delocalization term $\left(\Delta E_{del}^{HF}\right)$, and the repulsive Heitler–London exchange term $\left(E_{x}^{HL}\right)$. The energy decomposition analysis further anticipates that the $E_{el}^{(10)}$ contribution crucially originates from the charge–charge interactions even though the higher-order energy components associated with the dipole–dipole, dipole–quadruple, and quadruple–quadruple interactions are found to be important to determine the stabilizing factors for the π-π interactions. Thus, the obtained results substantiate the necessity of $\sum M_r$ and $\Delta\alpha$ together with the $\sum|\sigma_m|$ or $\sum|\sigma_p|$ parameters in the multiple regression analysis to evaluate the binding energies in extended π-systems in gas phase.

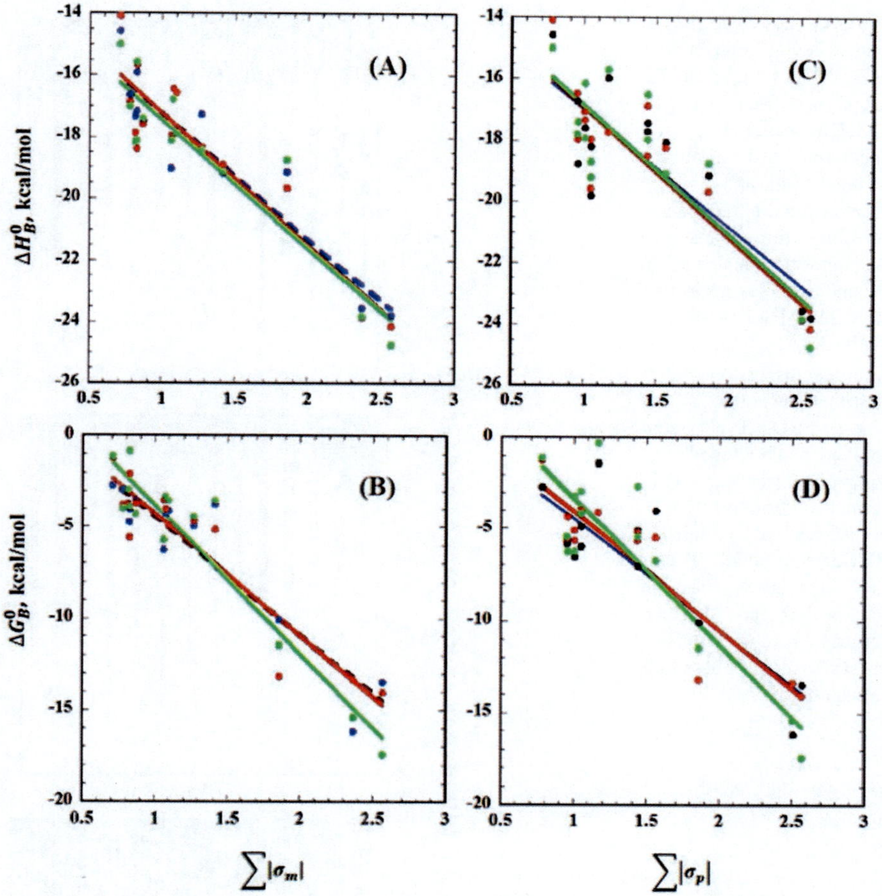

Fig. 3.18 Correlations of $\Delta H_B{}^0$ and $\Delta G_B{}^0$ with $\Sigma|\sigma_m|$ (panels **a** and **b**) $\Sigma|\sigma_p|$ (panels **c** and **d**) for the interactions of GR, 3BGR, and 3NGR with various m-, p- and several di- and tri-substituted nitrobenzene derivatives at the DFT/M06/cc-pVDZ level. Correlations in panel (**a**): GR... nitrobenzenes (blue): r = 0.93; 3BGR... nitrobenzenes (red): r = 0.92; 3NGR... nitrobenzenes (green): r = 0.92. Correlations in panel (**b**): GR... nitrobenzenes (blue): r = 0.93; 3BGR... nitrobenzenes (red): r = 0.94; 3NGR... nitrobenzenes (green): r = 0.95. Correlations in panel (**c**): GR... nitrobenzenes (blue): r = 0.85; 3BGR... nitrobenzenes (red): r = 0.90; 3NGR... nitro-benzenes (green): r = 0.86. Correlations in panel (**d**): GR... nitrobenzenes (blue): r = 0.86; 3BGR... nitrobenzenes (red): r = 0.92; 3NGR... nitrobenzenes (green): r = 0.95 (reprinted with permission from ref. [117] (https://pubs.acs.org/doi/10.1021/acsomega.7b01912), Copyright (2018) American Chemical Society; "Further permissions related to the material excerpted should be directed to the ACS")

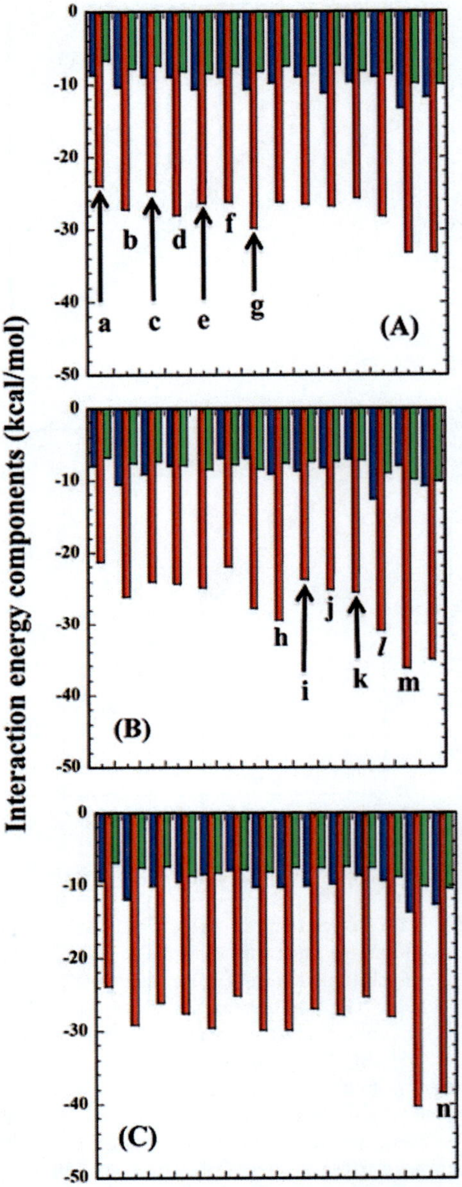

Fig. 3.19 Bar chart graphs comparing the contributions of $E_{el}^{(10)}$ and ΔE_{disp} in GR (panel **a**)/3BGR (panel **b**)/3NGR (panel **c**) interactions with various m- and several di- and tri-substituted nitrobenzene derivatives through energy decomposition analysis. The red and blue bars respectively represent ΔE_{disp} (MP2) and $E_{el}^{(10)}$ contributions. The empirical dispersion contributions at the M06/cc-pVDZ level (green bars) are also included for comparison (reprinted with permission from ref. [117] (https://pubs.acs.org/doi/10.1021/acsomega.7b01912), Copyright (2018) American Chemical Society; "Further permissions related to the material excerpted should be directed to the ACS")

3.6 Modeling Non-covalent Interactions in Bio-inspired Supramolecular Systems

Hierarchical molecular self-assembly, also called supramolecular polymerization, is prevalent in nature, and the molecular recognition pattern of such naturally occurring biomolecules can be exploited convincingly in vitro to generate functional nanomaterials with distinctive electronic, mechanical, and biological properties [118–123]. The potential applications of supramolecular assemblies in diverse fields including nanotechnology, biotechnology, and medicine necessitate a comprehensive understanding of the mechanism of molecular assembly [124–128]. Experimental techniques such as electron microscopy (scanning and transmission), nuclear magnetic resonance (NMR), X-ray diffraction (XRD), and in situ atomic force microscopy (AFM) fail to provide a detailed overview of the atomic interactions and the molecular driving forces that essentially promote the formation of macroscopic structures [129]. The design of tailored hierarchical soft materials using molecular dynamics simulations is also a formidable task, since a broad array of conformational sampling is necessary for the convergence of thermodynamic and structural quantities. The supramolecular polymerization is assisted by the delicate balance of attractive forces between the building blocks including hydrogen bonding and van der Waals interactions.

By performing atomistic MD simulations for a cylindrical nanofiber comprising 144 peptide amphiphile (PA) molecules (Fig. 3.20a) in water with physiological ion concentration and employing CHARMM force field, Schatz and co-workers have shown that the self-assembly of PAs into nanofiber is aided by the electrostatic interaction between the PAs and the sodium counterions as well as the van der Waals interaction between the PA units (Fig. 3.20b) [130]. Furthermore, the formation of β-sheets parallel to the fiber axis through the hydrogen-bonding

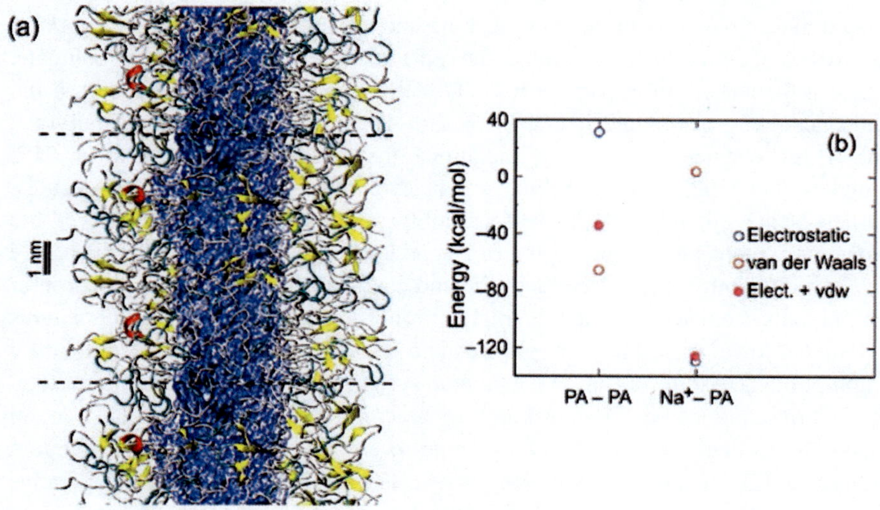

Fig. 3.20 (**a**) Snapshot of self-assembled PAs at 40 ns. The hydrophobic core is represented by a blue surface: R-helixes are in red, β-sheets are in yellow, turns are in cyan, and coils are in gray. (**b**) Intermolecular interaction energies between PAs are shown in the left column, and interaction energies between sodium ion and PA are shown in the right column. The sum of electrostatic (open blue circle) and van der Waals energies (open red circle) is shown as a filled purple circle (adapted with permission from ref. [130] Copyright (2011) American Chemical Society)

network is predicted to be the critical factor for the ordered structure of cylindrical nanofiber. In another study [131], the pertinency of plane-wave (PW)-based DFT computations in determining the driving forces for the experimentally perceived cooperativity in the hydrogen-bond-mediated supramolecular polymerization of $C3$-symmetrical trialkylbenzene-1,3,5-tricarboxamides (BTAs) has been assessed by analyzing the function of electrostatic interactions on the total cooperativity. The calculated interaction energy for the BTA dimer derived from PW-DFT using PBE functional is found to be –38.7 kJ/mol, which further enhances to –116 kJ/mol for the BTA oligomer holding seven monomers, and thereby leading to the total cooperative effect of 200%. The escalation of binding strength for the higher oligomers corroborates well with the shortening of hydrogen-bond length. The key interactions accounted for the cooperativity include long-range dipole-dipole interactions, short-range polarization, and resonance-assisted hydrogen bonding. The contribution of long-range dipole-dipole interactions for the self-assembled supramolecular polymer comprising seven BTA monomers is predicted to be 43% within a framework of pairwise dipole-dipole model. The remainder 71% of the interaction energy was assigned to the short-range polarization and resonance-assisted hydrogen bonding accompanied by electron redistribution along the BTA chain. The electron-density difference maps further manifest that the polarization of electron density alters substantially around the amide hydrogen bond with the augmentation of chain length of BTA oligomer, as delineated in Fig. 3.21.

Later, by performing classical MD simulations, Kang et al. [132] have demonstrated the impact of both the π-π stacking and the hydrogen bonding network on the growth of chiral filaments through the self-assembly of a peptide-drug conjugate, where a β-sheet forming peptide (CGVQIVYKK, or Tau) is conjugated with the anticancer drug camptothecin (CPT) by dint of disulfide linker (disulfylbutyrate, buSS). As revealed by the MD simulations, the π-π stacking between the CPT moieties, that emanates from the planar pentacyclic structure of the drug molecules, harnesses the initial phases of self-assembly process. The π-π stacking of the planar drug molecules of the neighboring peptide-drug conjugate is diverse in nature and exhibits both the displaced parallel and sandwiched configurations. The preassembled system accommodates a higher proportion of van der Waals interactions (31%) and lower contribution of electrostatic interactions (69%) between the drug amphiphiles (DAs) in contrast to the random system even though the calculated total interaction energies per DA are found to be comparable. Moreover, the fraction of intermolecular hydrogen bonds is found to escalate from 46% in the random system to 77% in the preassembled system, leading to a substantial cooperative rearrangement of the hydrogen bonding network prior to the commencement of nanofilament structure.

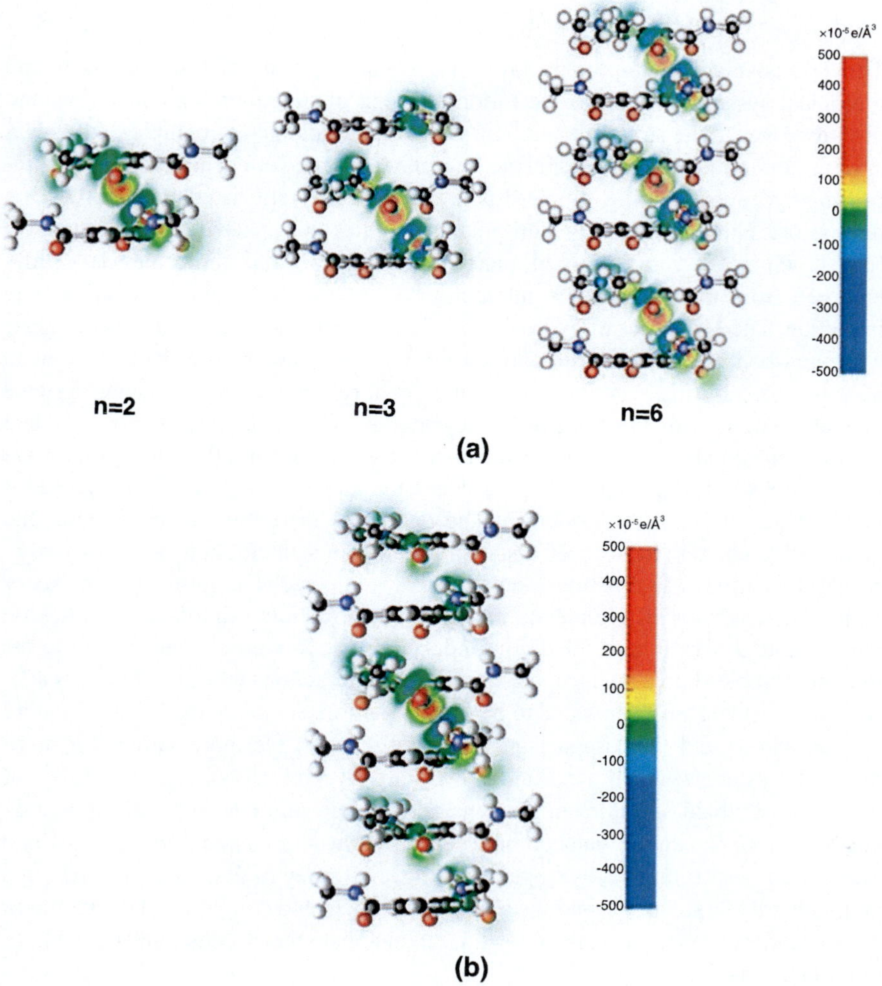

Fig. 3.21 Electron-density differences calculated (**a**) by subtracting the electron density of the individual monomers from the electron density of the constituting oligomer (dimer, trimer, and hexamer) and (**b**) by subtracting the electron density of two trimers within a hexamer from the electron density of the constituting hexamer. Values are in electrons/$Å^3$. Red regions denote accumulation of electron density, and blue regions correspond to depletion of electron density upon formation of the hydrogen-bond complex (reprinted with permission from ref. [131] Copyright (2010) American Chemical Society)

3.7 Conclusions

The non-covalent interactions play a significant role in diverse chemical and biological systems as well as the rational design of functional nanomaterials and drug discovery. The structure and conformational dynamics of complex molecular systems are influenced by the delicate interplay of non-covalent interactions encompassing hydrogen bonding, dipole-dipole interactions, steric repulsion, and London dispersion. The interaction of non-polar molecules is impacted by the dispersive forces, while the interaction of charged or highly polar molecules is chiefly emerged from the electrostatic interaction including Coulombic interaction and induction. The energetics of hydrogen-bonded systems, especially where polymeric hydrogen-bonding network are formed via O-H/N-H bonds, are demonstrated to be influenced by the specific low-frequency vibrations of the molecular systems in combination with the high-frequency modes. The reckoning of non-covalent interactions of halide ions with the surrounding solvent is turned out to be important to interpret the CTTS bands that are characteristics of charge-transfer spectra of halide-water clusters. As evident by the numerous theoretical investigations, the quantitative characterization of dispersion-dominated interactions such as the π-π and the cation-π interactions demand advanced correlated ab initio methods or empirical treatment. The understating of origin of cation-π interactions relies on the accurate description of the dipole-induced dipole as well as higher order terms like quadrupole-dipole, quadrupole-quadrupole interactions etc. The Hammett substituent constants are manifested to be critical parameters to comprehend the nature of interactions and their impact in predicting reaction thermodynamics for those molecular systems which are stabilized by the π-π interactions. The stability of the self-assembled structure in proteins and protein-mimetic materials is mainly attributable to the subtle balance between different non-covalent interactions like electrostatic, hydrophobic, hydrogen bonding, and van der Waals as revealed by the atomistic MD simulations; and the ability to estimate and comprehend the nature of non-covalent interactions is thus essential to elucidate the structure and function of hybrid materials.

The high-level wave function-based methods such as CCSD(T) or MP2 together with large basis sets like aug-cc-pVTZ and anharmonicity corrections are found to yield reliable predictions of weak and strong non-covalent interactions. However, the practical implementations of such approximate methods for large molecular systems are categorically hindered by their slow convergence with the basis set size and associated large basis set superposition errors. The dispersion-corrected DFT-based methods or range-separated functionals with a long-range correlation contribution from wave function methods followed by ZPE and BSSE corrections appear to be effective approach to trade-off between computational cost and accuracy.

Acknowledgments This work has been supported by NSF-CREST (Award No. 154774) and EPSCOR R-II (Award No. OIA - 1632899). One of the authors (S.R.) acknowledges the financial support by the Faculty of Chemistry of Wroclaw University of Science and Technology.

References

1. Pauling L (1931) The nature of the chemical bond. Application of results obtained from quantum mechanics and from theory of paramagnetic susceptibility to the structure of molecules. J Am Chem Soc 53:1367–1400
2. Coulson CA (1953) Valence. Clarendon, Oxford
3. Szabo A, Ostlund NS (1989) Modern quantum chemistry, introduction to advanced electronic structure theory. Dover, New York
4. Roothaan CCJ (1951) New developments in the molecular orbital theory. Rev Mod Phys 23:69–89
5. Møller C, Plesset MS (1934) Note on an approximation treatment for many-electron systems. Phys Rev 46:618–1622
6. Foresman JB, Frisch AE (2015) Exploring chemistry with electronic structure methods.3rd edn. Gaussian, Wallingford, CT
7. Hehre WJ, Radom L, Schleyer PvR, Pople JA (1986) Ab initio molecular orbital theory. Wiley-Interscience, New York
8. Helgaker T, Jørgensen P, Olsen J (2000) Molecular electronic-structure theory. Wiley, Chichester
9. Hohenberg P, Kohn W (1964) Inhomogeneous electron gas. Phys Rev 136:B864–B871
10. Kohn W, Sham LJ (1965) Self-consistent equations including exchange and correlation effects. Phys Rev 140:A1133–A1138
11. Stratmann RE, Scuseria GE, Frisch MJ (1998) An efficient implementation of time-dependent density-functional theory for the calculation of excitation energies of large molecules. J Chem Phys 109:8218–8244
12. Cossi M, Rega N, Scalmani G, Barone V (2003) Energies, structures, and electronic properties of molecules in solution with C-PCM solvation model. J Comput Chem 24:669–681
13. Kier LB, Aldrich HSA (1974) Theoretical study of receptor site models for trimethylammonium group interaction. J Theor Biol 46:529–541
14. Höltje H-D, Kier LB (1975) Nature of anionic or α-site of cholinesterase. J Pharm Sci 64:418–420
15. Dougherty DA (1996) Cation-p interactions in chemistry and biology: a new view of benzene, Phe, Tyr, and Trp. Science 271:163–168
16. Ma JC, Dougherty DA (1997) The cation–π interactions. Chem Rev 97:1303–1324
17. Minuox H, Chipot C (1999) Cation-π interactions in proteins: can simple models provide an accurate description? J Am Chem Soc 121:10366–10372
18. Kim KS, Tarakeswar P, Lee JY (2000) Molecular clusters of π-systems: theoretical studies of structures, spectra, and origin of interaction energies. Chem Rev 100:4145–4186
19. Dougherty DA, Stauffer DA (1990) Acetylcholine binding by a synthetic receptor: implications for biological recognition. Science 250:1558–1560
20. Kumpf RA, Dougherty DA (1993) A mechanism for ion selectivity in potassium channels: computational studies of cation–π interactions. Science 261:1708–1710
21. Gallivan JP, Dougherty DA (1999) Cation–π interactions in structural biology. Proc Natl Acad Sci USA 96:9459–9461
22. Xue Z, Suhm MA (2009) Probing the stiffness of the simplest double hydrogen bond: the symmetric hydrogen bond modes of jet-cooled formic acid dimer. J Chem Phys 131:054301-1–054301-4
23. Blandamer MJ, Fox MF (1970) Theory and applications of charge-transfer-to-solvent spectra. Chem Rev 70:59–93
24. Wheeler SE, Houk KN (2008) Substituent effects in the benzene dimer are due to direct interactions of the substituents with the unsubstituted benzene. J Am Chem Soc 130:10854–10855

25. Chen X, Chen B (2015) Macroscopic and spectroscopic investigations of the adsorption of nitroaromatic compounds on graphene oxide, reduced graphene oxide, and graphene nanosheets. Environ Sci Technol 49:6181–6189
26. Zhao N, Yang M, Zhao Q, Gao W, Xie T, Bai H (2017) Superstretchable nacre-mimetic graphene/poly(vinyl alcohol) composite film based on interfacial architectural engineering. ACS Nano 11:4777–4748
27. Schöllkopf W, Toennies JP (1996) The nondestructive detection of the helium dimer and trimer. J Chem Phys R E 1155–1158
28. Grisenti RE, Schöllkopf W, Toennies JP (2000) Determination of the bond length and binding energy of the helium dimer by diffraction from a transmission grating. Phys Rev Lett 85:2284–2287
29. Roy D, Marionski M, Maitra NT, Dannenberg JJ (2012) Comparison of some dispersion-corrected and traditional functionals with CCSD(T) and MP2 ab initio methods: dispersion, induction, and basis set superposition error. J Chem Phys 137:134109-1–134109-12
30. Rigby M, Smith EB, Wakeham WA, Maitland GC (1986) The forces between molecules. Clarendon, Oxford
31. Standard JM, Certain PR (1985) Bounds to two- and three-body long-range interaction coefficients for S-state atoms. J Chem Phys 83:3002–3008
32. Lennard-Jones JE (1924) On the determination of molecular fields. —II. From the equation of state of a gas. Proc R Soc Lond Ser A 106:463–477
33. Hill TL (1948) Steric effects. I. Van der Waals potential energy curves. J Chem Phys 16:399–404
34. Vedani A (1988) YETI: an interactive molecular mechanics program for small-molecule protein complexes. J Comp Chem 9:269–280
35. Tang KT, Toennies JP (2003) The van der Waals potentials between all the rare gas atoms from He to Rn. J Chem Phys 118:4976
36. Tang KT, Toennies JP (1984) An improved simple model for the van der Waals potential based on universal damping functions for the dispersion coefficients. J Chem Phys 80:3726–3741
37. Grimme S (2006) Semiempirical GGA-type density functional constructed with a long-range dispersion correction. J Comput Chem 27:1787–1799
38. Grimme S, Antony J, Ehrlich S, Krieg H (2010) A consistent and accurate ab initio parametrization of density functional dispersion correction (DFT-D) for the 94 elements H-Pu. J Chem Phys 132:154104-1–154104-18
39. Grimme S, Ehrlich S, Goerigk L (2011) Effect of the damping function in dispersion corrected density functional theory. J Comput Chem 32:1456–1465
40. Xantheas SS (1994) Ab initio studies of cyclic water clusters $(H_2O)_n$, n=1-6. II. Analysis of many-body interactions. J Chem Phys 100:7523–7534
41. Becke AD, Johnson ER (2005) A density-functional model of the dispersion interaction. J Chem Phys 122:154101-1–154101-9
42. Koide A (1976) A new expansion for dispersion forces and its application. J Phys B 9:3173–3183
43. Gora RW, Barkowiak W, Roszak S, Leszczynski J (2002) A new theoretical insight into the nature of intermolecular interactions in the molecular crystal of urea. J Chem Phys 117:1031–1039
44. Gora RW, Sokalski WA, Leszczynski J, Pett VB (2005) The nature of interactions in the ionic crystal of 3-pentenenitrile, 2-nitro-5-oxo, ion (−1) sodium. J Phys Chem B 109:2027–2033
45. Jeziorski B, Moszynski R, Szalewicz K (1994) Perturbation theory approach to intermolecular potential energy surfaces of van der Waals complexes. Chem Rev 94:1887–1930
46. Heßelmann A (2018) DFT-SAPT intermolecular interaction energies employing exact-exchange Kohn–Sham response methods. J Chem Theor Comput 14:1943–1959
47. Gutowski M, Piela L (1988) Interpretation of the Hartree-Fock interaction energy between closed-shell systems. Mol Phys 64:337–355

48. Larson JW, McMahon TB (1984) Gas-phase bihalide and pseudobihalide ions. An ion cyclotron resonance determination of hydrogen bond energies in XHY- species (X, Y = F, Cl, Br, CN). Inorg Chem 23:2029–2033
49. Desiraju GR, Steiner T (1999) The weak hydrogen bond: In structural chemistry and biology. International Union of Crystallography, Monographs on crystallography, 9. Oxford University Press, Oxford and New York
50. Lipinski CA (2004) Lead- and drug-like compounds: the rule-of-five revolution. Drug Discov Today Technol 1:337–341
51. Varadwaj PR, Varadwaj A, Marques HM, Yamashita K (2019) Significance of hydrogen bonding and other non-covalent interactions in determining octahedral tilting in the CH3NH3PbI3 hybrid organic-inorganic halide perovskite solar cell semiconductor. Sci Rep 9:1–29
52. Johnson ER, Keinan S, Mori-Sánchez P, Contreras-García J, Cohen AJ, Yang W (2010) Revealing non-covalent interactions. J Am Chem Soc 132:6498–6506
53. Kuo S-W (2018) Hydrogen bonding in polymer materials. Wiley-VCH, Weinheim
54. Pugliano N, Saykally RJ (1992) Measurement of quantum tunneling between chiral isomers of the cyclic water trimer. Science 257:1937–1940
55. Cruzan JD, Braly LB, Liu K, Brown MG, Loeser JG, Saykally RJ (1996) Quantifying hydrogen bond cooperativity in water: VRT spectroscopy of the water tetramer. Science 271:59–62
56. Liu K, Brown MG, Cruzan JD, Saykally RJ (1996) Vibration-rotation tunneling spectra of the water pentamer: structure and dynamics. Science 271:62–64
57. Liu K, Brown MG, Saykally RJ (1997) Terahertz laser vibration–rotation tunneling spectroscopy and dipole moment of a cage form of the water hexamer. J Phys Chem 101:8995–9010
58. Vernon MF, Krajnovich DJ, Kwok HS, Lisy JM, Shen YR, Lee YT (1982) Infrared vibrational predissociation spectroscopy of water clusters by the crossed laser-molecular beam technique. J Chem Phys 77:47–57
59. Gruenloh CJ, Carney JR, Arrington CA, Zwier TS, Fredricks SY, Jordan KD (1997) Infrared spectrum of a molecular ice cube: the S4 and D2d water octamers in benzene-(water)$_8$. Science 276:1678–1681
60. Buck U, Ettischer I, Melzer M, Buch V, Sadlej J (1998) Structure and spectra of three-dimensional $(H_2O)_n$ clusters, n= 8, 9, 10. Phys Rev Lett 80:2578–2581
61. Xantheas SS, Dunning Jr TH (1993) Ab initio studies of cyclic water clusters $(H_2O)_n$, n=1–6. I. Optimal structures and vibrational spectra. J Chem Phys 99:8774–8792
62. Fowler JE, Schaefer III HF (1995) Detailed study of the water trimer potential energy surface. J Am Chem Soc 117:446–452
63. Brudermann J, Melzer M, Buck U, Kazimirski JK, Sadlej J, Buch V (1999) The asymmetric cage structure of $(H_2O)_7$ from a combined spectroscopic and computational study. J Chem Phys 110:10649–10652
64. Combariza JE, Kestner NR, Jortner J (1994) Energy-structure relationships for microscopic solvation of anions in water clusters. J Chem Phys 100:2851–2864
65. Kim KS, Mhin BJ, Choi U-S, Lee K (1992) Ab initio studies of the water dimer using large basis sets: the structure and thermodynamic energies. J Chem Phys 97:6649–6662
66. Kim J, Kim KS (1998) Structures, binding energies, and spectra of isoenergetic water hexamer clusters: extensive ab initio studies. J Chem Phys 109:5886–5895
67. Kim J, Majumdar D, Lee HM, Kim KS (1999) Structures and energetics of the water heptamer: comparison with the water hexamer and octamer. J Chem Phys 110:9128–9134
68. Temelso B, Archer KA, Shields GC (2011) Benchmark structures and binding energies of small water clusters with anharmonicity corrections. J Phys Chem A 115:12034–12046
69. Markovich G, Pollack S, Giniger R, Cheshnovsky O (1991) Photoelectron spectroscopy of iodine anion solvated in water clusters. J Chem Phys 95:9416–9419
70. Markovich G, Pollack S, Giniger R, Cheshnovsky O (1994) Photoelectron spectroscopy of CI-, Br-, and 1- solvated in water clusters. J Chem Phys 101:9344–9353

71. Combariza JE, Kestner NR, Jortner J (1993) Microscopic solvation of anions in water clusters. Chem Phys Lett 203:423–428
72. Baik J, Kim J, Majumdar D, Kim KS (1999) Structures, energetics, and spectra of fluoride–water clusters $F^{-}(H_2O)_n$, n=1–6: ab initio study. J Chem Phys 110:9116–9127
73. Shi R, Wang P, Tang L, Huang X, Chen Y, Su Y, Zhao J (2018) Structures and spectroscopic properties of $F^{-}(H_2O)_n$ with n = 1–10 clusters from a global search based on density functional theory. J Phys Chem A 122:3413–3422
74. Arshadi M, Yamadgni R, Kebarle P (1970) Hydration of the halide negative ions in the gas phase. II. Comparison of hydration energies for the alkali positive and halide negative ions. J Phys Chem 74:1475–1482
75. Hiraoka K, Mizuse S, Yamabe S (1988) Solvation of halide ions with H_2O and CH_3CN in the gas phase. J Phys Chem 92:3943–3952
76. Fox MF, Hayon E (1977) Far ultraviolet solution spectroscopy of the iodide ion. J Chem Soc Faraday Trans 173:1003–1016
77. Serxner D, Dessent CEH, Johnson MA (1996) Precursor of the I_{aq}^{-} charge-transfer-to-solvent (CTTS) band in $I^{-}.(H_2O)_n$ clusters. J Chem Phys 105:7231–7234
78. Lehr L, Zanni MT, Frischkorn C, Weinkauf R, Neumark DM (1999) Electron solvation in finite systems: femtosecond dynamics of iodide·$(water)_n$ anion clusters. Science 284:635–638
79. Takahashi N, Sakai K, Tanida T, Watanabe I (1995) Vertical ionization potentials and CTTS energies for anions in water and acetonitrile. Chem Phys Lett 246:183–186
80. Majumdar D, Kim J, Kim KS (2000) Charge transfer to solvent (CTTS) energies of small $X^{-}(H2O)_{n=1-4}$ (X=F, Cl, Br, I) clusters: ab initio study. J Chem Phys 112:101–105
81. Kim J, Lee HM, Suh SB, Majumdar D, Kim KS (2000) Comparative ab initio study of the structures, energetics and spectra of $X^{-}(H2O)_{n=1-4}$ [F, Cl, Br, I] clusters. J Chem Phys 113:5259–5272
82. Kim J, Suh SB, Kim KS (1999) Water dimer to pentamer with an excess electron: ab initio study. J Chem Phys 111:10077–10087
83. Lee S, Kim J, Lee SJ, Kim KS (1997) Novel structures for the excess electron state of the water hexamer and the interaction forces governing the structures. Phys Rev Lett 79:2038–2041
84. Olbert-Majkut A, Akhokas J, Lundell J, Petterson M (2011) Raman spectroscopy of acetic acid monomer and dimers isolated in solid argon. J Raman Spectrosc 42:1670–1681
85. Cato Jr MA, Majumdar D, Roszak S, Leszczynski J (2013) Exploring relative thermodynamic stabilities of formic acid and formamide dimers—role of low-frequency hydrogen-bond vibration. J Chem Theory Comput 9:1016–1026
86. Copeland C, Menon O, Majumdar D, Roszak S, Leszczynski J (2017) Understanding the influence of low-frequency vibrations on the hydrogen bonds of acetic acid and acetamide dimers. Phys Chem Chem Phys 19:24866–24878
87. Tan TL, Goh KL, Ong PP, Teo HH (1999) Rovibrational constants for the ν_6 and $2\nu_9$ bands of HCOOD by Fourier transform infrared spectroscopy. J Mol Spectrosc 198:110–114
88. Bertie JE, Michaelian KH, Eysel HH, Hager D (1986) The Raman-active O–H and O–D stretching vibrations and Raman spectra of gaseous formic acid-d1 and -OD. J Chem Phys 85:4779–4789
89. Pettersson M, Lundell J, Khriachtchev L, Räsänen M (1997) IR spectrum of the other rotamer of formic acid, cis-HCOOH. J Am Chem Soc 119:11715–11716
90. Marushkevich K, Khriachtchev L, Lundell J, Domanskaya A, Räsänen M (2010) Matrix isolation and ab initio study of trans–trans and trans–cis dimers of formic acid. J Phys Chem A 114:3495–3502
91. Marushkevich K, Siltanen M, Räsänen M, Haloneu L, Khriachtchev L (2011) Identification of new dimers of formic acid: the use of a continuous-wave optical parametric oscillator in matrix isolation experiments. J Phys Chem Lett 2:695–699

92. Marushkevich K, Khriachtchev L, Räsänen M, Melqvuori M, Londell J (2012) Dimers of the higher-energy conformer of formic acid: experimental observation. J Phys Chem A 116:2101–2108
93. Senent ML (2001) Ab initio determination of the torsional spectra of acetic acid. Mol Phys 99:1311–1321
94. Maçôas EMS, Khriachtchev L, Pettersson M, Fausto R, Räsänen M (2003) Rotational isomerism in acetic acid: the first experimental observation of the high-energy conformer. J Am Chem Soc 125:16188–16189
95. Chocholoušová J, Vacek J, Hobza P (2003) Acetic acid dimer in the gas phase, nonpolar solvent, microhydrated environment, and dilute and concentrated acetic acid: ab initio quantum chemical and molecular dynamics simulations. J Phys Chem A 107:3086–3092
96. Roszak S, Gee RH, Balasubramanian K, Fried LE (2005) New theoretical insight into the interactions and properties of formic acid: development of a quantum-based pair potential for formic acid. J Chem Phys 123:144702-1–144702-10
97. Kollipost F, Larsen RW, Domanskaya AV, Nörenberg M, Shum MA (2012) Communication: the highest frequency hydrogen bond vibration and an experimental value for the dissociation energy of formic acid dimer. J Chem Phys 136:151101-1–151101-4
98. Goubet M, Soulard P, Pirali O, Asselin P, Réal F, Gruet S, Huet TR, Roy P, Georges R (2015) Standard free energy of the equilibrium between the trans-monomer and the cyclic-dimer of acetic acid in the gas phase from infrared spectroscopy. Phys Chem Chem Phys 17:7477–7488
99. Marrdyukov A, Sánchez-Garcia E, Rodziewicz P, Doltsinis NL, Sander W (2007) Formamide dimers: a computational and matrix isolation study. J Phys Chem A 111:10552–10561
100. Albrecht M, Rice CA, Suhm MA (2008) Elementary peptide motifs in the gas phase: FTIR aggregation study of formamide, acetamide, N-methylformamide, and N-methylacetamide. J Phys Chem A 112:7530–7542
101. Nicholas JB, Hay BP, Dixon DA (1999) Ab initio molecular orbital study of cation–π binding between the alkali-metal cations and benzene. J Phys Chem A 103:1394–1400
102. Sunner J, Nishizawa K, Kebarle P (1981) Ion-solvent molecule interactions in the gas phase. The potassium ion and benzene. J Phys Chem 85:1814–1820
103. Kim KS, Lee JY, Lee SJ, Ha T-K, Kim DH (1994) On binding forces between aromatic ring and quaternary ammonium compound. J Am Chem Soc 116:7399–7400
104. Kadlubanski P, Calderón-Mojica K, Rodriguez WA, Majumdar D, Roszak S, Leszczynski J (2013) Role of the multipolar Electrostatic interaction energy components in strong and weak cation–π interactions. J Phys Chem A 117(33):7989–8000
105. Meyer EA, Castellano RK, Diederich F (2003) Interactions with aromatic rings in chemical and biological recognition. Angew Chem Int Ed 42:1210–1250
106. Kryger G, Silman I, Sussman JL (1999) Structure of acetylcholinesterase complexed with E2020 (Aricept): implications for the design of new anti-Alzheimer drugs. Structure 7:297–307
107. Burley SK, Petsko GA (1985) Aromatic-aromatic interaction: a mechanism of protein structure stabilization. Science 229:23–28
108. McGaughey GB, Gagne M, Rappe AK (1998) π-stacking interactions: alive and well in proteins. J Biol Chem 273:15458–15463
109. Gung BW, Xue X, Reich HJ (2005) The strength of parallel-displaced arene–arene interactions in chloroform. J Org Chem 70:3641–3644
110. Sinnokrot MO, Sherrill CD (2006) High-accuracy quantum mechanical studies of π–π interactions in benzene dimers. J Phys Chem A 110:10656–10668
111. Sinnokrot MO, Sherrill CD (2004) Substituent effects in π–π interactions: sandwich and T-shaped configurations. J Am Chem Soc 126:7690–7697
112. Sinnokrot MO, Valeev EF, Sherrill CD (2002) Estimates of the ab initio limit for π–π interactions: the benzene dimer. J Am Chem Soc 124:10887–10893

113. Tsuzuki S, Uchimaru T, Mikami M (2006) Intermolecular Interaction between hexafluorobenzene and benzene: ab initio calculations including CCSD(T) level electron correlation correction. J Phys Chem A 110:2027–2033
114. Tsuzuki S, Honda K, Uchimaru T, Mikami M (2005) Ab initio calculations of structures and interaction energies of toluene dimers including CCSD(T) level electron correlation correction. J Chem Phys 122:144323
115. Tsuzuki S, Honda K, Uchimaru T, Mikami M, Tanabe K (2002) Origin of attraction and directionality of the π/π interaction: model chemistry calculations of benzene dimer interaction. J Am Chem Soc 124:104–112
116. Hunter CA, Sanders JKM (1990) The nature of π-π interactions. J Am Chem Soc 112:5525–5534
117. Khan MH, Leszczynska D, Majumdar D, Roszak S, Leszczynski J (2018) Interactions of substituted nitroaromatics with model graphene systems: applicability of Hammett substituent constants to predict binding energies. ACS Omega 3:2773–2785
118. McManus JJ, Charbonneau P, Zaccarelli E, Asherie N (2016) The physics of protein self-assembly. Curr Opin Colloid Interface Sci 22:73–79
119. Simon AJ, Zhou Y, Ramasubramani V, Glaser J, Pothukuchy A, Gollihar J, Gerberich JC, Leggere JC, Morrow BR, Jung C et al (2019) Supercharging enables organized assembly of synthetic biomolecules. Nat Chem 11:204–212
120. Ferrari S, Kahl G, Bianchi E (2018) Molecular dynamics simulations of inverse patchy colloids. Eur Phys J E Soft Matter Biol Phys 41:1–5
121. Eom K (2019) Computer simulation of protein materials at multiple length scales: from single proteins to protein assemblies. Multiscale Sci Eng 1:1–25
122. Zhang Z, Marson RL, Ge Z, Glotzer SC, Ma PX (2015) Simultaneous nano- and microscale control of nanofibrous microspheres self-assembled from star-shaped polymers. Adv Mater 27:3947–3952
123. Elemans JAAW, Rowan AE, Nolte RJM (2003) Mastering molecular matter. Supramolecular architectures by hierarchical self-assembly. J Mater Chem 13:2661–2670
124. Mandal D, Nasrolahi Shirazi A, Parang K (2014) Self-assembly of peptides to nanostructures. Org Biomol Chem 12:3544–3561
125. Eskandari S, Guerin T, Toth I, Stephenson RJ (2017) Recent advances in self-assembled peptides: implications for targeted drug delivery and vaccine engineering. Adv Drug Deliv Rev 110–111:169–187
126. Qiu F, Chen Y, Tang C, Zhao X (2018) Amphiphilic peptides as novel nanomaterials: design, self-assembly and application. Int J Nanomed 13:5003–5022
127. Frederix PWJM, Patmanidis I, Marrink SJ (2018) Molecular simulations of self-assembling bio-inspired supramolecular systems and their connection to experiments. Chem Soc Rev 47:3470–3489
128. Tuttle T (2015) Computational approaches to understanding the self-assembly of peptide-based nanostructures. Isr J Chem 55:724–734
129. Colombo G, Soto P, Gazit E (2007) Peptide self-assembly at the nanoscale: a challenging target for computational and experimental biotechnology. Trends Biotechnol 25:211–218
130. Lee O-S, Stupp SI, Schatz GC (2011) Atomistic molecular dynamics simulations of peptide amphiphile self-assembly into cylindrical nanofibers. J Am Chem Soc 133:3677–3683
131. Filot IAW, Palmans ARA, Hilbers PAJ, van Santen RA, Pidko EA, de Greef TFA (2010) Understanding cooperativity in hydrogen-bond-induced supramolecular polymerization: a density functional theory study. J Phys Chem B 114:13667–13674
132. Kang M, Zhang P, Cui H, Loverde SM (2016) π–π stacking mediated chirality in functional supramolecular filaments. Macromolecules 49:994–1001

Chapter 4
DNA Damage Radiosensitizers Geared Towards Hydrated Electrons

Magdalena Zdrowowicz, Lidia Chomicz-Mańka, Kamila Butowska, Paulina Spisz, Karina Falkiewicz, Anna Czaja, and Janusz Rak

Abstract Hypoxia, a hallmark of solid tumors, which account for ca. 80% of cancer cases, makes the cancer cells resistant toward ionizing radiation and as a consequence to radiotherapy, one of the most common modality employed against tumor. Oxygen present in significant amounts in the normoxic cells fixes DNA damage exerted by the radiation-formed hydroxyl radicals which leads to about three folds larger radiosensitivity of the normoxic cells compared to the hypoxic ones. The imparted reactivity of hydroxyl radicals ($^{\bullet}$OH) in the absence of oxygen and presence of equimolar to the $^{\bullet}$OH radicals number of solvated electrons under these conditions (under normoxia hydrated electrons (e_{hyd}) react with the dissolved oxygen leading to relatively nonreactive $O_2^{\bullet-}$ radicals) suggest the necessity of the employment of sensitizers sensitive to the hydrated electrons for an efficient radiotherapy. In the present chapter several groups of radiosensitizers geared towards hydrated electrons are thoroughly discussed. In *Introduction* we show how the hypoxic conditions influence cellular damage induced by ionizing radiation. Then we demonstrate that hydrated electrons do not have significant DNA damaging potential although they effectively bind to DNA (due to the subsequent protonation of the anions formed in result of electron attachment to DNA). Finally, we will draw the readers' attention to the need for using radiosensitizers reacting with hydrated electrons along with ionizing radiation. Introduction is followed by three sections devoted to the discussed classes of radiosensitizers (1) modified nucleosides, (2) oxygen mimetics and (3) metallic nanoparticles and metal complexes. Lastly, the review is concluded with a short summary, where the advantages and disadvantages of all the discussed classes of sensitizers are compiled.

M. Zdrowowicz · L. Chomicz-Mańka · K. Butowska · P. Spisz · K. Falkiewicz · A. Czaja · J. Rak (✉)
Faculty of Chemistry, University of Gdańsk, Gdańsk, Poland
e-mail: janusz.rak@ug.edu.pl

J. Leszczynski, M. K. Shukla (eds.), *Practical Aspects of Computational Chemistry V*, https://doi.org/10.1007/978-3-030-83244-5_4

4.1 Introduction

Cancer is the main reason of death in the developed countries [1] and radiotherapy belongs to one of the most common modalities for treating it in humans [1]. Indeed, more than 50% of cancer patients receive radiotherapy at certain stage of their treatment. Ionizing radiation (IR), a key component of radiotherapy, effectively damages cellular DNA when the concentration of oxygen assumes values typical for normal tissues (so called physoxia). On the other hand, solid tumors comprise ca. 80% of all diagnosed cases [2]. In this type of tumors low oxygen level, called hypoxia, is observed, and results from abnormal vascularization of tumor and increased metabolism of cancer cells (Fig. 4.1) [3]. First report on hypoxia in solid tumors, was published by Thomlinson and Gray in the middle of twentieth century [4]. They showed a relationship between hypoxia and resistance to radiation therapy. Similarly, in 1979, Brown demonstrated that poor response to radiotherapy is correlated with low oxygen levels in tumor microenvironment [5]. In fact, the efficacy of damaging the hypoxic cells with IR is three times lower than that of damaging the normoxic ones [6].

The action of radiotherapy is based on DNA damage, which if unrepaired leads to cellular death. It was demonstrated that the contribution of indirect effects (i.e. free radical processes triggered by radiolysis of water) to biological damage for sparsely

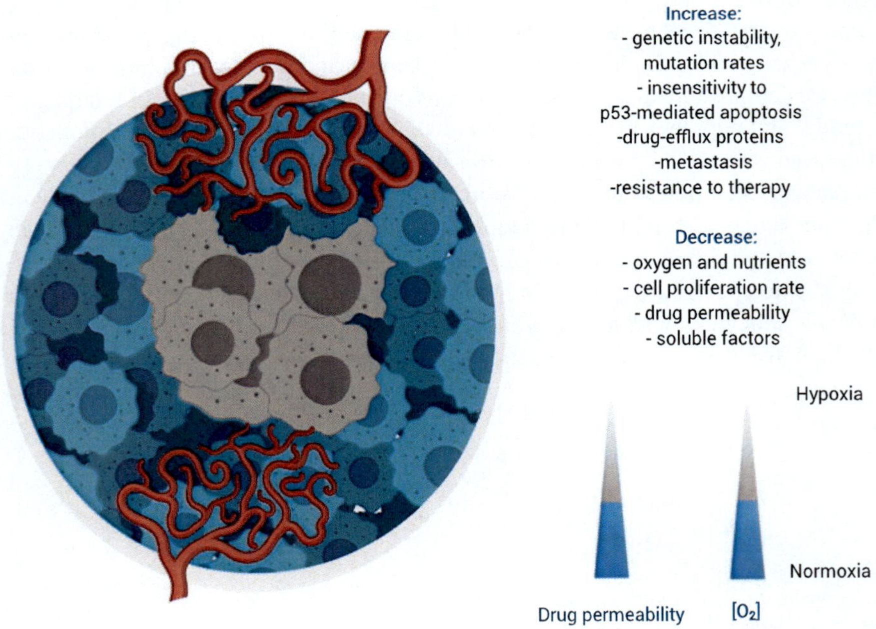

Fig. 4.1 Hypoxia in the tumor microenvironment. Hypoxia can limit radiotherapy and diffusion of chemotherapeutic agent into tumor, leading to radio- and chemoresistance.

ionizing radiation far exceeds that of its direct action [1]. The most abundant products of water radiolysis are hydroxyl radicals and secondary electrons. The latter species undergo very fast thermalization, becoming hydrated (e_{hyd}) within the picosecond time-scale. Until the seminal work of Sanche and coworkers [7], who demonstrated efficient generation of single and double strand breaks in plasmid DNA under ultra-high vacuum by electrons of energies well below ionization threshold of DNA, it was assumed that hydroxyl radicals are the main source of cellular DNA damage. Indeed, it has long been known that hydrated electrons do not influence the activity of biologically active DNA while hydroxyl radicals inactivate it inducing strand breaks [8, 9]. The assumption of major role of hydroxyl radical in DNA damage seems to be quite reasonable since the reaction between hydrated electrons and O_2 transforms hydrated electrons into the practically unreactive towards DNA $O_2^{\bullet-}$ radical anions [10]. The formation of $O_2^{\bullet-}$ happens before the electrons reach DNA and it is very probable under physoxia/normoxia. However, solid tumor cells, as was mentioned above, frequently suffer from the low concentration of oxygen. Thus, the scavenging properties of O_2 toward hydrated electrons, high radiosensitivity under normoxia and low radiosensitivity under hypoxia (the latter is explained in terms of oxygen dependent fixation of the initial damage induced by the hydroxyl radicals [11]) suggest that low energy electrons (LEEs) are not responsible for the biological effects of ionizing radiation. Indeed, it is well known from theory [12] and experiment [9] that hydrated electrons do not cause strand breaks in DNA but bind efficiently to nucleobases. Actually, it is well documented that hydrated electrons attach to nucleobases, nucleotides and DNA with almost diffusion controlled rate [13]. The resulting nucleobase anion radicals protonate which prevents formation of strand breaks, leading instead to the formation of ultimate molecular products such as dihydrothymine or dihydrocytosine [14, 15].

These circumstances suggest a strategy that could resolve the situation unfavorable from radiotherapy viewpoint. Since hypoxia imparts the action of hydroxyl radical one could work out a way leading to the "activation" of the second major product of water radiolysis i.e. hydrated electrons. Previous reports [16], demonstrating that even 0 eV electrons are able to induce strand breaks in DNA, emphasize the damaging potential that inhabits electron-DNA interactions. To overcome the disability of hydrated electrons to damage DNA in an aqueous solution, a specific type of radiosensitizer can be used. An exemplary compound is a derivative of thymidine, 5-bromo-2′-deoxyuridine (BrdU), which undergoes an ease phosphorylation in the cell, giving the respective triphosphate, and in this form incorporates into DNA during its replication or repair [17]. Electron attachment to the BrdU labelled DNA leads to the dissociation of the C5-Br bond in 5-bromouracil that leaves behind the reactive uracil radical which in turn transforms, in a series of secondary chemical reactions, into a strand break [18].

In the following we will focus on various types of radiosensitizers which utilize electrons for their activity. In the first section, we will describe BrdU and similar derivatives of nucleosides that after incorporation into DNA undergo efficient dissociative electron attachment (DEA) leading to reactive radicals in the DNA molecule and, in consequence, to DNA strand breaks. Here, we will discuss a computational model which enables potentially radiosensitizing derivatives to be proposed yet before their synthesis. The next section will be devoted to oxygen mimetics with a focus to nitroimidazolic derivatives which currently seem to be most promising group of this type of radiosensitizers. Besides their ability to "fix" DNA damage induced by the hydroxyl radicals their action triggered by the attachment of hydrated electrons will also be elaborated. Finally, DNA damage by the simultaneous action of ionizing radiation and metallic nanoparticles or metal complexes will be described. We will end up with a summary emphasizing that our future anticancer therapies should be supported by radiosensitizers that employ hydrated electrons which are one of the major products of water radiolysis and which are not utilized by classical radiotherapy. Future perspectives of this relatively new direction will be shortly discussed in the closing statements.

4.2 Nucleoside Derivatives: A Trojan Horse Approach to Radiotherapy

4.2.1 Modified Nucleosides as Radiosensitizers

Effective radiosensitizers should exhibit selectivity towards cancer cells. Based on this assumption, only two groups of sensitizers that can be used in clinical practice were distinguished: (1) hypoxic-cell radiosensitizers and (2) halogenated pyrimidines [19]. The first group of radiosensitizers enhances the radiation effects only in hypoxic cells. The hypoxia is characteristic only for cancer cells present in a tumor tissue [19]. The selectivity towards cancer cells is due to the difference in oxygenation levels between cancer and normal cells. In case of the second group of radiosensitizers, the selectivity results from cancer cells ability for rapid growth and uncontrolled division, which means a higher degree of pyrimidine derivatives incorporation into DNA of cancer cell, compared to healthy tissue. In other words, the halogenated analogs need the tumor tissue to be cycling faster than in case of the normal cells. The extent of radiosensitization increases with the amount of incorporated halo derivative. An excellent representative of this group is 5-fluorouracil (for structure see Fig. 4.2) which enhances the cellular response to ionizing radiation. This analog is metabolized to the monophosphate form and acts mainly by inhibition of thymidylate synthase [20]. The other one, gemcitabine (for its structure see Fig. 4.2), is a well-known radiosensitizer used in the treatment of many types of cancer. Its incorporation into the cells requires phosphorylation of the nucleoside by kinases to gemcitabine 5′-diphosphate. This metabolite inhibits activity of ribonucleotide reductase and decreases the pool of dGTP, dATP, dCTP, and dTTP. Furthermore, gemcitabine is converted to active triphosphate and, when incorporated into DNA, leads to the termination of a DNA chain [21, 22].

Modified nucleosides possess exceptional properties making them especially good agents for the radiation-induced cell killing. Their most unique property is

Fig. 4.2 Structures of modified nucleoside: 5-fluorouracil (FU) and gemcitabine (dFdC)

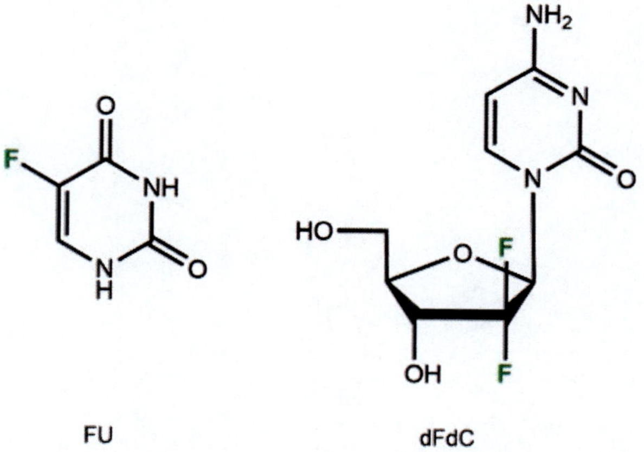

the ability (of at least some of them) to substitute native nucleosides in DNA without affecting its structure and biological function. For this reason, as mentioned above, modified nucleosides can be extensively incorporated into DNA during non-controlled proliferation of cancer cells—and in consequence—they are selective towards tumor. However, if the nucleosides are appropriately structurally modified, they can additionally engage hypoxia as a second selectivity factor. Such modification should rely on the introduction of substituents that increase nucleosides' sensitivity to degradation induced by solvated electrons, which are one of the major products of water radiolysis under hypoxic conditions [23] (in normoxia hydrated electrons are mostly scavenged by oxygen) and remain non-reactive to native DNA in water solution [24]. Thus, a modified nucleoside usually should have an electrophilic substituent, undergoing efficient DEA that leaves behind a nucleoside radical, which in secondary reactions is able to produce damage to DNA leading ultimately to cancer cell death [1].

4.2.2 *Nucleoside Derivatives as Trojan Horses*

At the cellular level, electrophilic nucleoside derivatives act like a "Trojan horse"—the lethal effects (DNA damage leading to cancer cells death) are produced only as a result of interactions between radiation (producing solvated electrons) and these compounds incorporated into DNA [1]. According to this approach, IR is the external stimulus that leads to DNA degradation (Fig. 4.3). This means that radiosensitizing derivatives (1) should be non-cytotoxic by themselves and (2) due to their structural similarity to their natural analogs, they should be incorporated into DNA structure during the replication or repair of the biopolymer.

The first assumption implies selection of analog, which after introduction into a cell will not cause its death. Cytotoxicity of nucleosides analogs results predominantly from their interfering with synthesis of nucleic acids by altering

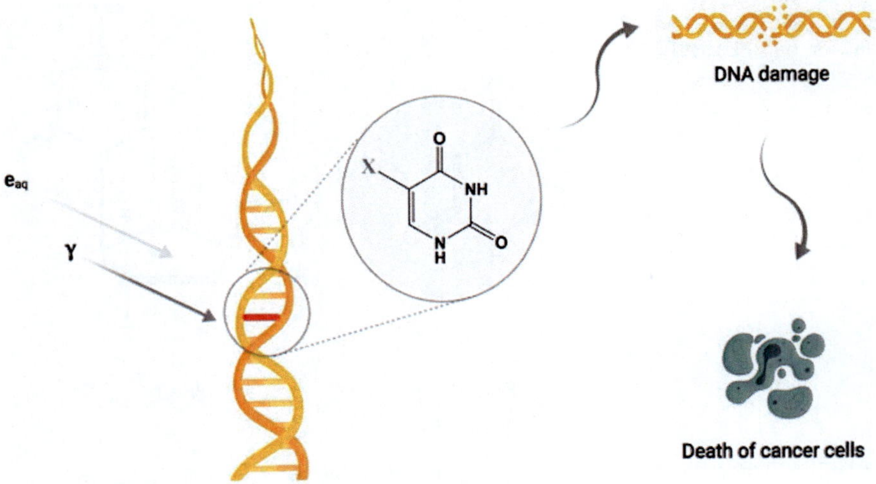

Fig. 4.3 Trojan horse approach to radiotherapy with the use of radiosensitizing nucleosides derivatives (X-electrophilic substituent, undergoing efficient dissociative electron attachment)

DNA structure, interfering with various enzymes involved in biosynthesis and repair mechanisms of DNA or by modifying the metabolism of native nucleosides/nucleotides [25]. Nevertheless, many of electrophilic base-modified nucleosides show low cytotoxicity towards cancer cells. With the use of MTT viability test, it has been shown that the modified nucleosides considered as radiosensitizers e.g., 5-selenocyanato-2′-deoxyuridine (SeCNdU) [26], 5-trifluoromethanesulfonyl-2′-deoxyuridine (OTfdU) [26], 5-iodo-4-thio-2′-deoxyuridine (ISdU) [27] or 5-bromo-4-thio-2′-deoxyuridine (BrSdU) [28] are characterized by relatively low cytotoxicity towards both cancer and normal cells. The second requirement, i.e. the necessity of introducing the modified nucleosides into cellular DNA, is not a trivial task and seems to be a limiting step of the Trojan horse approach. There are several possible methods of cellular DNA labelling: (1) using the substituted nucleosides themselves, (2) via the substituted nucleobases and (3) using prodrugs. In the first method, a modified nucleoside must easily penetrate cellular membrane, be phosphorylated in cytoplasm by the specific kinases and, as a triphosphate, attached to a newly synthesized DNA or enter the biopolymer during its repair (with the involvement of DNA polymerases, e.g. via BER repair mechanism). Labelling DNA using a nucleobase exploits the reserve pathway of nucleotide biosynthesis, where the attachment of phosphoribosyl pyrophosphate to the nucleobase, leading to the nucleoside 5′-monophosphate, is catalysed by a phosphoribosyltransferase. The other option is chemical conversion of nucleosides into a prodrugs e.g., lipophilic cylcoSal pronucleotides which will hydrolyze in cytoplasm to nucleoside monophosphates [29]. Since the first phosphorylation step is assumed to be the most difficult one, the cycloSal pronucleotides is described as a universal way of the modified nucleoside incorporation into the genome [30].

Despite potential difficulties during incorporating of nucleoside derivatives into DNA, there are several reports addressing the effectiveness of this process. Uracil analogs, notably 5-bromo- and 5-iodo-2′-deoxyuridine (IdU), can be used by a cell for DNA biosynthesis almost as easily as natural ones. For instance, it was demonstrated that 5-iodo-2′-deoxyuridine was effectively incorporated (17% after 14 days of continuous infusion) into human granulocytes [31]. It has been also proved that 4-thiothymidine [32] as well as halogenated derivatives of 4-thio-2′-deoxyuridine—5-iodo-4-thio-2′-deoxyuridine and 5-bromo-4-thio-2′-deoxyuridine—are incorporated into the DNA of cultured human cells [33].

Since the human kinases play a critical role in the activation of important biologically-active nucleoside analogs, the structural and kinetic studies on the activity of these enzymes towards various nucleoside derivatives have been carried out [34]. For instance, Jagiello et al. [35] presented the quantitative structure-activity relationship (QSAR) model allowing to identify and understand the molecular properties of modified nucleosides related to the hTK1 kinase activity. This model is based on only two molecular features: the ability of the modified nucleoside to intermolecular interactions with the enzyme and the shape of the molecule determined by atom substitutions. Its use allows to predict the activity of hTK1 kinase towards a new nucleoside analog before its synthesis [35].

4.2.3 *Electron-Induced Degradation of Modified Nucleosides: Experimental Studies*

One of the best known radiosensitizers of Trojan horse type is 5-bromouracil (BrU). This is a highly electron affinic molecule, prone to dissociative electron attachment—the process that leads to a reactive uracil-5-yl radical via the unstable 5-bromouracil anion. The promising characteristics of the LEE-induced damage to DNA by the halo derivatives of native nucleosides has been demonstrated many times. For instance, the radiosensitizing properties of all four bromonucleosides were studied by Sanche's and Rak's groups, by bombarding a thin layer of single stranded trinucleotides TXT (where X = BrdU, BrdC, BrdA or BrdG) with low energy electrons under ultra-high vacuum [36, 37]. The comparison of the results obtained for the native and brominated trimers suggests that the main pathways of LEE-induced TXT degradation are related to the release of bromide anions [36]. Moreover, a 2–3-fold increase in the sensitivity of TXT oligonucleotides has also been demonstrated compared to the susceptibility of their native counterparts to electron attachment [37]. These findings were confirmed within the radiolytic studies of the TXT trimers irradiated in water containing a hydroxyl radical scavenger. Indeed, it has been shown that solvated electrons are unable to damage native DNA, whereas DNA labelled with bromoderivatives are vulnerable to damage. Strand breaks and abasic sites have been identified as the main products of brominated trimers radiolysis [18]. The radiolytic studies on electron-induced damage to trimeric oligonucleotides labelled with iodinated pyrimidines showed that electron attachment to the iodo bases leads mainly to the formation of single strand breaks. The worth mentioning fact is that the iodinated trimers are two-fold more sensitive to solvated electrons than the brominated ones, which can be explained by the barrier-free dissociation of the iodinated base anions [38].

Besides the halo derivatives, several new C5-substituted pyrimidine analogs have been proposed in the context of efficient electron-induced degradation [39, 40]. One of them is 5-thiocyanato-2′-deoxyuridine (SCNdU) [41]. Using low-temperature ESR and steady-state radiolysis at ambient temperature coupled with mass spectrometry, it has been shown that the electron attachment to SCNdU leads to two parallel reactions producing quite different products (Fig. 4.4a). In one path, the C5-S bond is broken and the secondary dU$^{\bullet}$ is formed, giving dU as a stable product. On the other path, the S-CN bond cleavage in the thiocyanate substituent produces the dU-S$^{\bullet}$ radical that ultimately gives a stable dimer (dU-S-S-dU). This radiation chemical study of electron addition to SCNdU establishes SCNdU as a potential radiosensitizer that could cause strand breaks, both intra- and interstrand DNA crosslinking as well as DNA-protein crosslinking via the formation of S-S dimers [41].

A similar approach was adopted for SeCNdU. It has been proved, that dU-Se$^{\bullet}$ is the primary product of dissociative electron attachment to SeCNdU. Electron binding to the studied derivative results in its degradation leading to two major products: the dU-Se-Se-dU dimer and $dUSeO_2H$ (product of the reaction between dU-Se$^{\bullet}$ and oxygen, produced during radiolysis) (Fig. 4.4b) [26]. In case of the next proposed sensitizer—OTfdU—radiolysis of water solution containing $^{\bullet}$OH radical scavenger leads to the formation of two stable products, dU and OHdU (Fig. 4.4c). The formation of dU is more favorable which suggests that the main transient product induced by dissociative electron attachment is dU$^{\bullet}$. On the other hand, the presence of OHdU among the radiolysis products shows also the involvement of $^{\bullet}$OdU radical in the DEA process. The extent of decay of the studied analogs induced by X-rays has been determined (30% for SeCNdU and 16% for OTfdU) and compared with the yield obtained for well-known radiosensitizer—BrdU (17%) [26].

Other promising candidates for Trojan horse type radiosensitizer seem to be the derivatives of 4-thio-2′-deoxyuridine. It has been demonstrated that ISdU sensitized effectively DNA to IR. The radiolysis of deoxygenated water solution of ISdU containing $^{\bullet}$OH radical scavenger was performed. The experimental results (LC-MS) showed the formation of five major products of degradation induced by X-rays (Fig. 4.4d). One of the major products of ISdU irradiation, SdU, results from the DEA process. The others are oxidation products—ISOdU or IdU, and the dimers—$(ISdU)_2$ and ISdU-SdU. The radiolitic decay of ISdU is almost 1.5-fold more effective compared to BrdU. At the cellular level, ISdU can effectively lead to X-ray induced death, which has been shown by clonogenic assay. Flow cytometry analysis (the histone H2A.X phosphorylation assay), on the other hand, demonstrated that ISdU pretreatment sensitizes cancer cells to ionizing radiation, at least in part, by

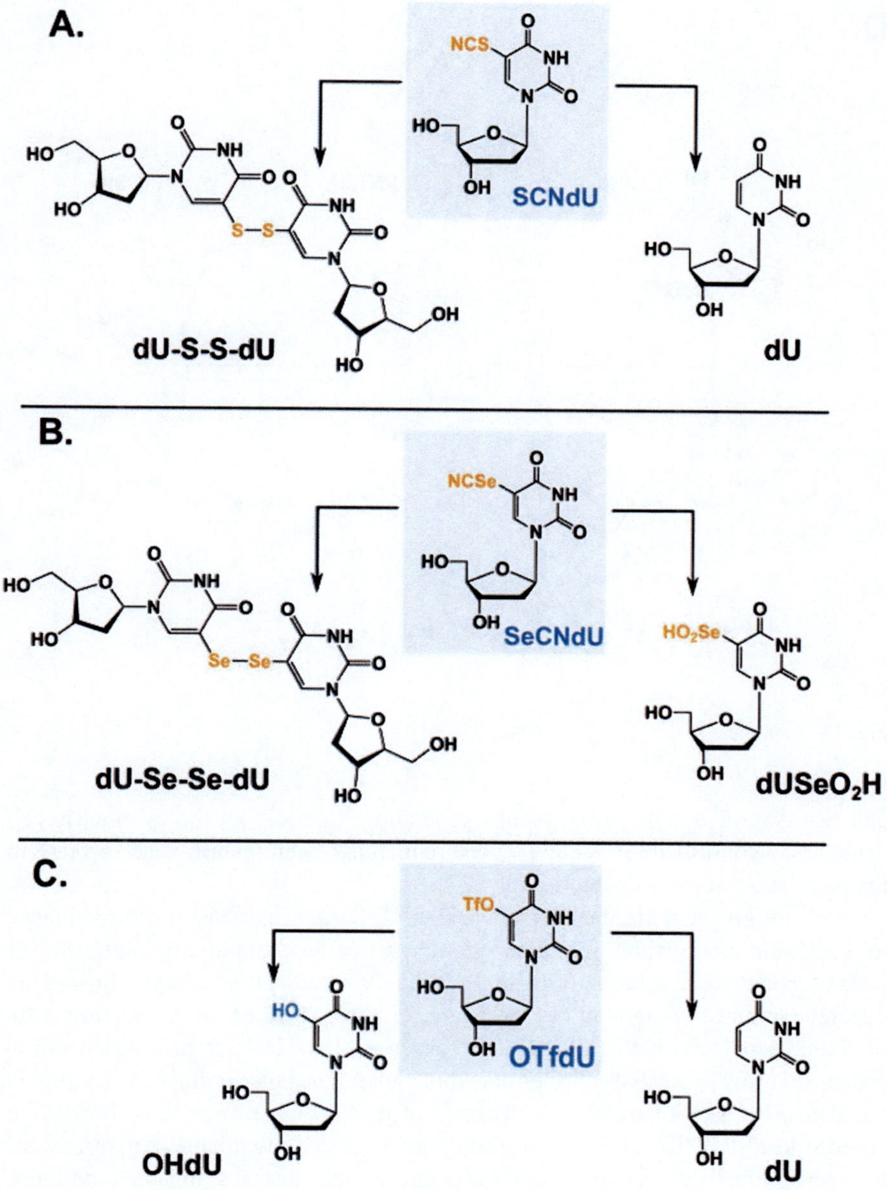

Fig. 4.4 Experimental determined products of X-ray radiolysis of (**a**) 5-thiocyanato-2′-deoxyuridine (SCNdU), (**b**) 5-selenocyanato-2′-deoxyuridine (SeCNdU), (**c**) 5-trifluoromethanesulfonyl-2′-deoxyuridine (OTfdU) and (**d**) 5-iodo-4-thio-2′-deoxyuridine (ISdU)

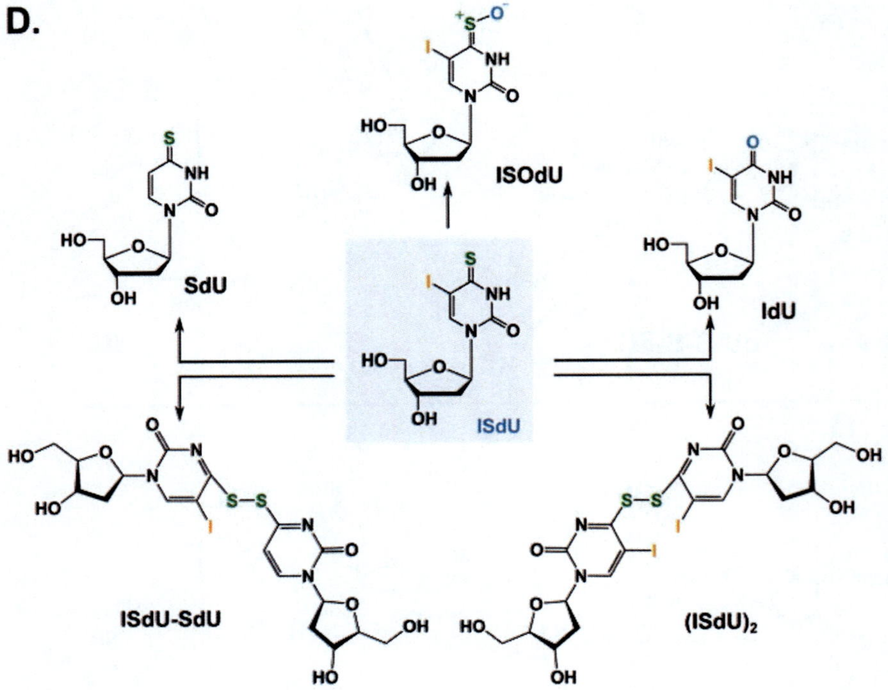

Fig. 4.4 (continued)

enhancement of double strand breaks generation. In addition, the cytometric cell death assay confirmed the X-ray induced reduction of cell viability and increase in the population of early apoptotic cells [27].

Interestingly, very similar derivative, BrSdU, despite its structural resemblance to ISdU did not exhibit radiosensitization properties. Stationary radiolysis of deoxygenated water solution of BrSdU with an $^{\bullet}$OH radicals scavenger showed the decomposition of compound but the characteristic pattern of DEA and formation of its expected product—SdU was not observed. The lack of radiosensitization properties of BrSdU was confirmed at cellular level. The pulse radiolysis experiment confirmed the lack of transient absorption of dU-S$^{\bullet}$ radical in case of BrSdU (in contrast to ISdU). The above-mentioned results were well-explained by difference between the activation barriers for ISdU and BrSdU that accompanies the elimination of the halogen anion from the XSdU$^{\bullet-}$ anion radical. Unfortunately, this fact proved the lack of radiosensitizing properties of BrSdU, but also confirmed the crucial role of DEA process in the mechanism of radiosensitization of DNA damage by the modified nucleosides [28].

The design and characterization of the above-mentioned new potent radiosensitizers with the use of computational chemistry methods are presented in the next section (see Sect. 4.3).

4.3 Computational Studies on Nucleoside Radiosensitizers

4.3.1 5-Bromo-2′-Deoxyuridine

Although 5-brominated uridine was synthesized more than hundred years ago [42], first significant wave of interest in 5-bromouracil (BrU) falls in the 1950s, when the developed methods of its synthesis already existed. The attempts of incorporation of BrU into bacterial DNA and first toxicity tests were reported [43–47] at the time. Still on this wave of interest began the works on BrU as a potential sensitizer, firstly to ultraviolet irradiation [48, 49], and soon later to X-ray radiation [50–52]. Through decades, BrU, also in its nucleoside form (5-bromo-2′-deoxyuridine) enjoyed variable popularity, rising with the works showing its radiosensitizing properties in in vitro and in vivo experiments [53–56]. It was demonstrated that the presence of BrU in a DNA strand increases three-fold the susceptibility of cells to be killed by high energy radiation [57]. Experimental studies on BrU as a radiosensitizer have reached the stage of clinical trials several times [58–61]. However this promising compound has not been successfully introduced into common clinical practice. The transition from in vivo research to clinical trials has brought the main problem with achievement of therapeutic level of BrdU in cancer cells. It is bounded with dehalogenation of BrdU by the liver—more than 90% of BrdU administered intravenously is deactivated in liver within one hour [62]. To reduce impact of dehalogenation, administration of the drug (like intra-arterial or intratumor directly), omitting the liver and avoiding systemic circulation can be introduced [54]. Additionally, it was observed (in vitro firstly [63]) that the uptake of BrdU into the cancer cell nuclei can be also increased by pretreatment with small amounts of antimetabolites—methotrexate, 5-fluoro-2′-deoxyuridine or 5-fluorouracil—inhibitors of thymidylate synthetase. In presence of those antimetabolites, native thymine biosynthesis is inhibited, so the dividing cancer cells are forced to incorporate BrdU—the available thymidine analog—into their DNA [54]. Side effects bounded with uptake of modified uracils by normal cells are separate problem. For instance, Sano et al. in their BAR therapy (BrdU-antimetabolite-continuous intra-arterial infusion radiation therapy) of brain tumors mentioned two side effects: onychomadesis of the nails and more frequent epilation and radiodermatitis of irradiated scalp, probably due to radiosensitization of the skin cells [54]. Groves et al. in their work about high-dose BrdU sensitized radiotherapy of glioblastoma multiforme combined with chemotherapy described substantial myelosuppressive and dermatologic toxicity caused by BrdU doses [60]. Moreover, none of them found a significant extended survival in patients treated with BrdU during radiotherapy.

Despite the clinical problems with BrU, the deep understanding of molecular mechanism of action of this derivative seems to be crucial for future radiosensitizers development. Computations joined the experimental studies on BrU at the beginning of the twenty-first century, when Wetmore et al. and Li et al. analyzed at the DFT level the process of electron induced dehalogenation of a set of 5-substituted

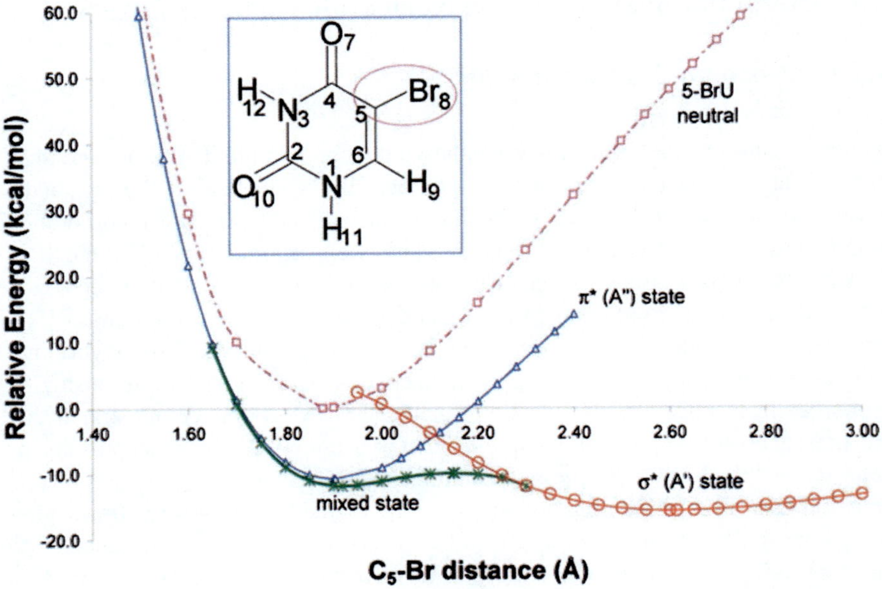

Fig. 4.5 Potential energy surfaces (PES) of 5-bromouracil and its anion radical along the C5-Br bond. The energy of the optimized neutral molecule is set as reference. The C5-Br bond is circled in the structure. Reprinted with permission from [65]. Copyright (2002) American Chemical Society

halouracils [64, 65]. They showed that attachment of excess electron to BrU leads to the formation of a stable anion radical, $BrU^{\bullet-}$. Such anion radical is then prone to produce, with a low kinetic barrier (related to a transition from π^* to σ^* state), the Br^- anion···uracil-5-yl radical $U^\bullet$ complex (of ca. 2.6 Å C5-Br bond length, see Fig. 4.5). In effect, dissociative electron attachment (DEA) process leads, in the case of BrU, to the formation of reactive $U^\bullet$ radical, which, if present in a DNA strand, can initiate the damage process. Hence, Wetmore et al. and Li et al. theoretically described BrU DEA process proposed former by Abdoul-Carime et al. on the basis of their experimental study [66].

It was also calculated [65] that the value of electron affinity (EA) increases as follows:

$$U - yl^\bullet \gg BrU > ClU > FU \gg U \quad (4.1)$$

Although the adiabatic EA value is the highest for uracil of all DNA/RNA bases, the halogenated derivatives, according to their EA values, are more effective electron acceptors than natural nucleobases, which means they can act as "electron traps". Another important conclusions are that the highest EA for uracil-5-yl radical indicates its reactive nature, which seems to be important for further consideration of uracil derivatives labelled DNA degradation mechanisms.

4.3.2 Crucial Characteristic of Electron Attachment Process

The analysis of excess electron–biomolecule interaction should start by defying electron affinity and ionizing potential (IP), which are part of Mulliken's definition of electronegativity [67]. According to Mulliken, electronegativity of neutral particle is:

$$\chi = \frac{1}{2}(I + A) \tag{4.2}$$

where I is its ionizing potential and A is electron affinity. Simplifying, electron affinity [68] characterizes molecule's ability to attach electron and create its anionic form:

$$X + \mathrm{e}^- \rightarrow \mathrm{X}^- + \mathrm{EA} \tag{4.3}$$

Numerical value of EA is defined as the difference between energy of the neutral moiety and its corresponding anion. On the other hand, IP characterizes the minimum energy which should be supplied to the system in order to detach an electron:

$$X + \mathrm{IP} \rightarrow \mathrm{X}^+ + \mathrm{e}^- \tag{4.4}$$

To discuss electron–molecule interactions during excess electron attachment we can provide additional nomenclature. Electron stability of anion radicals, $X^{\bullet-}$, formed after electron attachment to the neutral X moiety can be described by three parameters correlated with electron affinity: adiabatic electron affinity (AEA), vertical attachment energy (VAE) and vertical detachment energy (VDE) [69, 70]. All these parameters are calculated as the difference in energy between the respective species (E—enthalpy or G—Gibbs free energy) of X and $X^{\bullet-}$ (Fig. 4.6):

- For AEA value, X and $X^{\bullet-}$ states are in their optimal geometries;
- For VAE value, both are in the neutral (X) geometry;
- For VDE value, both are in the anion radical ($X^{\bullet-}$) geometry.

$$AEA = E_X - E_{X^{\bullet-}} \tag{4.5}$$

$$VAE = E_X - E_{X^{\bullet-}}\ (in\ the\ X\ geometry) \tag{4.6}$$

$$VDE = E_X\ \left(in\ the\ X^{\bullet-}\ geometry\right) - E_{X^{\bullet-}} \tag{4.7}$$

Analyzing the plot of electron energy against molecular geometry (Fig. 4.6) one can state that to obtain electronically stable anion radical $X^{\bullet-}$ via electron attachment

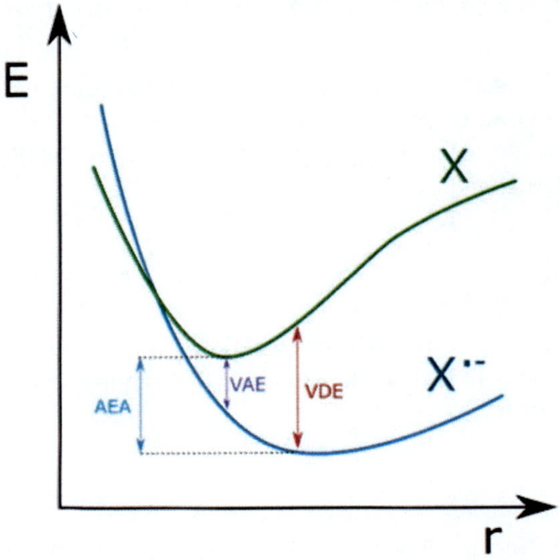

Fig. 4.6 AEA, VAE and VDE parameters definition. E states for electron energy, while r—molecule's geometry

to the neutral X moiety, this process should decrease system energy. For AEA, VDE and VEA parameters it means they should be positive and of relatively high values. The positive and high values of electron affinity prove that electron attaches willingly to a molecule. Therefore, the values of AEA, VAE and VDE, calculated for potentially radiosensitizing nucleobase should be larger in comparison to those parameters calculated for native nucleobases, and are the initial characteristics, which are tested during the modeling of radiosensitizers [71].

4.3.3 Bromonucleobases

As was discussed in the previous paragraph, 5-bromouracil is one of the best studied modified nucleobase of proven radiosensitizing properties, which has not found its place in clinical practice, though. As an analog of thymine, it can be incorporated to DNA and its electron affinity is larger than that of the native nucleobase (2.6 eV for BrU compared to 2.3 eV for U) [64]. It is confirmed that the formed via electron attachment stable $BrU^{\bullet-}$ anion radical is able to produce genotoxic uracil-5-yl radical ($U^{\bullet}$) in the DEA process. Although it has not found application in clinical practice (see the previous paragraph), BrU with its electron-accepting properties seems to be a good starting point on the way of searching for new radiosensitizers. Substitution of H5 hydrogen atom in uracil with bromine atom leading to radiosensitizer suggests bromine derivatives of other native nucleobases could be worth of considering as the potential radiosensitizers.

Based on this concept, Chomicz et al. analyzed at the DFT level the electron affinity and DEA mechanism for four methylated bromonucleobases (methyl group

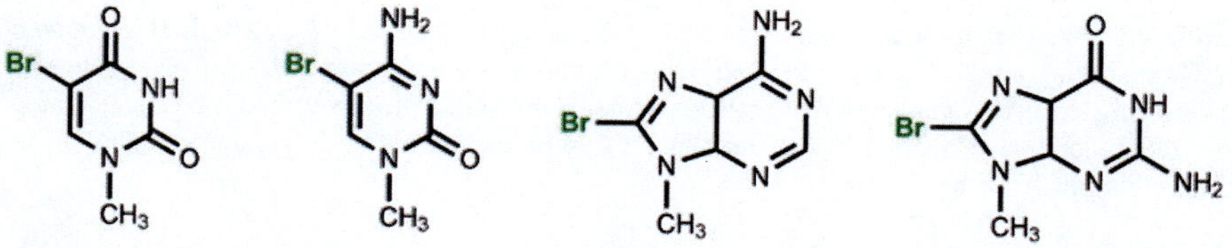

Fig. 4.7 Geometrical structures of bromonucleobases

was used to mimic the *N*-glycosidic bond): 8-bromo-9-methyladenine (BrA), 8-bromo-9-methylguanine (BrG), 5-bromo-1-methylcytosine (BrC) and 5-bromo-1-methyluracil (BrU) for comparison (Fig. 4.7) [72].

Based on the respective EA and VDE values (AEA, VEA, VDE) it was determined that all brominated nucleobases (BrX) are able to form stable valence anion radicals ($BrX^{\bullet-}$) and BrU is the one which attaches an excess electron most willingly. Electrophilic character of BrX can be ordered according to the following series (see Table 4.1):

$$BrG < BrA < BrC < BrU \tag{4.8}$$

According to the AEA values of halogenated nucleobases [72], 5-bromopyrimidines occurred to be better electron acceptors than 8-bromopurines. The ease of electron attachment (in an aqua solution or the gas phase) has the same order as that depicted by Eq. (4.8) (cf. Table 4.1).

On the other hand, the DEA process—dehalogenation of BrX producing reactive nucleobase radicals, $X^{\bullet}$, although kinetically and thermodynamically possible for all analyzed derivatives, occurred to be easier for the modified purines, with no activation barrier for breaking the C8-Br bond in case of BrA (for details, see Table 4.1).

As mentioned above DFT work took into account reaction environment at the PCM (Polarizable Continuum Model) level only, there was a question, if the explicit model of surrounding water would affect the DEA mechanism. Thus, ab initio molecular dynamics (AIMD) was used to better understand and describe electron attachment to BrX [73]. Due to the AIMD calculations we described the process of localization of an electron on DNA bases which allowed to notice some differences in behavior between brominated purines and pyrimidines. The results of AIMD modeling [73] were similar to the previous DFT studies [72] and showed that the excess electron readily attach to the neutral structure of BrX creating the anion radical ($BrX^{\bullet-}$). In such anion radical, $BrX^{\bullet-}$, the Br-X bond dissociates (with low activation barrier for pyrimidines and barrier-free in case of purines) which transforms the system to an anion Br^- – radical $X^{\bullet}$ complex, that finally dissociates

Table 4.1 Electron affinity's values (in eV) and dissociative electron attachment thermodynamic (ΔG) and kinetic (ΔG^{*}) data (in kcal/mol) for chosen uracil/uridine derivatives—potential radiosensitizers. AEA values in free enthalpy scale. If not stated differently, ΔG, ΔG^{*} calculated as a difference between anion radical and complex anion radical free enthalpy and calculations in aqua solution (PCM model)

Compound	Electron affinity			DEA			Ref.
	AEA	VDE	VAE	ΔG	ΔG^{*}	Mechanism	
BrU[a]	2.48	2.74	1.90	–8.00	2.54	$U^{\bullet} + Br^{-}$	[72]
BrC[a]	2.26	2.51	1.65	–14.44	2.15	$C^{\bullet} + Br^{-}$	
BrA[a]	–	2.91	1.31	–	Barrier-free	$A^{\bullet} + Br^{-}$	
BrG[a]	1.55	2.04	1.00	–29.45	0.41	$G^{\bullet} + Br^{-}$	
CNU[a]	2.83	3.18	–	40.2	41.2	$U^{\bullet} + CN^{-}$	[39]
SCNU[a]	2.70	3.07	–	–3.7	3.4	$U^{\bullet} + SCN^{-}$	
NCSU[a]	2.73	3.86	–	10.5	19.5	$U^{\bullet} + NCS^{-}$	
NCOU[a]	2.40	2.78	–	10.5	20.1	$U^{\bullet} + NCO^{-}$	
OCNU[a]	2.62	3.07	–	–19.3	2.5	$U^{\bullet} + OCN^{-}$	
SHU[a]	2.26	2.66	–	6.8	10.4	$U^{\bullet} + SH^{-}$	
NNNU[a]	2.38	2.77	–	10.6	12.5	$U^{\bullet} + NNN^{-}$	
NOOU[a]	3.55	3.81	–	26.8	29.3	$U^{\bullet} + NOO^{-}$	
SCNdU[b,c]	2.64	–	–	–1.60	8.69	$dU^{\bullet} + SCN^{-}$	[41]
				–16.21	4.07	$dU\text{-}5\text{-}S^{\bullet} + CN^{-}$	
NCU[a]	2.7	3.0	–	12.9	25.6	$U^{\bullet} + NC^{-}$	[40]
NOU[a]	3.9	4.0	–	40.7	41.0	$U^{\bullet} + NO^{-}$	
OHU[a]	2.2	2.6	–	27.7	28.0	$U^{\bullet} + OH^{-}$	
CCHU[a]	2.5	2.9	–	57.3	58.8	$U^{\bullet} + CCH^{-}$	
$NHNO_2U$[a,d]	3.1	4.5	–	27.0	29.2	$U^{\bullet} + NHNO_2^{-}$	
			–	–25.6	–0.3	$UNH^{\bullet} + NO_2^{-}$	
SeCNU[a,d]	2.5	2.9	–	0.4	6.5	$U^{\bullet} + SeCN^{-}$	
	2.5	2.9	–	–14.4	–0.2	$USe^{\bullet} + CN^{-}$	
SOFU[a,d]	4.4	–	2.4	–	–	$USO^{\bullet} + F^{-}$	
CCl_3U[a,d]	4.7	–	2.1	–	–	$UCCl_2^{\bullet} + Cl^{-}$	
SClU[a,d]	4.9	–	3.2	–	–	$US^{\bullet} + Cl^{-}$	
IOU[e]	4.0	–	2.6	–	–	$IU + O^{\bullet -}$	
$5FCH_2dUDP$[f]	2.45	2.95	1.84	–16.35	1.11	$CH_2dUDP^{\bullet} + F^{-}$	[86]
$5ClCH_2dUDP$[f]	3.84	–	1.93	–96.70[g]	–	$CH_2dUDP^{\bullet} + Cl^{-}$	
$5BrCH_2dUDP$[f]	3.84	–	1.99	–86.35[g]	–	$CH_2dUDP^{\bullet} + Br^{-}$	
5BrdUDP-dADP[f]	2.50	3.02	1.95	–	–	–	
$5FCH_2dUDP$-dADP[f]	2.42	2.93	1.79	–16.80	1.92	$CH_2dUDP^{\bullet}$-dADP + F^{-}	
$5ClCH_2dUDP$-dADP[f]	3.82	–	1.88	–95.97[g]	–	$CH_2dUDP^{\bullet}$-dADP + Cl^{-}	
$5BrCH_2dUDP$-dADP[f]	3.78	–	1.95	–84.73[g]	–	$CH_2dUDP^{\bullet}$-dADP + Br^{-}	

(continued)

Table 4.1 (continued)

Compound	Electron affinity			DEA			Ref.
	AEA	VDE	VAE	ΔG	ΔG*	Mechanism	
OTfdU[b]	2.7	–	–	–51.4	1.5	$dUO^{\bullet} + Tf^-$	[26]
				–17.6	8.6	$dU^{\bullet} + OTf^-$	
SeCNdU[b]	2.6	–	–	–13.2	1.2	$dUSe^{\bullet} + CN^-$	
				5.5	6.4	$dU^{\bullet} + SeCN^-$	
CH_2CNU[b]	2.3	–	–	3.2	19.7	$dUCH_2^{\bullet} + CN^-$	
				43.7	44.0	$dU^{\bullet} + CH_2CN^-$	
BrdU[b]	2.4	–	–	–3.9	5.3	$dU^{\bullet} + Br^-$	
ISU[e]	–	–	–	–31.6	0.6	$(ISU)_2$	[27]
	–	–	–	–107.0	2.8	SU	
	–	–	–	–43.2	15.3	ISOU	
	–	–	–	–4.9	24.5	IOSU	
	–	–	–	–45.0	1.8	IU	
	–	–	–	–42.2	2.8	ISU-SU	
BrSU[e]				–2.5	6.2	$SU^{\bullet} + Br^-$	[28]

[a]B3LYP/6-31++G(d,p)
[b]M06-2X/6-31++G(d,p)
[c]C3′-exo-front conformation, values for another conformations available
[d]Available data calculated at the M06-2X/6-31++G(d,p) and MPWB1K/6-31++G(d,p) levels, too
[e]B3LYP/DGDZVP++, available data calculated at the M06-2X/DGDZVP++ and MPWB1K/DGDZVP++ levels, too
[f]B3LYP-D3/6-31++G(d,p), gas phase
[g]$\Delta Gt = G({}^{\bullet}dUCH_2DP/{}^{\bullet}dUCH_2DP\text{-}dADP) + G(X^-) - G(5XCH_2dUDP/5XCH_2dUDP\text{-}dADP)$

to a separated bromine anion Br^- and the reactive nucleobase radical $X^{\bullet}$ (see Eq. 4.9).

$$BrX + e^-_{hyd} \rightarrow BrX^{\bullet -} \rightarrow \left[Br^- \cdots X^{\bullet}\right] \rightarrow Br^- + X^{\bullet} \tag{4.9}$$

Barrier-free dehalogenation found for purines may suggest that they could be more effective sensitizers than pyrimidines. On the other hand, electron localizes on pyrimidines faster than on purines (7–9 fs for pyrimidines vs. 12–15 fs for purines), probably because of the higher values of electron affinity of the former bases. Concluding, all the analyzed bromonucleobases can easily attach an excess electron and undergo the DEA process, producing the reactive nucleobase radical. Hence, all of them could be considered as potential radiosensitizers.

In order to elucidate the mechanism of electron induced degradation of DNA labeled with brominated nucleobases, the fate of the nucleobase radicals was examined at the DFT level for the chosen nucleotides having geometries characteristic for double-stranded DNA. Possible electron induced degradation pathways were tracked both for brominated purine [74, 75] and pyrimidine [76, 77] nucleotides. Thus, the radical reactions were studied on the following purine diphos-

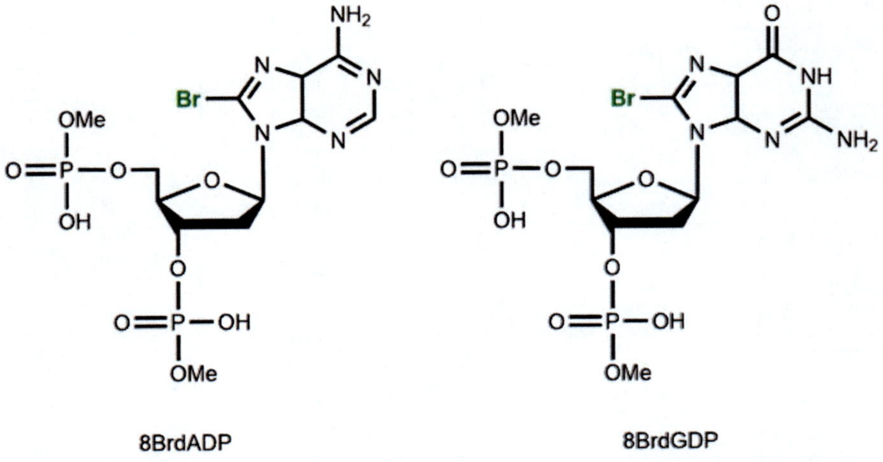

Fig. 4.8 Brominated purines dinucleotides models along with their abbreviated names

phates: 8-bromo-2′-deoxyadenosine 3′,5′-diphosphate (8BrdADP) and 8-bromo-2′-deoxyguanosine 3′,5′-diphosphate (8BrdGDP, see Fig. 4.8). Their geometries were excised from the structure of the B-DNA double helix and H8 replaced by the bromine atom while the C8-Br distance set to the value characteristic for 8-brominated purines [74].

For both adenosine and guanosine derivatives, electron attachment induces barrier-free breaking of the C8-Br bond (as was calculated for bromonucleobases [72, 73]) which leads to the anion radical complex [dXDP–Br]$^{\bullet-}$, and finally to the isolated bromine anion and purine's ring centered, dXDP$^{\bullet}$, radical (see Fig. 4.9). The latter radical can be thermodynamically stabilized by detaching a hydrogen atom from geometrically accessible positions, C2′, C3′ or C5′, of deoxyribose to form the sugar-centered radicals: rad2′, rad3′ or rad5′ (Fig. 4.9). Formation of rad2′ seemed to be a "dead-end" route, as further hydrogen atom transfers, leading subsequently to the breakage of sugar-phosphate chain, are kinetically forbidden (kinetic barriers as high as 33–41 kcal/mol [74]). When it comes to stabilization process for rad3′ radical of adenosine and guanosine, breaking the phosphodiester bond and a ketone derivative formation were postulated. An analogous reaction for the rad 5′ radical may lead to the formation of an aldehyde derivative, but that route is kinetically unfavorable [74, 75]. Moreover, the most advantageous route within the analyzed pathways occurred to be the cyclization reaction. The cycloX mutation can be easily produced from the rad5′ radical as a result of an attack on the C8 position of the respective purine (alternative radical stabilization, see Fig. 4.9). The cyclic product, 5′,8-cyclo-2′-deoxyadenosine, was observed experimentally in late seventies as one of the main products of γ-radiolysis of 2′-deoxyadenosine aqueous solutions in the absence of oxygen [78]. Further research on the cyclic purine lesions were critically reviewed by Chatgillialoglu et al. in 2010 [79], so they will not be discussed in details here. This is worth of emphasizing that the cyclopurines can block DNA

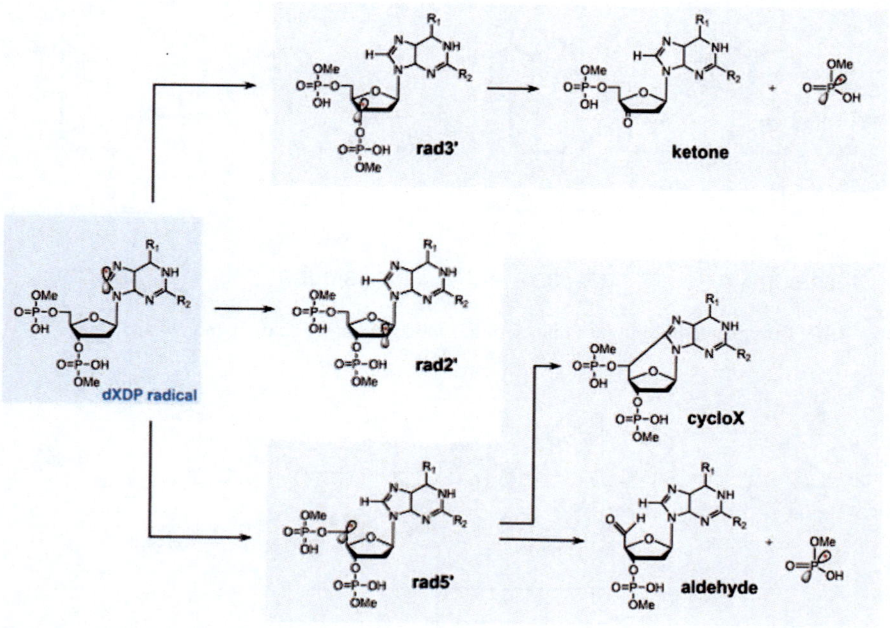

Fig. 4.9 Degradation paths for debrominated purine nucleotide radical (adenosine, R_1: $-NH_2$, R_2: $-H$; guanosine, R_1: $=O$, R_2: $-NH_2$)

replication and transcription [80]. In summary, brominated purines incorporated into DNA strand might sensitize this biomolecule to radiation, but rather via cyclopurine mutations than the formation of strand-breaks.

The same methodology and model construction were used to study electron induced degradation processes in the brominated pyrimidines [76, 77]. Except the 5-bromosubstituted derivatives, 6-substituted nucleosides were also taken under consideration, giving four diphosphate nucleotide models: 5- and 6-bromo-2′-deoxyuridine 3′,5′-diphosphates (5BrdUDP and 6BrdUDP, see Fig. 4.10) as well as 5- and 6-bromo-2′-deoxycytidine 3′,5′-diphosphates (5BrdCDP and 6BrdCDP, Fig. 4.10).

The electron induced debromination of all studied derivatives occurs easily, with a small kinetic barrier (ca. 3 kcal/mol). However, the stabilization of the rad5 radicals, formed via debromination (see Fig. 4.11), by the hydrogen atom transfer from deoxyribose to pyrimidine's ring is difficult because of steric hindrance. Strong deformation of rad5 is required to transform it into rad3′ (kinetic barriers as high as 67 kcal/mol for 5BrdCDP and 76 kcal/mol for 5BrdUDP degradation) or rad5′ (kinetic barriers in kcal/mol: 53 for 5BrdCDP and 44 for 5BrdUDP [76, 77]). Therefore, it was suggested, that the radiation induced formation of strand breaks in DNA labeled with 5BrdX could be explained by using an extended model of DNA, including at least two neighboring nucleotides [81].

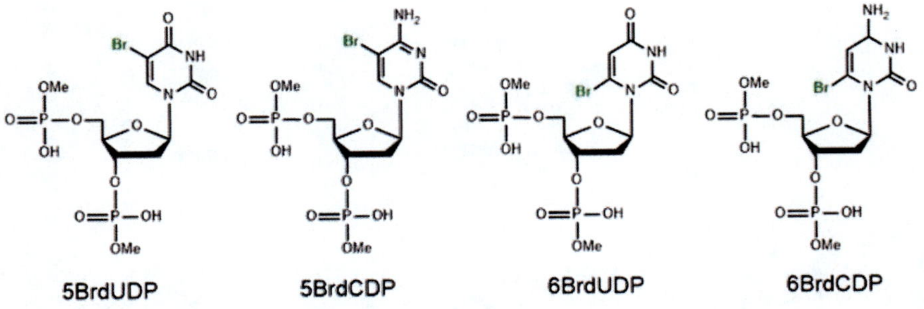

Fig. 4.10 Brominated pyrimidines dinucleotides models along with their abbreviated names

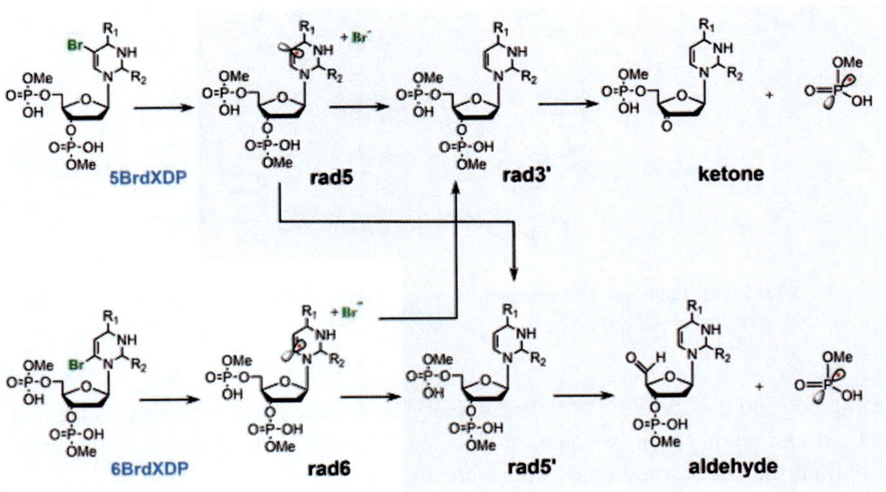

Fig. 4.11 Degradation paths for debrominated pyrimidine nucleotide radicals (uridine, R_1: =O, R_2: =O; cytosine, R_1: –NH_2, R_2: =O)

High activation barriers found for rad5→ rad3′ and rad5 → rad5′ transformations as well as pyrimidine geometries suggest that 6-substituted bromopyrimidines could lead to a DNA strand break more likely than the 5-substituted ones. Therefore, the isomer of rad5, the rad6 radical, which could form as a result of electron attachment to the 6-bromopyrimidine was studied as well [76, 77]. The 6-bromoderivatives turned out to be even more sensitive to the excess electron than the 5-bromoderivatives. The rad6 radical is formed as a result of the barrier-free breakage of the C6-Br bond. The lack of steric hindrance, that is present in the rad5 radicals, causes that rad6 stabilizes much easier by detaching hydrogen from the C3′ or C5′ position of the sugar moiety. Such produced rad3′ or rad5′ radicals could then undergo the breakage of the phosphodiester bond (leading to DNA strand break), producing stable ketone (C3′ site) or aldehyde (C5′ site) derivatives. Therefore, 6-bromopyrimidines could be considered as potential DNA radiosensitizers. However, despite promising computational results obtained for 6BrdX derivatives, they will probably not be used as radiosensitizers due to their insufficient stability in aqueous solutions [82].

Summing up, the computational studies on brominated nucleobases led to conclusions that theoretically all of the described above compounds could work as potential radiosensitizers, and promising nucleobase derivatives should possess at least two properties: (1) high electron affinity to work as an electron trap, (2) ease to undergo DEA process which produces a reactive nucleobase centered radical.

4.3.4 5-Substituted Uracils as Potential Radiosensitizers

Obtaining a reactive radical inside the DNA strand after electron attachment is crucial to effective sensitization of this biopolymer to radiation. Based on this assumption, an easy and fast computational approach was propose for initial prescreening of radiosensitizing properties of substituted uracils [39]. Although the uracil molecule could be modified at various positions, only 5-position substitution was taken under consideration in the above mentioned study, as this site of the pyrimidine ring is not involved into the formation of hydrogen bonds within DNA double helix and it is susceptible to chemical modifications.

5-substituted uracil radiosensitizers should have higher electron affinity, compared to the native nucleobases, which enables them to act as "electron traps" and to catch hydrated electrons generated during water radiolysis. To be sure that the chosen compound will have high electron-withdrawing properties, only substituents with positive values of the Hammet's inductive constant σ_1, falling in the range from 0.3 (thiol function) to 0.63 (strong electrophile nitro group) [83], were selected [39]. Secondly, the size of the substituent bonded to uracil moiety should be as small as possible to fit into DNA double helix as a substituted uracil replaces thymine, i.e. 5-methyluracil. Another criterion for selecting the substituents was related to the ease of decomposition of the respective anion radical. The latter criterion results from the fact that an effective nucleoside radiosensitizer is expected to easily undergo DEA process to produce reactive genotoxic $U^{\bullet}$ radical inside the DNA strand. Thus, to preselect compounds which could be further degraded to $U^{\bullet}$ during the DEA process, the substituents with relatively weak bond (in comparison to the C5-Br bond) with uracil moiety were chosen (see Fig. 4.12).

For those compounds the electronic stability of anions was analyzed at the DFT level (AEA and VDE values; for details see Table 4.1). The results confirmed that all those derivatives could act as an electron trap if introduced into DNA strand. However, further computational study on the DEA process (providing the kinetic—ΔG^* and thermodynamic—ΔG barriers for breaking the bond between uracil and its substituent in the substituted uracil radical anion) revealed [39] that only two of the modified uracils, i.e. SCNU and OCNU, behave similarly to BrU, willing to produce uracil $U^{\bullet}$ radical due to electron attachment (see Fig. 4.12).

To confirm the quantum chemical predictions, photoelectron spectroscopy (PES) was then employed. Two derivatives, differing in their radiosensitizing properties

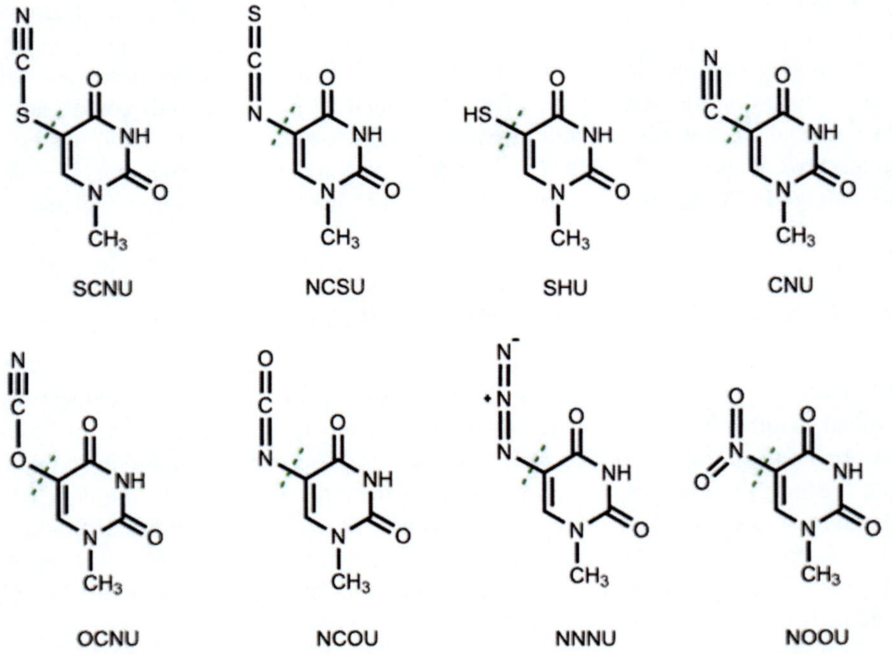

Fig. 4.12 5-substituted 1-methyl-uracil derivatives studied with corresponding name abbreviations. Dotted lines indicate bonds considered to be broken during DEA process. Adapted with permission from [39]. Copyright (2013) American Chemical Society

were chosen for the experimental part of studies: SCNU, expected to undergo DEA process with producing $U^•$ radical (potential sensitizer) and CNU, believed to produce thermodynamically stable $CNU^{•-}$ anion radical after electron attachment (no sensitizing properties). The results of PES experiment, in which anionic $CNU^{•-}$ form for CNU and SCN^- (and no $SCNU^{•-}$) for SCNU were detected, confirmed the reliability of the computational approach [39].

It also turned out that the described above simple computational tool is not always able to predict the outcome of DEA process even for seemingly simple uracil/uridine derivatives. ESR experiment carried out for 5-thiocyanato-2′-deoxyuridine (SCNdU) revealed that the main electron induced degradation path leads to the breakage of the S-C bond inside the substituent rather than between the substituent and uracil, which was predicted by that simple DFT model [41]. It is worth of mentioning that this finding does not exclude SCNdU from the radiosensitizer group. Produced via path B the U-5-$S^•$ radical (see Fig. 4.13) was experimentally found to dimerize to dU-S-S-dU (see Fig. 4.4a). Therefore, it was suggested that the formation of such radical within DNA strand could result in the inter- and intrastrand crosslink damage. Moreover, the side product of the B path, the cyanide anion (CN^-), is known for its cytotoxicity related to the efficient blocking of cellular respiration [84].

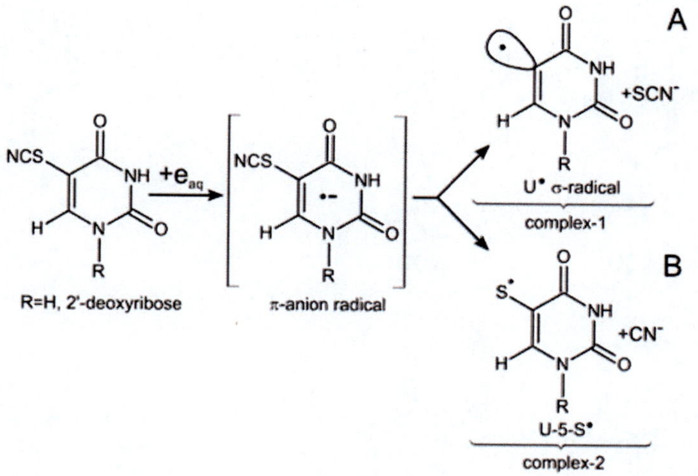

Fig. 4.13 Formation of U• (ca. 10% yield) and uracil-5-thiyl radical (U-5-S•, ca. 90% yield) by dissociative electron attachment to 5-thiocyanatouracil (SCNU) or 5-thiocyanato-2′-deoxyuridine (SCNdU). [41]—Reproduced by permission of the PCCP Owner Societies

The possibility of various DEA paths, leading not only to U• but also to another types of radicals, prompted to consider electron attachment induced dissociation not only between the substituent and uracil moiety but also within the substituent itself [40]. Based on previous experience [39], another set of electrophilic 5-substituted uracils were computationally studied (see Fig. 4.14). For the first group (A: NCU, NOU, OHU and CCHU, see Fig. 4.14) chemically available was only electron induced uracil-substituent bond breaking. Group B consists of compounds for which electron attachment induces barrier-free breakage of a bond inside the substituent (spontaneously broken bonds marked in Fig. 4.14). For two compounds: $NHNO_2U$ and SeCNU both degradation routes were available—path A leading to U•, as well as path B leading to U-5-•NH and U-5-Se•. Path B was found to be kinetically more favorable [40]. Completely different behavior was observed for IOU derivative, as DEA process led in this case to the production of 5-iodouracil (IU) and oxygen radical ($O^{\bullet -}$) as a side product [40]. Careful analysis of kinetics and thermodynamics of the electron induced degradation processes enabled to select three most promising radiosensitizing agents: SeCNU, the source of cyanide anions able to block cellular respiration, SClU, as the source of U-5-S• believed to produce crosslinks in DNA, and IOU as a prodrug, as it may transform into a radiosensitizing IU. This computational study led to the further experimental research on selenocyano- modified uridine SeCNdU [26, 85].

The above described computational approach was also used by Wang et al. to analyze radiosensitizing properties of 5-(halomethyl) uridines: 5-XCH_2U, where X = F, Cl, or Br [86]. They found out that all those uridine derivatives are likely to attach an excess electron and undergo DEA process leading to breaking inter-substituent X-C bond with the formation of the U-5-$^{\bullet}CH_2$ radical. Thus, they

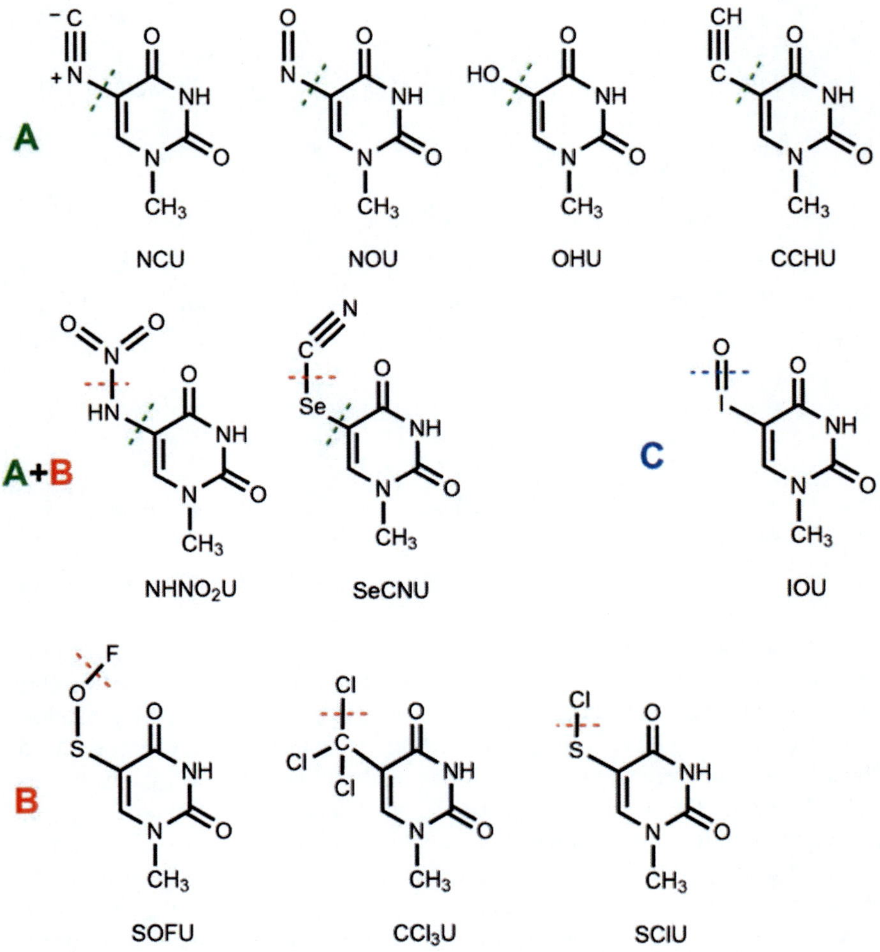

Fig. 4.14 Structures of 5-substituted uracil derivatives (UXY). The bonds broken due to electron attachment are marked with dotted lines. Group **a**: U-XY bond breaking was considered only, groups **b** and **c**: consider breaks to the UX-Y bond during degradation. The $NHNO_2OU$ and SeCNU compounds belong to both groups [40]. Taken with permission from Ref. [40] Copyright (2016) (John Wiley and Sons)

suggested that halomethylated uridines could be potential radiosensitizers inducing crosslink damage in 5-XCH_2U labeled DNA, and their sensitizing abilities are increasing as follows: 5-FCH_2U < 5-$ClCH_2U$ ≈ 5-$BrCH_2U$.

Similarly, it was demonstrated [87] that various pyrimidine nucleosides with azido modification at the C5-site of the pyrimidine base (Fig. 4.15) bind radiation-produced electrons forming the highly unstable azide anion radical. The latter species undergoes prompt release of nitrogen molecule resulting in the nitrene radical which after swift protonation is converted into the damaging $RNH^{\bullet}$ radical.

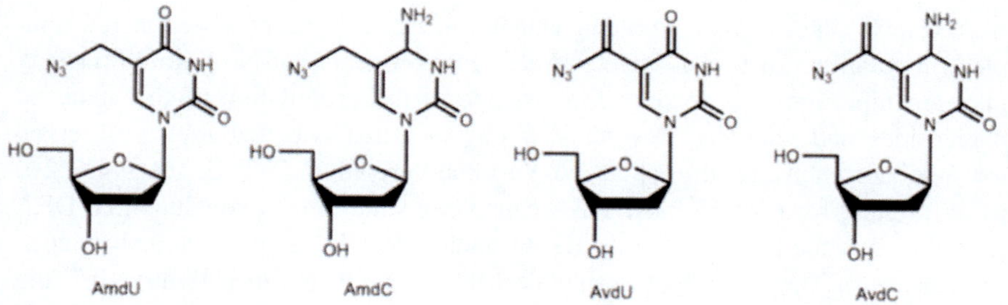

Fig. 4.15 Structures of potentially radiosensitizing azido-modified pyrimidine nucleosides [87]

It was shown that 5-azidomethyl-2′-deoxyuridine (AmdU)—see Fig. 4.15—reveals significant radiosensitizing properties against EMT6 breast cancer cells.

4.3.5 A Need to Expand the Computational Model for Difficult Derivatives

As it was emphasized in the previous paragraphs, relatively easy and cheap DFT calculations [39, 40] allow the pre-selection of potentially radiosensitizing compounds via analysis of the mechanisms of their DEA processes. On the other hand, such simplified methodology cannot anticipate all possible experimental problems, related, for example, to reaction environment like water. Continuous PCM model admittedly allows to take into account electrostatic stabilization exerted by water, but remains ineffective for the description of hydrogen bonds. The solution could be obviously employment of more complex and time-consuming methods, AIMD or QM/MM, that take into account explicit water molecules and reaction environment [73, 81].

On the other hand, unexpected experimental results can trigger in-depth analysis of computational results—and vice versa. So was in the case of SCNdU radiosensitizer for which water environment changed dissociative electron attachment route in the ESR experiment, as was described in the previous paragraph [41]: •S-dU radical was mainly observed experimentally instead of 5-dU•, predicted by the computations on DEA to SCNU in the gas phase [39]. Another interesting case was related to water impact in the reactions triggered by the excessive electron attachment to BrU and studied using the ESR method [88]. In that case computations helped to explain the formation of radiolysis products by suggesting that the dU• radical is able to produce •OH radicals in reaction with water. Thus, additional damaging properties of dU• could be bonded with the producing of genotoxic radicals in the close vicinity to the labeled DNA strand. Another problem related to the presence of water in biological systems which should not be ignored while searching for new radiosensitizers, is that new radiosensitizing compounds may undergo hydrolysis losing their sensitizing properties.

The next experimental problem, which can be easily overlooked in too simple computational model is that in water environment acid-base equilibria may play an important role, leading for instance to the protonation of the analyzed nucleosides and changing degradation routes in that way (as it was observed for 5-trifluoromethanesulfonyl 2′-deoxyuridine derivative, OTfdU, see Fig. 4.16) [26]. Initially, for OTfdU, basic DEA calculations allowed to conclude that DEA should lead to the formation of the dU-O$^{\bullet}$ radical and Tf^{-} anion (kinetic barrier of 1.5 kcal/mol). The second less probable path run with the formation of dU$^{\bullet}$ and OTf^{-} (barrier 8.6 kcal/mol). However, irradiation of solutions containing OTfdU gave different results. Namely, both products were formed, while the second pathway was found more efficient. It allowed to conclude that simple DEA model does not include important processes. Differences between degradation products obtained experimentally and predicted by calculations were finally explained by the possibility of anion radical protonation or neutral form deprotonation and the influence of pK_a on the DEA process.

Another similar case is the ISdU study [27]. During radiolysis, the compound effectively undergoes the DEA process with the formation of a number of stable products whose mechanisms of formation were proposed by calculations. Also here, besides the expected DEA product—SdU, a number of other products—dimers or oxidation products—are formed. To explain the reaction mechanisms the presence of *tert*-butoxyl radicals (*t*-BuO$^{\bullet}$, the product of *t*-BuOH reaction with OH$^{\bullet}$) and H_2O_2 (water radiolysis product) in the environment were taken into account. The calculations allowed to conclude that the apparent decrease in the efficiency of SU formation during radiolysis in solution is affected by the SU$^{\bullet}$ radical consumption during the formation of ISU-SU dimer.

Those two examples of potential radiosensitizers discussed above—OTfdU and ISdU—are the exemplars of possible problems that might be encountered while characterizing potential radiosensitizers. In both cases the calculations allowed for better understanding and explanation of the experimental results and highlighted a need to take into account a number of aspects of the DEA process, not only electron affinity or Hammet's constant of substituents.

Fig. 4.16 Structures of modified 2′-deoxyuridines considered as potential radiosensitizers: 5-trifluoromethanesulfonyl 2′-deoxyuridine (OTfdU) and 5-iodo-4-thio-2′-deoxyuridine (ISdU)

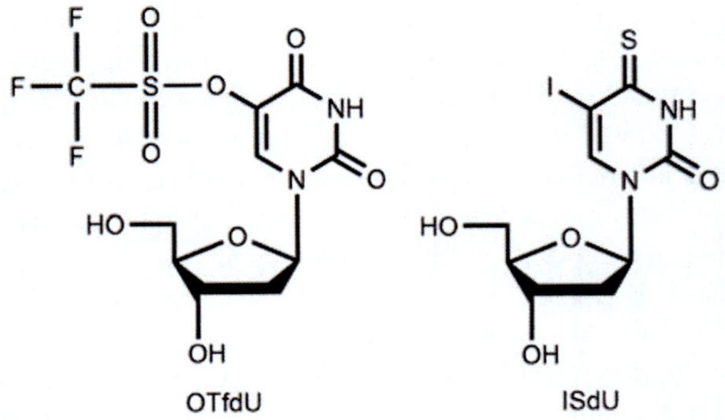

4.4 Oxygen Mimetics

In the 1960s, a concept of joining electron affinity of a compound with its radiosensitizing efficiency has been extensively explored [89]. Such radiosensitizers were to be imitated (mimic) oxygen in hypoxic cells (Fig. 4.17). The best known and effective class of oxygen mimetics are nitroimidazoles [90].

Nitroimidazole antibiotics display main role in clinical use to "fix" DNA damage induced by ionizing radiation. Adams et al. noted that significant parameter for radiosensitization and toxicity is the electron affinity of 2- and 5-nitroimidazoles [91]. Thus, 2-nitroimidazole—misonidazole—with higher electron affinity is more effective than 5-nitroimidazole—metronidazole. However, clinical use of misonidazole is limited, because it turned out to be neurotoxic [92]. To minimize neurotoxicity of 2- and 5-nitroimidazoles, their side chains were modified by more polar groups [93, 94] (Fig. 4.18). Amide analog of 2-nitroimidazole named etanidazole, demonstrated more hydrophilic character, hence a slower uptake by neural tissues compared to misonidazole [95]. Despite lesser toxicity, etanidazole offered no benefit to patient in clinical trials [93].

Another 2-nitro sensitizer with the reduced neurotoxicity, doranidazole (Fig. 4.18), showed radiosensitization in hypoxia. Doranidazole has been evaluated in I/II phase of clinical trials with promising results in chemoradiotherapy treatment of non-small-cell lung cancer [96]. The most efficient sensitizer seems to be 5-nitroimidazole analog called nimorazole. It is well tolerated and has been clinically used in the cancer treatment in Denmark. Currently, nimorazole has been investigated in III phase of clinical studies to use in radiotherapy in head and neck cancers [97]. Despite of the fact that nitroimidazolic compounds have been under consideration for years, it is important to note that molecular mechanism of their action still remains unclear.

One of such mechanisms, proposed by Wardman et al. [98], assumes that nitro derivative induces DNA damage in a manner similar to oxygen. A possible pathway leading to DNA strand break and cell death is based on the reaction of the hydroxyl radical with a DNA base, forming carbon-centred radicals. In the

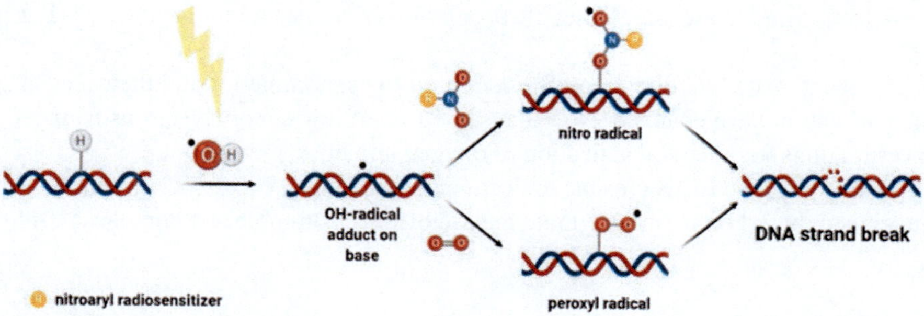

Fig. 4.17 Mechanism of oxygen and oxygen-mimetics lead to DNA strand break

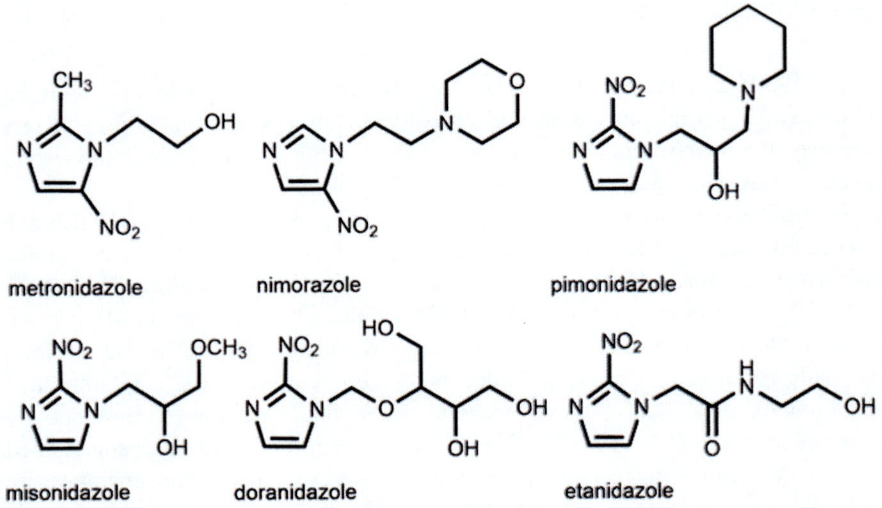

Fig. 4.18 Clinically investigated nitroimidazole radiosensitisers

next step, oxygen/nitroimidazole can be added to a nucleobase, forming a long-lived peroxyl/nitro radical. In either case, the long-lived radical might abstract hydrogen atom from a neighbouring sugar moiety leading ultimately to a heterolytic phosphodiester bond breakage (Fig. 4.19). The sugar radical-cation, a product of heterolytic phosphodiester bond cleavage, is able to react with other molecules, e.g. water, forming stable products [10]. On the other hand, nitroimidazole radiosensitizers with high electron affinity have been identified as hypoxia-activated prodrugs (HAPs) or hypoxia-selective cytotoxins (HSCs) [99]. In the presence of oxygen, HAPs are reduced by one-electron reductase forming nitro radical anions, which undergo oxidation to reproduce the nitroimidazole prodrug. In hypoxic conditions, nitroimidazole prodrug radical is fragmented by disproportionation or undergoes subsequent reduction leading to various products capable of attacking cellular macromolecules, such as DNA [100].

On the other hand, potentially active drug species may be produced directly by two-electron reductase generating nitroso, hydroxylamine and amine species (Fig. 4.20).

In recent years, the direction of research on oxygen mimetics and their mechanism of action have changed. Nowadays, the role of low energy electrons is under investigations for potential activation of oxygen mimetics.

In the gas phase LEEs are able to decompose molecules via dissociative electron attachment based on an intermediate metastable electron-molecule complex [101].

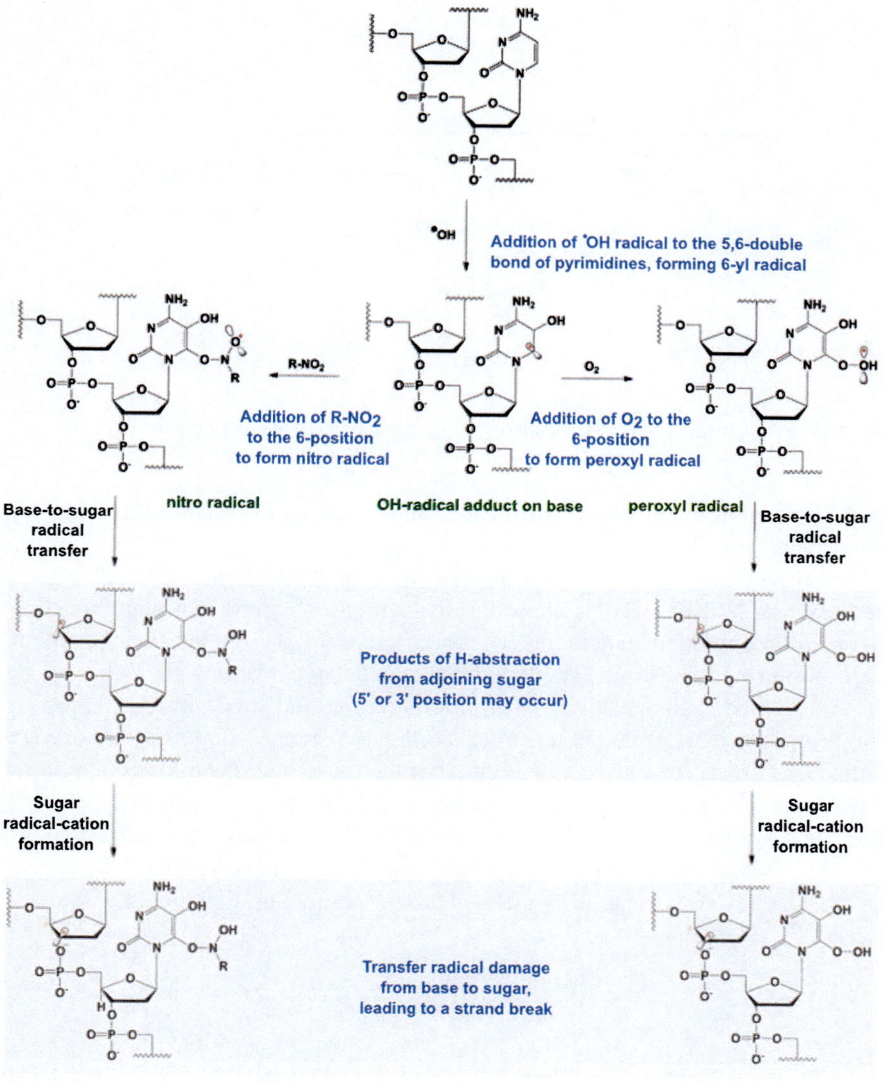

Fig. 4.19 Possible mechanism for strand break formation by oxygen-mimetics

Overall, DEA is a two-step reaction:

$$e^- + M \rightarrow \left[M^{\bullet -}\right]^{\#} \rightarrow [M\text{–}X]^- + X \tag{4.10}$$

$$\left[M^{\bullet -}\right]^{\#} \rightarrow M^{(*)} + e^- \tag{4.11}$$

where M is a molecule, $[M^{\bullet -}]^{\#}$ assigns the metastable intermediate transitory anion formed by Franck-Condon transition and X is a radical or neutral fragment [102].

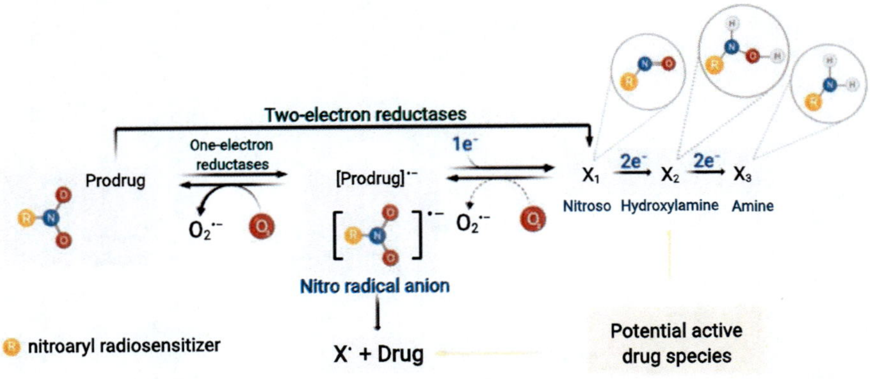

Fig. 4.20 Mechanism of hypoxia-activated prodrug by one- and two-electron reduction

Transitory anion may lose the extra electron, recovering the neutral molecule, by autodetachment.

In 2012, theoretical calculations showed breaking of four chemical bonds in one-step reaction by LEEs [103]. These results, demonstrated efficient bond breaking in organic compound and in the next years, shed light on the attachment of LEEs to nitroimidazole. In fact, experimental studies performed by Tanzer et al. [104] exhibited high sensitivity of 4-nitroimidazole and its derivative 1-methyl-4-nitroimidazole to LEEs in energies of 0–8 eV range. Both nitroimidazolic compounds create the same ionic products in the crossed electron-molecular beam experiment and detected by mass spectrometry, however ion yields are different. They observed the loss of H atom (only for 4-nitroimidazole), expressed as Eq. (4.12), cleavage of the C–NO_2 bond and loss of neutral $^{\bullet}OH$. The cleavage of the C–NO_2 bond in 4-nitroimidazole occurs through two DEA reactions:

$$e^- + M \rightarrow \left[M^{\bullet -}\right]^{\#} \rightarrow [M\text{–}H]^- + H^{\bullet} \tag{4.12}$$

$$e^- + M \rightarrow \left[M^{\bullet -}\right]^{\#} \rightarrow [M\text{–}NO_2]^{\bullet} + NO_2{}^- \tag{4.13}$$

$$e^- + M \rightarrow \left[M^{\bullet -}\right]^{\#} \rightarrow [M\text{–}NO_2]^- + NO_2{}^{\bullet} \tag{4.14}$$

Kossoski et al. showed that generation of $NO_2{}^-$ goes by indirect dissociation mechanism, where coupling of π^* state and the repulsive $\sigma^*{}_{CN}$ state occurs [105].

The loss of a neutral $^{\bullet}OH$ unit for 4-nitroimidazole is related to the cleavage of two bonds, N–O and C–H or N–H ($e^- + M \rightarrow [M^{\bullet -}]^{\#} \rightarrow [M\text{–}OH]^- + {}^{\bullet}OH$), while the formation of $[M\text{–}OH]^-$ is quenched in the methylated compounds.

Methylated 4-nitroimidazole reactivity is completely blocked in the energy below 2 eV, but target molecule reactivity is operative in higher LEEs energies.

This phenomenon has been interpreted via the formation of vibrational Feshbach resonance (VFR) [106]. Additionally, the excess electron in VFRs is weakly bonded in the field of vibrationally excited molecule. It is worth of mentioning that nitroimidazole isomers and their derivatives have different dipole moment, i.e. 4-nitroimidazole has a dipole moment around 7.2 D, compared to only 3.24 D for 5-nitroimidazole. For the methylated compound large dipole moment (above critical minimum value to bind an excess electron) may account for the formation of VFR and increase efficiency of DEA, affecting this way the ion intensity [107].

Further observations of 4-nitroimidazole, 1-methyl-4-nitroimidazole and 1-methyl-5-nitroimidazole compounds confirmed bond cleavage with NO_2^- formations, loss of neutral $^{\bullet}$H and $^{\bullet}$OH, but also propose possible pathway of complete degradation of the nitroimidazole derivatives leading to CN^- formation [104]. The CN^- generation from DNA bases has been reported by Märk et al. already in 2004, but at higher energies [108]. For nitroimidazolic molecules, complete degradation with the formation of CN^- or CNO^- is viable at 0 eV via reactions [109]:

$$e^- + M \rightarrow \left[M^{\bullet -}\right]^{\#} \rightarrow CN^- + CO_2 + N_2 + {}^{\bullet}CH_3 \tag{4.15}$$

$$e^- + M \rightarrow \left[M^{\bullet -}\right]^{\#} \rightarrow CNO^- + CO + N_2 + {}^{\bullet}CH_3 \tag{4.16}$$

As mentioned above, 4- and 5-nitroimidazole and their methylated derivatives demonstrated similar fragmentation pathways. In 2017, Ribar et al. [109] suggested that in 4-nitroimidazole single bond cleavage related to the loss of neutral $H^{\bullet}$ at position N1 occurs. Most remarkably, vertical electron affinity (VEA) of the dipole-bound state is comparable to that of the valence-bound anion for 4-nitroimidazole. This is why the N1-H bond is broken due to electron attachment for 4-nitroimidazole via dipole-bound anion formation (VEA of dipole bound state eq. 158 meV vs. 160–170 meV for valence bound anion) while the fragmentation pathway for 2-nitroimidazole, leading to the loss of H_2O (see Eq. 4.17), proceeds via the valence anion since the VEA of its dipole bound state amounts to only 43 meV.

$$e^- + M \rightarrow \left[M^{\bullet -}\right]^{\#} \rightarrow [M\text{–}H_2O]^{\bullet -} + H_2O \tag{4.17}$$

Another pathway not observed for the degradation of 4- and 5-nitroimidazole is the loss of HONO:

$$e^- + M \rightarrow \left[M^{\bullet -}\right]^{\#} \rightarrow [M\text{–}HONO]^{\bullet -} + HONO \tag{4.18}$$

These results suggest that 2-nitroimidazole could be more effective radiosensitiser because of dominant losses of $^{\bullet}$OH, $^{\bullet}$ON, H_2O and another variety of radicals and radical anions which could damage biological molecules. Recently, Meißner et al. observed for 2-nitroimidazole additional simple bond cleavages, which lead to

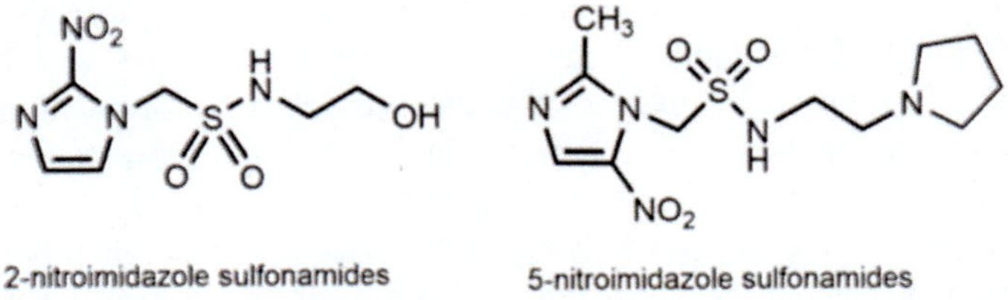

Fig. 4.21 Examples of nitroimidazole sulfonamides

the loss of the neutral CH_3 and CH_2NO_2 [110]:

$$e^{-} + M \rightarrow \left[M^{\bullet -}\right]^{\#} \rightarrow [M\text{–}CH_3]^{-} + CH_3 \tag{4.19}$$

$$e^{-} + M \rightarrow \left[M^{\bullet -}\right]^{\#} \rightarrow [M\text{–}CH_2\text{–}NO_2]^{-} + CH_2NO_2 \tag{4.20}$$

Despite radiosensitising properties of 2-nitroimidazole, its neurotoxicity is too large to be used in clinic [95]. So, in 2019, Denifl et al. focused on nimorazole, which is clinically used since 1990 [111]. The DEA formation of NO_2^- from nimorazole is the only dissociative process observed in the electron energy range of 2–4 eV. A competitive to the DEA process is so called associative attachment (AA), i.e. the formation of intact anion without further dissociation [112]. They also studied electron attachment to nimorazole in water clusters ($M(H_2O)_n$) and found out that DEA is reduced for solvated system in favour of AA. It confirms assumption about energy dissipation to the environment in solution [113]. As a consequence, presence of water lead to significant quenching of DEA to nimorazole as compared to the gas phase. It is worth of emphasizing, that intact anion formation seems to be crucial for radiosensitizing activity of these compounds. Indeed, nitroimidazolic species have to be present in the time of irradiation for therapeutic effects [98].

In recent years increasing interest in novel nitroimidazole derivatives has been observed. One of the strategies employed to propose new compounds is transformation of nitroimidazoles into alkylsulfonamides, which were obtained by modification of appropriate side chains [114]. This approach was reported first time in 2014 and has a major impact on the physicochemical properties of the nitroimidazolic compounds. First of all, sulfonamide chain in respective derivatives with extra hydroxyl or amine groups leads to the compounds of increased solubility and electron affinity (Fig. 4.21). Second, some of the studied nitroimidazole sulfonamides derivatives had selective toxicity against hypoxic cells.

On the other hand, Hay et al. modified 2- and 5-nitroimidazole sulfonamides by esterification to phosphate ester obtaining radiosensitizers called phosphate prodrugs (Fig. 4.22) [115]. Phosphate modifications showed better in vivo activity comparing to nitroimidazole sulfonamides and what is most important their electron affinity is significantly higher.

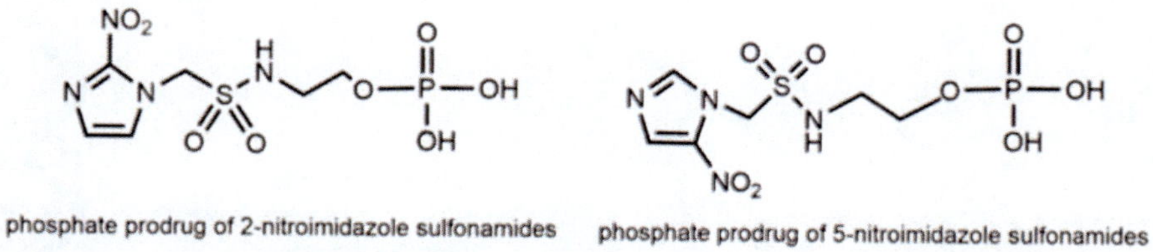

Fig. 4.22 Examples of phosphate prodrug of nitroimidazole sulfonamides

4.5 Metallic Nanoparticles and Metal Complexes

4.5.1 Possible Mechanisms of Radiosensitization

Metal chemistry offers a wide range of biologically-active molecules [116]. Many of them are considered as anti-cancer agents. Besides being effective single-agent chemotherapeutics, metallic nanoparticles and complexes are explored as potent radiosensitizers [117].

Research done in recent years indicates that heavy-metal nanomaterials with high atomic number appear to be especially promising radiosensitizers. One of their exceptional features is ability to absorb, scatter and emit radiation energy. Furthermore, the metallic nanoparticles possess favorable kinetic profiles of drug exposure to the tissues, satisfying chemical stability, high biocompatibility and low toxicity. In addition, fast distribution and enhanced uptake in tumor tissue are characteristic for nanoscale materials. Ionizing radiation can interact with metallic nanoparticles and cell structures in several ways. In general, from the viewpoint of physical processes, the radiation-enhancing action is related to absorption of ionizing radiation and emission of secondary electrons. Such phenomena as Auger electrons production, Compton or coherent scattering and photoelectric effect, lead to enhancement of local dose of ionizing radiation in tumor site and in consequence, to direct or indirect (mostly via reactive oxygen species (ROS)) DNA damage (Fig. 4.23) [118–120].

The biological mechanism of radiosensitization by metallic nanoparticles has been studied intensively. There are several theories explaining the synergistic cooperation between ionizing radiation and metallic nanoparticles. One of them is related to cell cycle regulation [121]. It is believed that enhancement of radiosensitivity is achieved when nanoparticles can alter cell cycle phases and lead to cells accumulation in the G2/M phase, and reduction of cells population in the G0/G1 phase [122, 123]. Other theory assumes that metallic nanomaterials inhibit the repair of DNA damage, thereby increasing their pool [124]. Another one says that nanoparticles are able to mediate cell signalling and can act via mechanism known as bystander effect wherein directly irradiated cells transmit damaging signals to non-irradiated cells [125, 126].

The other metal-based radiosensitizing agents are metal complexes. There are several possible modes of their combined action with radiation at the molecular

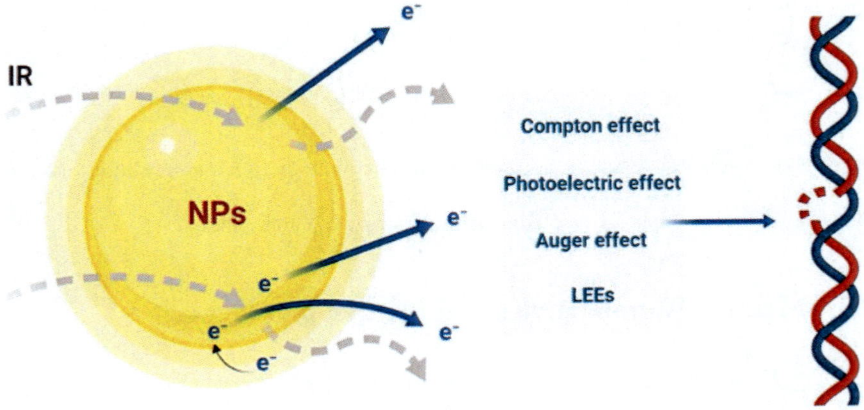

Fig. 4.23 Possible action modes of the metallic nanoparticles (NPs) in radiosensitization. Absorption of ionizing radiation (IR) and subsequent phenomena related to secondary electrons production or scattering lead to enhancement of local dose of ionizing radiation in tumor site and in consequence, to direct or indirect DNA damage

level. Essentially, it is believed that the most explored example of radiosensitizing metal complexes, cisplatin, dissolves in the cellular environment and binds to DNA structure leading to potentiation of its sensitivity to low energy electrons and •OH radicals [127, 128]. Other interpretations emphasize the role of effective dissociation of cisplatin by low energy electrons [129, 130] or increased generation of secondary electrons on Pt atoms and their further interaction with DNA leading to the enhancement of bond cleavages [131].

The cellular effects of synergy between radiation and cisplatin are mostly related to the binding of cisplatin to DNA. It is not surprising that the resulting Pt–DNA adducts act by distorting the structure of DNA, inhibiting replication and transcription [116]. The radiosensitizing effect of cisplatin at cellular level is explained by (1) enhancement of DNA damage caused by additional immediate species created directly by radiation, or/and by (2) interference with repair by non-homologous end-joining (NHEJ) when Pt–DNA lesions occur near radiation-generated double strand breaks [116, 132, 133].

4.5.2 On the Role of Low Energy Electrons in the Radiosensitization of DNA by Metallic Nanoparticles and Complexes

Since low energy electrons are one of the major secondary products formed in water environment by high energy radiation, interest in them is constantly increasing. The special properties of LEEs, at energies less than 30 eV, make them a good target in cancer treatment. Furthermore, interaction between high energy radiation and atoms with high atomic number e.g., gold nanoparticles, which are source of LEEs, leads to synergic and desired effect in radiotherapy [134]. A lot of experiments have been performed to understand molecular interactions of LEEs with small individuals (e.g., O_2, H_2O) and large biomolecules (e.g., DNA) as well. The latter class of interactions is very important for the potential application of metallic nanoparticles and metal complexes in radiosensitization. Previous experiments were usually performed on the gaseous isolated targets or under vacuum [135, 136]. In such conditions the primary mechanism of biomolecules damage caused by electrons is a resonance mechanism. An unoccupied orbital captures LEEs and forms transient anions (TA), then undergoing DEA process, which can lead to various damages of DNA e.g., crosslinks or single/double strand breaks. On the other hand, after anion deionization, the site, where electron was attached, can stay in a dissociative electronically excited state leading also to DNA damage [135, 137].

One of the nanoparticle types, which have potential radiosensitizing properties are gold nanoparticles (GNPs). They lead to the enhancement of DNA damage. Xiao et al. and Zheng et al. indicated that most of electrons emitted from GNPs possess energy in a range of 0–30 eV [135, 138, 139]. Furthermore, Zheng et al. suggested that radiosensitization of DNA by GNPs could result from (1) an enhancement of local absorption of ionizing radiation (IR), which leads to an increase of short-range secondary electrons pool, that can damage DNA, and/or (2) an increase of DNA sensitivity to damages induced by LEEs [140].

Ligands, which stabilize the binding with GNPs are thiolated organic molecules (e.g., oligonucleosides, peptides etc.). In these complexes, the thiol ligand self-assembles on the gold surface through the Au–S covalent bond. Ligands allow entering into the cell but their role on the mechanism of action of thiolated GNPs radiosensitization is not clear [139]. Other species, which can coordinate to GNPs, are alkane ligands which lead to the increase of photoexcitation and photoemission of nanoparticles. On the other hand, gadolinium (Gd) chelating dithiolated pentetic acid (DTDTPA), which should lead to the increase of X-ray absorption and the amount of LEEs, do not exhibit any radiosensitization effect [139].

Xiao et al. conducted an experiment using GNPs and three GNPs complexes: (1) with thiolated undecane ($S–C_{11}H_{23}$), (2) with DTDTPA and (3) with DTDTPA:Gd chelating ligands. Dry films of plasmid DNA pGEM-3zF(−), DNA with GNPs and DNA with three aforementioned GNPs complexes, were bombarded with 60 keV electrons. These experiments confirmed that GNPs caused increase in the absorption of IR, which affects the production of LEEs. However, coating of GNPs led to the decrease amount of LEEs and in consequences, to reduction of double and single strand breaks formation. It has been proved that the binding of GNP without coating to DNA leads to increase of DSB and SSB by a factor of 2.3, while in case of GNPs coated with $S–C_{11}H_{23}$, reduction of this factor to 1.6 was observed. The binding of GNPs via the relatively long DTDTPA linkers results in essentially complete attenuation of LEE's activity because of their short effective range. For this reason, it has been concluded that GNPs should be bonded to DNA of cancer cells by a relatively short linker [139].

Other type of molecules which sensitizes DNA to IR by LEEs are platinum (Pt) compounds, e.g. cisplatin [135]. Zheng et al. conducted an experiment, where they studied the influence of LEEs on damage to DNA linked with cisplatin. They prepared a dry film of pure plasmid DNA and a DNA–cisplatin complex and bombarded them with electrons in the range of 1 eV to 60 keV. DNA damage was determined by count of SSBs and DSBs. The results shown that enhancement factor (EFs) for DNA–cisplatin complex increases from 1.3 to 4.4 depending on (1) the dissociation of bond caused by the formation of transient anions and (2) the ratio of cisplatin/plasmid. In this experiment Zheng et al. proved that in DNA–cisplatin complex, SSBs and DSBs are induced by LEEs and are substantially enhanced when cisplatin is covalently bonded to DNA [141].

Bao et al. and Luo et al. measured conformational damage in pure plasmid DNA and DNA–cisplatin complex, in dependence of energy of LEEs (2–20 eV). Except a strong resonance in pure DNA at 5 and 10 eV they observed two additional signals at 13.6 and 17.6 eV. The presence of these two additional resonances can explain the increase of damage in cisplatin–DNA complex [142, 143]. Bao et al. and Rezaee et al. claimed that, when Pt-analogs bound two guanines on the opposite strands. Transient anion, formed from the Pt-adduct would have additional electron delocalized equally between two unoccupied σ^* molecular orbital (Fig. 4.24). This led to the bond breaks between cisplatin and two guanines. As a result two guanine radicals are created, which may detach hydrogens from the DNA backbone [137, 143].

Besides inter-strand interactions between cisplatin and DNA, its intra-strand binding was also observed. Indeed, Mantri et al. [144] explained, by using quantum chemical molecular simulations, preference of cisplatin bifunctional binding to the adjacent nucleobases in the AGA fragment of DNA. After initial platination at the central guanine, cisplatin prefers binding with the 5′-adenine. Both adducts with cisplatin, (AG and GA) have similar energies, but the adduct of AG is slightly more favored. It is caused by higher reaction barrier for the GA closure over the alternative AG binding motif, which is connected with a difference between the two forming transition states. The presence of a hydrogen bond between the axial ammine ligand of cisplatin and the phosphate backbone stabilizes the transition state of AG adduct. This interaction is not observed in the GA adduct. It means that the binding preference of AG adduct over the GA one is largely under kinetic control [144].

Another experiment was performed by Behmand et al. [145]. A water solutions of oligonucleotide (TTTTTGTTGTTT) with or without cisplatin in the presence of a hydroxyl radical scavenger were irradiated. The results showed that oligonucleotide interacted with solvated electrons, which led to the damage of thymine and to the break of bonds between cisplatin and one or both guanines. They proposed two possible mechanisms explaining this process. The first one comprised an attachment of solvated electron to thymine base and transfer of this electron from base to base up to the guanine site, where cisplatin released it via DEA. The second, assumed that solvated electron directly induces detachment of cisplatin from guanine—also via DEA [145, 146].

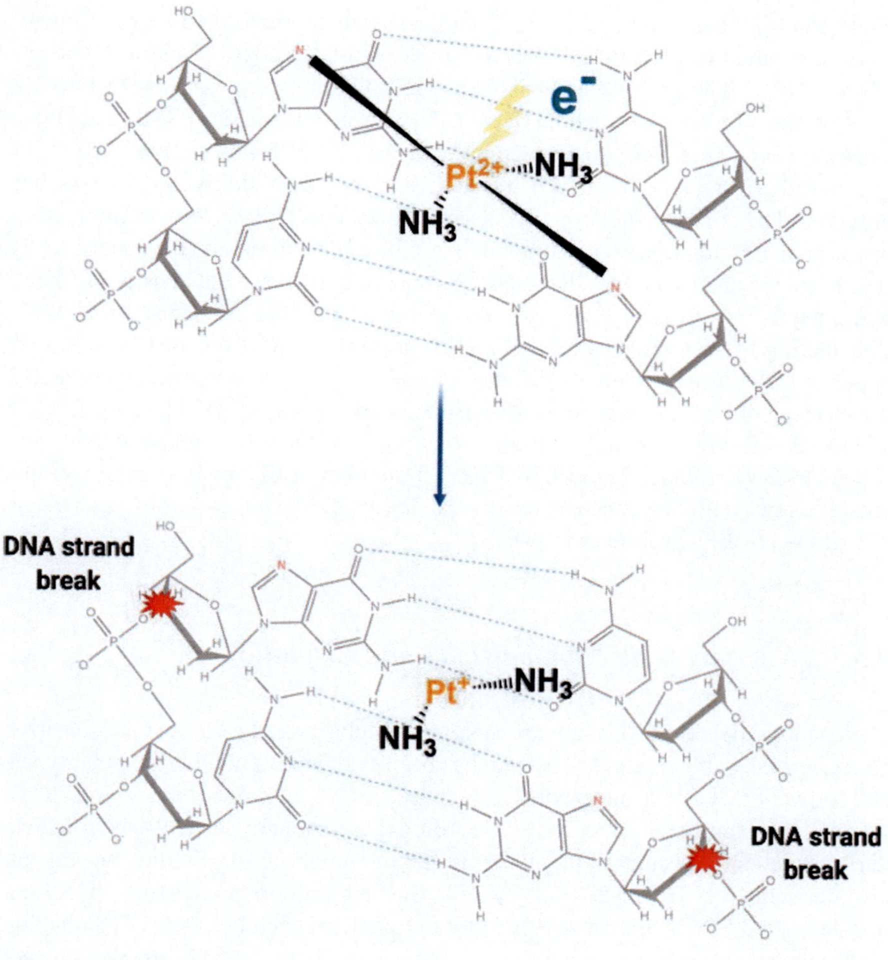

Fig. 4.24 Possible mechanism of DNA damage induced by single electron, when cisplatin is bound by two guanines on opposite strands

Another attempt to explain a molecular mechanism of LEEs interaction with cisplatin–DNA was undertaken by Dong et al. [136]. They prepared five-monolayer films of plasmid DNA (pGEM-3Zf (–), 3197 bp)–cisplatin complex and bombarded them with electrons with energies of 4.6 and 9.6 eV. They also observed the increase of EFs for SSBs, DSBs and crosslinks. Their EFs are the largest measured so far. The molecular mechanism of this process is based on that already proposed by Bao et al. and Rezaee et al. [134, 143].

Since, it has been shown that GNPs and cisplatin are radiation enhancers, it was intriguing if the electrostatic binding of GNPs with DNA–cisplatin complex could lead to a superadditive effect. Zheng and Sanche performed an experiment, where dry films of plasmid DNA and complexes of (1) DNA–cisplatin (1:2), (2) DNA–GNP (1:1) and (3) DNA–cisplatin–GNP (1:2:1) were exposed to 60 keV electrons under ultra-high vacuum (UHV). They measured yields of SSBs and DSBs generation, for pure DNA and its complexes. EFs of SSBs formation is equal to 2, 2.5 and 3 for (1), (2) and (3), respectively and superadditive effect was not observed. The opposite situation was observed in case of DSBs. Here, an impressive increase of EFs for DNA–cisplatin–GNP complex was observed (EF equal to 7.5) [147]. This is due to the fact that GNPs significantly increase the density of LEEs, while cisplatin causes the lowering of the energy barrier for DSBs generation [148]. Considering the above properties, single or multiple interactions of LEEs on two opposite DNA strands, within a distance of 10 base pairs, could affect significantly the increase of DSBs formed in DNA–cisplatin–GNP complex [137].

Described above experiments suggest that cisplatin and GNPs are good radiosensitizers to DNA damage induced by LEEs. Introducing LEEs to anticancer therapy creates a few challenges, such as selective delivery of molecules—which are source of LEEs—or LEEs dosimetry [134].

4.5.3 Not only Gold Nanoparticles and Cisplatin

Gold nanoparticles and cisplatin are the most widely studied metallic radiosensitizers, nevertheless the research spectrum has been expanded to include the other types and forms of metallic nanoparticles and complexes.

The first example is super magnetic iron oxide nanoparticles (SPIONs), which due to excellent biocompatibility undergo cellular uptake [149]. Studies carried out on three cell lines (Caco-2, MCF-7, 3T3) showed that citrate-coated SPION and malate-coated SPION enhance sensitivity to radiation, even by 300%, by inducing higher concentration of reactive oxygen species in tumor cells. Moreover, coated SPIONs do not show any cytotoxic effect on cells unexposed to X-ray radiation. Furthermore, super magnetism of the iron oxide nanoparticles enables the area of tumor occurrence through heterogeneous external magnetic field to be accessed [149, 150].

Another example are bimetallic core-shell gold-platinum polyethylene glycol (PEG)-grafted nanoparticles (Au:Pt–PEG NPs), which induce an 90% increase of double-stranded breaks (DSBs) compared to 34% increase using monometallic gold PEG-grafted nanoparticles (Au–PEG NPs). Radiosensitizing properties of core-shell mono- and bimetallic nanoparticles were investigated by evaluating a gamma-induced damage in plasmid DNA. The presence of dimethyl sulfoxide (DMSO) resulted in an 80% DSBs reduction demonstrating a significant role of hydroxyl radicals in damage. Monte Carlo simulations confirmed that bimetallic Au:Pt–PEG NPs radiosensitizers are more effective than monometallic Au–PEG NPs. Considering only the direct effects caused by primary photons and secondary electrons, a radiation enhancement by 45% for the bimetallic nanoparticles has been demonstrated, which remains in very good agreement with experimental results (42%) [151].

Moreover, Porcel et al. [152] also showed that platinum nanoparticles (NP–Pt) might behave as a radiation enhancer. Examination of the SSBs and DSBs was possible by the irradiation of DNA loaded with NP–Pt with X-rays tuned to the L_{III} and M_{III} resonant energy platinum shell (11,556 eV and 2649 eV, respectively) and with non-resonant radiation (11,536 eV and 2638 eV, respectively). Damages were induced in the presence of nanoparticles showing similar efficiency for radiation with or without electron shells resonance, which suggests that the non-resonant mechanism was mainly responsible for DNA damages enhancement. These studies suggested that LEEs can increase DNA damage by a non-resonant process mediated by secondary electrons, such as photoelectrons [152]. Alternative research demonstrated, that the enhanced damage of the plasmid DNA film deposited on a tantalum foil, such as SSBs, DSBs and crosslinks, are caused by X-ray-induced secondary electron emission from tantalum. The authors also noticed that the hydration level of DNA has a significant effect on the enhancement factor. When the number of water molecules per nucleotide increased from 6 to 21, the EFs increased for electron induced SSBs and DSBs, but decreased for crosslinks formation [153].

The electron induced processes in radiosensitization by metallic complexes are not widely studied, except of few cases. With the use of both, experimental and theoretical methods, the interaction of electrons with bis(pentamethylcyclopentadienyl)titanium(IV) dichloride ($Cp^*{}_2TiCl_2$) and difluoride ($Cp^*{}_2TiF_2$) was investigated. The research included the measurements of partial cross sections for DEA and the electron ionization (EI) mass spectra of isolated molecules. The fragmentation pattern in EI of these molecules is similar, while DEA to $Cp^*{}_2TiCl_2$ leads to strong fragmentation, mainly to Cl^- anion, and DEA for $Cp^*{}_2TiF_2$ to the creation of a stable parent anion. Due to such electron responses, these organometallic compounds can be promising radiosensitizers [128].

It has been demonstrated that not only cisplatin but also other platinum complexes possess the radiosensitizing properties. Rezaee et al. [154] bombarded pure plasmid DNA and three complexes of plasmid DNA with cisplatin, carboplatin or oxaliplatin, with 10 keV or 10 eV electrons, under UHV. Electrophoretic determination of the DNA forms, including nicked circular, supercoiled and linear, resulted from irradiation, allows to analyze the quantity of SSBs, DSBs. EFs in the yields of SSBs, DSBs and interduplex cross-links induced by 10 keV and 10 eV electrons in the presence of cisplatin, carboplatin, and oxaliplatin have been presented. When carboplatin, cisplatin, and oxaliplatin are bound to DNA, irradiation with electrons of 10 eV cause significant increase of DSB yields by factors of 3.1, 2.5, and 2.4, respectively. Similarly, the yields of interduplex cross-links induced by 10 eV electrons are enhanced as well by factors of 2.2, 3.1, and 4.1 in the presence of cisplatin, carboplatin, and oxaliplatin, respectively [154].

Dissociative electron attachment to platinum(II) bromide ($PtBr_2$) was examined at the range of LEEs energies (0–10 eV) in the gas phase. The study showed that Br^- was the only anion detected at 0.4, 1.2 and 7 eV. Significant temperature dependence of resonance was observed at 0.4 eV and the high ion current intensity for this value was explained by the formation of Br^- from HBr generated in the oven. Calculations allowed to link the 1.2 eV resonance to DEA reaction (Br^- + PtBr) and 7 eV resonance to DEA reaction leading to the formation of Pt, Br and Br^- fragments [155].

4.6 Summary

Efficient radiotherapy requires employment of a radiosensitizer sensitive to hydrated electrons besides ionizing radiation. Indeed, most tumors are hypoxic which imparts hydroxyl radical reactivity towards DNA. Moreover, in the absence of oxygen hydrated electrons are the second major product of water radiolysis. It is, however, well known that biological action of solvated electrons is negligible. Although in aqueous solutions native DNA binds solvated electrons, due to subsequent protonation of resulting anions, the electron attachment process does not lead to DNA strand breaks. In order to resolve this unfavorable from radiotherapy viewpoint situation, one may use substances that "activate" solvated electrons in the sense of DNA damage.

Three groups of radiosensitizers whose activity is related to electrons were discussed in this chapter. We started from the well-known bromo- and ododerivatives of nucleobases. Electron attachment to these species triggers efficient, low-barrier elimination of the halide anion and formation of the respective radical. Other modifications where the 5-substituent increases the electron affinity of the derivative and these modifications undergo low-barrier or barrier-free dissociation were also discussed. If such a modification of nucleobase is incorporated into the DNA molecule, then the reactive radical, being a product of DEA, may in the secondary chemical reactions lead to a strand break. A computational model that enables proposal of potentially radiosensitizing nucleosides/nucleobases was discussed. We demonstrated that it already predicted several new nucleosides/nucleobases which should work as the DEA sensitive systems. Among other, one could list here: SeCNdU, OTfdU, ISdU, BrSdU, SCNU, OCNU. Although the discussed computational model turned out to predict properly the behavior of the characterized compounds within the radiolytic studies, sometimes it failed completely. We suppose that the major drawback of that computational approach is related to the employment of the PCM model of water which treats water as a continuous medium characterized by a given dielectric constant. Indeed, the anions formed due to attachment of a hydrated electron are prone to protonation (which cannot be described at the PCM level) and if the DEA barrier for the intact anion is high enough, protonation becomes competitive to the DEA process, completely changing its thermodynamic and kinetic characteristics. Therefore, the most obvious development of such computational model should take into account the fact that water is not a continuous solvent. For these troublesome cases the AIMD or QM/MM approach, where at least several water molecules are treated quantum mechanically, should be used.

Another important aspect of radiosensitizing nucleobase derivatives is connected to their phosphorylation and incorporation into DNA. In fact, to induce DNA damage after electron attachment these molecules have to be present in DNA. Therefore, they have to be good substrate for nucleoside kinases and in the form of triphosphate for DNA polymerases. Computational QM/MM models that will be able to describe phosphorylation of nucleosides and their subsequent incorporation into dsDNA are being worked out in our laboratory.

Another type of radiosensitizers are oxygen mimetics. One of the most thoroughly studied group of such substances are derivatives of nitroimidazole (NIs). The studies that began in 1970s resulted in the mechanism in which the oxygen mimetic reacts with the radical formed after attachment of $^{\bullet}$OH to a nucleobase in DNA. However, recent studies demonstrate that NIs form a series of resonances in the gas phase leading even to complete degradation of those molecules that is accompanied with the formation of various radical species. On the other hand, if electron attachment process to NIs proceeds in water, the excess of energy is transferred to the environment and the primary product is an intact anion of nitroimidazolic compound (associative electron attachment). This anion can be protonated under physiological conditions and as a neutral radical can bind to DNA which may finally lead to a strand break. Studies trying to work out less toxic NIs have been recently executed.

Last but not least we discussed metallic nanoparticles and metal complexes. Metals have much larger absorption cross sections (compared to biological tissue) and therefore have the potential to increase the effectiveness of the radiation, increasing the amount of reactive species and damage they cause [156]. Therefore, a rational design of metal-based drugs is a promising strategy to discover potent radiation enhancers. Metallic nanoparticles and metal complexes can operate by different non-exclusive modes to exert their radiosensitizing effects on cancer cells. The increasing interest on electron induced processes in radiosensitization by metal-based agents is observed. Studies on the mechanisms of radiosensitization by metallic nanoparticles and complexes at physical and biological level are crucial for designing the perfect metal-based radiosensitizers.

Although still much has to be done to understand completely the processes triggered by radiosensitizers discussed in the current chapter the direction of further studies seems to be well defined. Since hypoxia is widespread in solid tumors, focusing on hydrated electron damage to DNA is an attractive idea that should result in a substantial increase of the efficiency of radiotherapy. Therefore, a search for new, better DEA sensitive nucleosides, oxygen mimetics binding electrons and metal complexes leading to electron-induced DNA damage is undoubtedly worth of our efforts.

Acknowledgements This work was supported by the Polish National Science Centre under Grant No. 2014/14/A/ST4/00405.

References

1. Rak J, Chomicz L, Wiczk J, Westphal K, Zdrowowicz M, Wityk P, Żyndul M, Makurat S, Golon Ł (2015) J Phys Chem B 119:8227
2. Visvader JE, Lindeman GJ (2008) Nat Rev Cancer 8:755
3. Sharma A, Arambula JF, Koo S, Kumar R, Singh H, Sessler JL, Kim JS (2018) Chem Soc Rev 48:771
4. Thomlinson RH, Gray LH (1955) Br J Cancer 9:539
5. Brown JM (1979) Br J Radiol 52:650
6. Daşu A, Denekamp J (1998) Radiother Oncol 46:269
7. Boudaïffa B, Cloutier P, Hunting D, Huels MA, Sanche L (2000) Science 287:1658
8. O'Neill P, Helden EM (1993) Adv Radiat Biol 17:53
9. Nabben FJ, Kerman JP, Loman H (1982) Int J Radiat Biol 42:23
10. von Sonntag C (2006) In: Free-radical-induced DNA damage and its repair. A chemical perspective. Springer, Berlin, p 180
11. Bryan T, Oronsky BT, Susan J, Knox SJ, Scicinski J (2011) Trans Oncol 4:189
12. Kohanoff J, McAllister M, Tribello GA, Gu B (2017) Condens Matter 29:383001
13. Scholes G (1978) In: Effects of ionizing radiation on DNA. Physical, chemical and biological aspects. Springer, Berlin, p 153
14. Wang W, Sevilla MD (1994) Radiat Res 138:9–17
15. Falcone JM, Becker D, Sevilla MD, Swarts SG (2005) Radiat Phys Chem 72:257–264
16. Panajotovic R, Martin F, Cloutier P, Hunting D, Sanche L (2006) Radiat Res 165:452
17. Goz B (1977) Pharmacol Rev 29:249
18. Westphal K, Wiczk J, Miloch J, Kciuk G, Bobrowski K, Rak J (2015) Org Biomol Chem 13:10362
19. Hall EJ, Giaccia AJ (2006) In: Radiosensitizers and bioreductive drugs. Lippincott Williams&Wilkins, Philadelphia, PA, pp 419–431
20. Gaba N (2011) In: Radioprotectors and radiosensitizers. Lambert Academic, Saarbrucken
21. Doyle TH, Mornex F, McKenna WG (2001) Clin Cancer Res 7:226
22. El-Naggar M, Omar M, Elgeriany A, Peters GJ, Mostafa A, Shehata S (2016) JCMT 2:188
23. von Sonntag C (1987) In: The chemical basis of radiation biology. Taylor and Francis, London
24. Cecchini S, Girouard S, Huels MA, Sanche L, Hunting DJ (2015) Biochemistry 44:1932
25. Galmarini CM, Mackey JR, Dumontet C (2002) Lancet Oncol 3:415
26. Makurat S, Zdrowowicz M, Chomicz-Mańka L, Kozak W, Serdiuk IE, Wityk P, Kawecka A, Sosnowska M, Rak J (2018) RSC Adv 8:21378
27. Makurat S, Spisz P, Kozak W, Rak J, Zdrowowicz M (2019) Int J Mol Sci 20:1308
28. Spisz P, Zdrowowicz M, Makurat S, Kozak W, Skotnicki K, Bobrowski K, Rak J (2019) Molecules 22:40
29. Meier C (2002) Mini-Rev Med Chem 2:219
30. Meier C, Balzarini J (2006) Antivir Res 71:282
31. Belanger K, Klecker RW, Rowland J, Kinsella TJ, Collins JM (1986) Cancer Res 46:6509
32. Gemenetzidis E, Shavorskaya O, Xu YZ, Trigiante G (2013) J Dermatol Treat 24:209
33. Brem R, Zhang X, Xu YZ, Karran P (2015) J Photochem Photobiol B Biol 145:1
34. Hazra S, Ort S, Konrad M, Lavie A (2010) Biochemistry 49:6784
35. Jagiello K, Makurat S, Pereć S, Rak J, Puzyn T (2018) Struct Chem 29:1367
36. Polska K, Rak J, Bass AD, Cloutier P, Sanche L (2012) J Chem Phys 136:075101
37. Park Y, Polska K, Rak J, Wagner JR, Sanche L (2012) J Phys Chem B 116:9676
38. Westphal K, Skotnicki K, Bobrowski K, Rak J (2016) Org Biomol Chem 14:9331
39. Chomicz L, Zdrowowicz M, Kasprzykowski F, Rak J, Buonaugurio A, Wang Y, Bowen KH (2013) J Phys Chem Lett 4:2853
40. Makurat S, Chomicz-Mańka L, Rak J (2016) ChemPhysChem 17:2572

41. Zdrowowicz M, Chomicz L, Żyndul M, Wityk P, Rak J, Wiegand TJ, Hanson CG, Adhikary A, Sevilla MD (2015) PCCP 17:16907
42. Levene PA, La Forge FB (1912) Ber Dtsch Chem Ges 45:608
43. Prusoff WH (1954) SEBM 85:564
44. Zamenhof S, Griboff G (1954) Nature 174:307
45. Scholler J, Gordon M, Sternberg SS (1956) SEBM 93:124
46. Price TD, Hudson PB, Hinds HH, Darmstadt RA, Zamenhof S (1956) Nature 178:684
47. Zamenhof S, De Giovanni R, Greer S (1958) Nature 181:827
48. Greer S, Zamenhof S (1957) Am Chem Soc Abstr (131st meeting) 3C
49. Greer S (1960) J Gen Microbiol 22:618
50. Larkiewicz Z, Szybalski W (1960) Biochem Biophys Res Commun 2:413
51. Stahl FW, Crasemann JM, Okun L, Fox E, Laird C (1961) Virology 13:98
52. Freifelder D, Freifelder DR (1966) Mutat Res 3:177
53. Zimbrick JD, Ward JF, Myers Jr LS (1969) Int J Radiat Biol 16:505
54. Sano K, Hoshino T, Nagai M (1968) J Neurosurg 28:530
55. Doiron A, Yapp DT, Olivares M, Zhu JX, Lehnert S (1999) Cancer Res 59:3677
56. Dextraze ME, Wagner JR, Hunting DJ (2007) Biochemistry 46:9089
57. Brust D, Feden J, Farnsworth J, Amir C, Broaddus WC, Valerie K (2000) Cancer Gene Ther 7:778
58. Coleman N, Mitchell JB (1999) J Clin Oncol 17:1
59. Dabaja BS, McLaughlin P, Ha CS, Pro B, Meyers CA, Seabrooke LF, Wilder RB, Kyritsis AP, Preti HA, Yung WK, Levin V, Cabanillas F, Cox JD (2003) Cancer 98:1021
60. Groves MD, Maor MH, Meyers C, Kyritsis AP, Jaeckle KA, Yung WK, Sawaya RE, Hess K, Bruner JM, Peterson P, Levin VA (1999) Int J Radiat Oncol Biol Phys 45:127
61. Prados MD, Seiferheld W, Sandler HM, Buckner JC, Phillips T, Schultz C, Urtasun R, Davis R, Gutin P, Cascino TL, Greenberg HS, Curran Jr WJ (2004) Int J Radiat Oncol Biol Phys 58:1147
62. Kriss JP, Maruyama Y, Tung LA, Bond SB, Révész L (1963) Cancer Res. 23:260
63. Sano K, Hoshino T, Nagai M (1966) Nippon Acta Neuroradiol 7:4
64. Wetmore SD, Boyd RJ, Eriksson LA (2001) Chem Phys Lett 343:151
65. Li X, Sanche L, Sevilla MD (2002) J Phys Chem A 106:11248
66. Abdoul-Carime H, Huels MA, Bruning F, Illenberger E, Sanche L (2000) J Chem Phys 113:2517
67. Cole LA, Perdew JP (1982) Phys Rev A 25:1265
68. Pritchard HO (1953) Chem Rev 52:529
69. Chomicz-Mańka L, Wityk P, Golon Ł, Zdrowowicz M, Wiczk J, Westphal K, Żyndul M, Makurat S, Rak J (2015) In: Handbook of computational chemistry. Springer, Dordrecht
70. Meunier M, Quirke N, Binesti D (1999) Mol Simulat 23:109
71. Li X, Cai Z, Sevilla MD (2002) J Phys Chem A 106:1596
72. Chomicz L, Rak J, Storoniak P (2012) J Phys Chem B 116:5612
73. Wieczór M, Wityk P, Czub J, Chomicz L, Rak J (2014) Chem Phys Lett 595–596:133
74. Chomicz L, Leszczynski J, Rak J (2013) J Phys Chem B 117:8681
75. Chomicz L, Furmanchuk A, Leszczynski J, Rak J (2014) Phys Chem Chem Phys 16:6568
76. Golon Ł, Chomicz L, Rak J (2014) Chem Phys Lett 612:289
77. Chomicz L, Golon Ł, Rak J (2014) Phys Chem Chem Phys 16:19424
78. Mariaggi N, Cadet J, Téoule R (1976) Tetrahedron 32:2385
79. Chatgilialoglu C, Ferreri C, Terzidis MA (2011) Chem Soc Rev 40:1368
80. Mazouzi A, Vigouroux A, Aikeshev B, Brooks PJ, Saparbaev MK, Morerab S, Ishchenko AA (2013) Proc Natl Acad Sci USA 110:E3071
81. Wityk P, Wieczór M, Makurat S, Chomicz-Mańka L, Czub J, Rak J (2017) J Chem Theory Comput 13:6415
82. Falkiewicz K, Kozak W, Zdrowowicz M, Spisz P, Chomicz-Mańka L, Torchała M, Rak J (2020, in preparation)
83. Hansch C, Leo A, Taft RW (1991) Chem Rev 97:165

84. Hall AH, Dart R, Bogdan G (2007) Ann Emerg Med 49:806
85. Sosnowska M, Makurat S, Zdrowowicz M, Rak J (2017) J Phys Chem B 121:6139
86. Wang S, Zhang M, Liu P, Xie S, Cheng F, Wang L (2018) Chem Phys Lett 692:374
87. Wen Z, Peng J, Tuttle PR, Ren Y, Garcia C, Debnath D, Rishi S, Hanson C, Ward S, Kumar A, Liu Y, Zhao W, Glazer PM, Liu Y, Sevilla MD, Adhikary A, Wnuk SF (2018) Org Lett 20:7400
88. Chomicz L, Petrovici A, Archbold I, Adhikary A, Kumar A, Sevilla MD, Rak J (2014) Chem Commun 50:14605
89. Adams GE, Cooke MS (1969) It J Radiat Biol 15:457
90. Brown JM (1975) Radiat Res 64:633
91. Adams GE, Clarke ED, Flockhart IR et al (1979) Int J Radiat Biol Relat Stud Phys Chem Med 35:133
92. Dische S, Saunders MI, Flockhart IR, Lee ME, Anderson P (1979) Int J Radiat Oncol Biol Phys 5:851
93. Eschwege F, Sancho-Garnier H, Chassagne D, Brisgand D, Guerra M, Malaise EP, Bey P, Busutti L, Cionini L, N'Guyen T, Romanini A, Chavaudra J, Hill C (1997) Int J Radiat Oncol Biol Phys 39:275
94. Overgaard JG, Eriksen M, Nordsmark M (2005) Lancet Oncol 6:757
95. Brown JM, Yu NY, Brown DM, Lee WW (1981) Int J Radiat Oncol Biol Phys 7:695
96. Nishimura Y, Nakagawa K, Takeda K, Tanaka M, Segawa Y, Tsujino K, Negoro S, Fuwa N, Hida T, Kawahara M, Katakami N, Hirokawa K, Yamamoto N, Fukuoka M, Ariyoshi Y (2007) Int J Radiat Oncol Biol Phys 69:786
97. Thomson D, Yang H, Baines H, Miles E, Bolton S, West C, Slevin N (2014) Clin Oncol 26:344
98. Wardman P (2007) Clin Oncol 19:397
99. Cole S, Stratford IJ, Adams GE, Fielden M, Jenkins TC (1990) Radiat Res 124:S38
100. Wilson WR, Hay MP (2011) Nat Rev Cancer 11:393
101. Langer J, Balog R, Stano M, Abdoul-Carime H, Illenberger E (2004) Int J Mass Spectrom 233:267
102. Klar D, Ruf MW, Hotop H (2001) Int J Mass Spectrom 205:93
103. Davis D, Vysotskiy VP, Sajeev Y, Cederbaum LS (2012) Angew Chem 51:8003
104. Tanzer K, Feketeová L, Puschnigg B, Scheier P, Illenberger E, Denifl S (2015) J Phys Chem A 119:6668
105. Kossoski F, Varella MTDN (2017) J Chem Phys 147:164310
106. Tanzer K, Feketeová L, Puschnigg B, Scheier P, Illenberger E, Denifl S (2014) Angew Chem 53:12240
107. Desfrancois C (1995) Phys Rev A At Mol Opt Phys 51:3667
108. Denifl S, Ptasinska S, Probst M, Hrusak J, Scheier P, Märk TD (2004) J Phys Chem A 108:6562
109. Ribar A, Fink K, Probst M, Huber SE, Feketeová L, Denifl S (2017) Chem Eur J 23:12892
110. Meißner R, Feketeová L, Illenberger E, Denifl S (2019) Int J Mol Sci 20:3496
111. Overgaard JG, Hansen HS, Lindelov B, Overgaared M, Jorgensen K, Rasmusson B, Berthelsen A (1991) Radiother Oncol 20:143
112. Meißner R, Kočišek J, Feketeová L, Fedor J, Fárník M, Limão-Vieira P, Illenberger E, Denifl S (2019) Nat Commun 10:2388
113. Bald I, Langer J, Tegeder P, Ingólfsson O (2008) Int J Mass Spectrom 277:4
114. Bonnet M, Hong CR, Gu Y, Anderson RF, Wilson WR, Pruijn FB, Wang J, Hicks KO, Hay MP (2014) Bioorg Med Chem 22:2123
115. Bonnet M, Hong CR, Gu Y, Wong WW, Liew LP, Shome A, Wang J, Gu Y, Stevenson RJ, Qi W, Anderson RF, Pruijn FB, Wilson WR, Jamieson SMF, Hicks KO, Hay MP (2018) J Med Chem 61:1241
116. Gill MR, Vallis KA (2019) Chem Soc Rev 48:540
117. Gasser G, Ott I, Metzler-Nolte N (2010) J Med Chem 54:3
118. Wang H, Mu X, He H, Zhang XD (2018) Trends Pharmacol Sci 39:24

119. Carter JD, Cheng NN, Qu Y, Suarez GD, Guo T (2007) J Phys Chem B 111:11622
120. Regulla DF, Hieber LB, Seidenbusch M (1998) Radiat Res 150:92
121. Roa W, Zhang X, Guo L, Shaw A, Hu X, Xiong Y, Gulavita S, Patel S, Sun X, Chen J, Moore R, Xig JZ (2009) Nanotechnology 20:375101
122. Pawlik TM, Keyomarsi K (2004) Int J Radiat Oncol Biol Phys 59:928
123. Liu Y, Chen W, Zhang P, Jin X, Liu X, Li P, Li F, Zhagn H, Zou G, Li Q (2016) Radiother Oncol 119:544
124. Kievit FM, Stephen ZR, Wang K, Dayringer CJ, Sham JG, Ellenbogen RG, Silber JR, Zhang M (2015) Mol Oncol 9:1071
125. Calatayud M, Asin L, Tres A, Goya GF, Ibarra MR (2016) Curr Nanosci 12:372
126. Rostami A, Toossi MT, Sazgarnia A, Soleymanifard S (2016) Rad Environ Biophys 55:461
127. Alizadeh E, Thomas OM, Sanche L (2015) Annu Rev Phys Chem 66:379
128. Langer J, Zawadzki M, Farnik M, Pinkas J, Fedor J, Kočišek J (2018) Eur Phys J D 72:112
129. Kopyra J, Koening-Lehmann C, Bald I, Illenberger E (2009) Angew Chem Int Ed Engl 48:7904
130. Lu QB, Kalantari S, Wang CR (2007) Mol Pharm 4(4):624
131. Xiao F, Luo X, Fu X, Zheng Y (2013) J Phys Chem B 117:4893
132. Boeckman HJ, Trego KS, Turchi JJ (2005) Mol Cancer Res 3:277
133. Sears CR, Cooney SA, Chin-Sinex H, Mendonca MS, Turchi JJ (2016) DNA Repair 40:35
134. Rezaee M, Hill RP, Jaffray DA (2017) Radiat Res 188:123
135. Sanche L (2016) Radiat Phys Chem 128:36
136. Dong Y, Zhou L, Tian Q, Zheng Y, Sanche L (2017) J Phys Chem C 121:17505
137. Sanche L (2009) In: Radical and radical ion reactivity in nucleic acid chemistry. Wiley, Hoboken, NJ, pp 239–295
138. Zheng Y, Hunting DJ, Ayotte P, Sanche L (2008) Radiat Res 169:19
139. Xiao F, Zheng Y, Cloutier P, He Y, Hunting D, Sanche L (2011) Nanotechnology 22:465101
140. Zheng Y, Cloutier P, Hunting DJ, Sanche L (2008) J Biomed Nanotechnol 4:469
141. Zheng Y, Hunting DJ, Ayotte P, Sanche L (2008) Phys Rev Lett 100:198101
142. Luo X, Zheng Y, Sanche L (2014) J Chem Phys 140:04B606
143. Bao Q, Chen Y, Zheng Y, Sanche L (2014) J Phys Chem C 118:15516
144. Mantri Y, Lippard SJ, Baik MH (2007) J Am Chem Soc 129:5023
145. Behmand B, Wagner JR, Sanche L, Hunting DJ (2014) J Phys Chem B 118:4803
146. Behmand B, Wagner JR, Hunting D, Marignier JL, Mostafavi M, Sanche L (2015) J Phys Chem B 119:9496
147. Zheng Y, Sanche L (2009) Radiat Res 172:114
148. Casta R, Champeaux JP, Sence M, Moretto-Capelle P, Cafarelli P (2015) Phys Med Biol 60:9095
149. Klein S, Sommer A, Distel LV, Neuhuber W, Kryschi C (2012) Biochem Biophys Res Comm 425:393
150. Klein S, Sommer A, Distel LV, Hazemann JL, Kroener W, Neuhuber W, Muller P, Proux O, Kryschi C (2014) J Phys Chem B 118:6159
151. Salado-Leza D, Traore A, Porcel E, Dragoe D, Muñoz A, Remita H, García G, Lacombe S (2019) Int J Mol Sci 20:5648
152. Porcel E, Kobayashi K, Usami N, Remita H, Le Sech C, Lacombe S (2011) J Phys Conf Ser 261:012004
153. Cai Z, Cloutier P, Hunting D, Sanche L (2006) Radiat Res 165:365
154. Rezaee M, Hunting DJ, Sanche L (2013) Int J Radiat Oncol Biol Phys 87:847
155. Tanzer K, Pelc A, Huber SE, Śmiałek MA, Scheier P, Probst M, Denifl S (2014) Int J Mass Spectrom 365:152
156. Hambley TW (2007) Dalton Trans 43:4929

Chapter 5
Application of Computational Chemistry for Contaminant Adsorption on the Components of Soil Surfaces

Glen R. Jenness, Levi A. Lystrom, Harley R. McAlexander, and Manoj K. Shukla

Abstract Computational chemistry has seen an explosion of methods and techniques over the last 30 years. In particular, the area of understanding surface adsorption has seen radical changes in methodology and systems of interest. It is our goal in the current chapter to review these techniques. Moreover, we will demonstrate how they can be used to understand the interaction of components of arid soils on potential environmental contaminants through a review of our groups research efforts on understanding munitions. Finally, we present an overview of new areas of research that will deliver the next generation of computational and theoretical tools.

5.1 Introduction

How chemicals react in the natural environment is an important consideration in determining which compounds make it to production. However, the natural environment is extremely heterogeneous and is comprised of numerous components that all interact with each other [1]. This makes the determination of how chemicals interact within the environment extremely difficult.

In the last decade, our group has risen to the forefront of understanding the fate and transport of munitions in the environment [2–16]. We have engaged both traditional computational chemistry methods, while also developing newer techniques for the prediction of chemical properties. Our work has revealed

G. R. Jenness (✉) · H. R. McAlexander · M. K. Shukla
Environmental Laboratory, US Army Engineer Research and Development Center, Vicksburg, MS, USA
e-mail: glen.r.jenness@usace.army.mil

L. A. Lystrom
Environmental Laboratory, US Army Engineer Research and Development Center, Vicksburg, MS, USA

Oak Ridge Institute for Science and Education, Oak Ridge, TA, USA

J. Leszczynski, M. K. Shukla (eds.), *Practical Aspects of Computational Chemistry V*, https://doi.org/10.1007/978-3-030-83244-5_5

several important factors in how these compounds are affected by the environment, including their solvation, interactions between various soil components, optical spectra, and degradation (*vide supra*).

It is the goal of the current study to examine the methodologies that we have used and developed in solving this problem. Here, we will be focusing primarily on the interaction of munitions with arid soil components. In Sect. 5.2, we present an overview of the periodic DFT method and include topics such as density of states analysis, non-local functionals, and self-interaction error corrections. For those familiar with these methods who wish to just see applications, one can jump to Sect. 5.3. In Sect. 5.3, we demonstrate how these methods can be applied to our problem at hand (i.e., understanding the fate and transport of munitions in soils). We present a number of case studies, including electrochemical properties, adsorption, charge transfer, Lewis acidity, and transport. In Sect. 5.4, we present up and coming techniques, including an examination of our latest work on eliminating empiricism in density functional tight binding (DFTB), and artificial intelligence/machine learning algorithms. Finally, we present our conclusions in Sect. 5.5.

5.2 Density Functional Theory (DFT)

5.2.1 *Preliminaries*

We begin our discussion with the time-independent Schrödinger equation,

$$\mathcal{H}\psi = E\psi, \tag{5.1}$$

where $\mathcal{H}$ is the Hamiltonian operator, E is the energy of the system, and ψ is the wave-function (in older literature one will find this is referred to as a state function). Equation (5.1) is an eigenvalue problem wherein the eigenvalues are the energy E, and the eigenvectors ψ on the right hand side do not change from the left hand side. Ideally, $\mathcal{H}$ contains all the interactions occurring within and on the system. For a system with N electrons, M nuclei, and no external potentials the Hamiltonian takes on the form,

$$\mathcal{H} = -\sum_{i=1}^{N} \frac{1}{2}\nabla_i^2 - \sum_{A=1}^{M} \frac{1}{2M_A}\nabla_A^2 - \sum_{i=1}^{N}\sum_{A=1}^{M} \frac{Z_A}{r_{iA}} + \sum_{i=1}^{N}\sum_{j>1}^{N} \frac{1}{r_{ij}} + \sum_{A=1}^{M}\sum_{B>1}^{M} \frac{Z_A Z_B}{r_{AB}}. \tag{5.2}$$

Solving the differential equation in Eq. (5.1) will produce a wave-function from which all physical quantities can be calculated from (given the appropriate quantum mechanical operator). However, in practice this equation lacks an exact solution due to the presence of the r_{ij}^{-1} term in Eq. (5.2), which results in the wave-function, ψ, to take on an unknown form [17, 18]. Consequently, we need to employ a number of approximations in order to solve Eq. (5.1).

First, we assume that the operator in Eq. (5.1) is simply a sum of one-electron terms,

$$\mathcal{H} = \sum_{i=1}^{N} h_i = \sum_{i=1}^{N} -\frac{1}{2}\nabla_i^2 - \sum_{A=1}^{M} \frac{Z_A}{r_{iA}}. \tag{5.3}$$

Thus, the wave-function is simply a product of one-electron wave-functions,

$$\psi = \prod_{i=1}^{N} \phi_i, \tag{5.4}$$

where ϕ_i is the solution to the one-electron Hamiltonian h_i. The total energy is then just a summation of one-electron energies,

$$E = \sum_{i=1}^{N} \epsilon_i. \tag{5.5}$$

Equation (5.3) neglects the r_{ij}^{-1} electron-electron interaction terms in Eq. (5.2); in order to account for these interactions, we introduce a second approximation in which we treat a single electron as acting in an *average* potential that arises from the other $N-1$ electrons. This potential is called the Hartree potential $\left(V_i^{\mathrm{H}}\{j\}\right)$, and is appended to Eq. (5.3) [17, 18],

$$h_i = -\frac{1}{2}\nabla_i^2 - \sum_{A=1}^{M} \frac{Z_A}{r_{iA}} + V_i^{\mathrm{H}}\{j\}. \tag{5.6}$$

However, in order to calculate $V_i^{\mathrm{H}}\{j\}$, we first need to know the set of wave-functions $\phi_{\{j\}}$, which we cannot know without first solving for ϕ_i. In other words, we need our answer in order to get our answer! This paradox is solved by our third approximation: the introduction of the basis set. Here, we express our initial wave-function ϕ_i as a linear combination of N ancillary functions,

$$\phi_i = \sum_{j=1}^{N} a_j \varphi_j. \tag{5.7}$$

In accordance with the variational principal, the larger N is (i.e., the larger the basis set), the more variationally complete the wave-function is and thus, the closer it is to the "exact" wave-function. Naturally, a variety of forms can be used here; however,

Slater and Gaussian forms are common,

$$\varphi(r, \zeta, n) = \begin{cases} r^{n-1}e^{-\zeta r} & \text{Slater} \\ r^{\ell}e^{-\zeta r^2} & \text{Gaussian.} \end{cases} \tag{5.8}$$

Here, r is the distance between two atoms, ζ is a pre-determined coefficient, n is the principal quantum number, and ℓ is the angular quantum number. By using pre-determined coefficients a_j and exponents ζ, and adjusting them iteratively through numerical minimization, we can circumvent the paradox outlined above. Such a procedure is called the self-consistent field (SCF) method [17, 18]. In the above discussion, we assumed that each atom i would receive their own basis function, with the molecular wave-function being a sum of these basis sets; this is referred to as the linear combination of atomic orbitals (LCAO) style basis set. In addition to the forms in Eq. (5.8), LCAO basis sets can also be generated numerically; here a finite-difference grid is used to construct the radial component of the wave-function [19].

Above, we introduced the Hartree potential $\left(V_i^{\mathrm{H}}\{j\}\right)$ without much fanfare. This term is rather complicated, and involves integrals over four basis function centers. Included within the Hartree potential is the Coulomb integral (denoted by J),

$$J = \int d\mathbf{r}_1 d\mathbf{r}_2 |\phi_a(\mathbf{r}_1)|^2 r_{12}^{-1} |\phi_b(\mathbf{r}_2)|^2. \tag{5.9}$$

Given that $|\phi_b(\mathbf{r}_2)|^2$ is equal to the electron density ρ, the integral in Eq. (5.9) represents the classical interaction of two charge densities interacting. Additionally, the Hartree potential includes the exchange integral, K,

$$K = \int d\mathbf{r}_1 d\mathbf{r}_2 \phi_a^*(\mathbf{r}_1)\phi_b(\mathbf{r}_1) r_{12}^{-1} \phi_b^*(\mathbf{r}_2)\phi_a(\mathbf{r}_2). \tag{5.10}$$

This integral has no classical component, and arises from the quantum phenomena of electrons being indistinguishable. This method is called the Hartree-Fock method, and it scales as $\mathcal{O}(N^4)$, where N is the number of basis functions. Moreover, the approximations employed result in an energy that is *under-predicted*, with the difference being referred to as the correlation energy [17]. While there exist several techniques for correcting for electron correlation (e.g., CCSD(T) [20], MP2 [21], etc.), these all increase the computational resources required.

This leads us to our final approximation: the density functional. Here, we postulate the existence of a functional V that contains all the information in regards to electron exchange and correlation (for this reason we call it the exchange-correlation functional, or V_{XC}). We refer to this quantity as a functional due to its dependence on the density, which in turn is dependent on position (in other words, a functional is a function of a second function). This allows us to re-write Eq. (5.6) as,

$$h_i^{\mathrm{KS}} = -\frac{1}{2}\nabla_i^2 - \sum_{A=1}^{M}\frac{Z_A}{r_{iA}} + \int \frac{\rho(\mathbf{r}')}{|\mathbf{r}_i - \mathbf{r}'|}d\mathbf{r}' + V_{\mathrm{XC}}[\rho]. \tag{5.11}$$

The newly defined Hamiltonian, now called the Kohn-Sham Hamiltonian (as denoted by the superscript KS), is in principal much easier to calculate. Moreover, given the exact V_{XC} we can re-produce the exact multi-electron Hamiltonian! Unfortunately, we do not know the exact V_{XC} with the exception of a few test systems (e.g., uniform electron gas). This has lead to a variety of density functionals being proposed, all with varying degrees of accuracy. Therefore, the choice of functional is dependent on the system in question, and the required accuracy. Due to the usage of a density functional, this method is referred to as density functional theory, or DFT. While not exact, it does feature a scaling of $\mathcal{O}(N^3)$ and does capture some electron-electron correlation without the need for higher level perturbation theory techniques [18, 22–24]. It has found a widespread success within the computational and theoretical chemistry fields, with implementation in a variety of open-source (e.g., `PSI4` [25], `GPAW` [26] etc.) and commercial (e.g., `Gaussian` [27], `VASP` [28–31], etc.) programs.

5.2.2 *Bloch Function*

We now consider a system with periodicity, i.e., a system with lattice vectors $(\mathbf{a}_1, \mathbf{a}_2, \mathbf{a}_3)$ whose atomic positions are,

$$[x_i, y_i, z_i] = [x_i, y_i, z_i] + (n_1\mathbf{a}_1 + n_2\mathbf{a}_2 + n_3\mathbf{a}_3) \tag{5.12}$$

where $[x_i, y_i, z_i]$ are the xyz-coordinates for atom i, and the set $\{n_{j=1-3}\}$ are arbitrary integers defining how many times the unit cell is replicated. This leads to Bloch's theorem which states that the wave-function of a periodic system is a plane-wave times a periodic function, $u_{\mathbf{k}}(\mathbf{r})$,

$$\phi_{\mathbf{k}}(\mathbf{r}) = e^{i\mathbf{k}\cdot\mathbf{r}}u_{\mathbf{k}}(\mathbf{r}), \tag{5.13}$$

where the vector $\mathbf{k}$ is referred to as reciprocal or k-space vector. What about the choice of $u_{\mathbf{k}}(\mathbf{r})$? Using a similar method as we did in Eq. (5.8), we expand $u_{\mathbf{k}}(\mathbf{r})$ in terms of ancillary functions,

$$u_{\mathbf{k}}(\mathbf{r}) = \sum_{\mathbf{G}} c_{\mathbf{G}} e^{i\mathbf{G}\cdot\mathbf{r}} u_{\mathbf{k}} \tag{5.14}$$

where $\mathbf{G}$ is the reciprocal space vectors $m_1\mathbf{b}_1 + m_2\mathbf{b}_2 + m_3\mathbf{b}_3$ with the vectors $\mathbf{b}_i$ being defined in terms of the real space unit cell vectors $\mathbf{a}_i$,

$$\mathbf{b}_i = 2\pi \frac{\mathbf{a}_{i+1} \times \mathbf{a}_{i+2}}{\mathbf{a}_i \cdot (\mathbf{a}_{i+1} \times \mathbf{a}_{i+2})}. \tag{5.15}$$

The $c_{\mathbf{G}}$ terms in Eq. (5.14) are coefficients to be determined during the calculation. In order to keep calculations reasonable, we need to impose a limit on the integers m_i. As the kinetic energy E can be written as,

$$E \propto |\mathbf{k} + \mathbf{G}|^2, \tag{5.16}$$

then we can specify an upper limit for the kinetic energy, which in turns limits how many basis functions we have! Thus, our wave-function can be written as,

$$\phi_{\mathbf{k}}(\mathbf{r}) = \sum_{|\mathbf{G}+\mathbf{k}|<G_{\text{cut}}} c_{\mathbf{k}+\mathbf{G}} e^{i(\mathbf{G}+\mathbf{k})\cdot\mathbf{r}}. \tag{5.17}$$

As this basis set is comprised of plane-waves, it is called a plane-wave basis set. While one could define the coefficients $c_{\mathbf{k}+\mathbf{G}}$ before hand, it is typical to just generate random numbers for them, and hold the electron density constant for the first few SCF steps. Additionally, instead of having a basis set wherein both coefficients and exponents have to be defined beforehand and have an arbitrary size, plane-wave basis sets are defined purely in terms of the energy cutoff employed. This makes the plane-wave method easy to judge in terms of accuracy [22]. However, the presence of plane-waves requires the use of Fourier transforms (which are notoriously hard to parallelize) and the enforcement of 3-dimensional periodicity, even for systems where such periodicity would be artificial. Thus, various alternatives for the choice of $u_{\mathbf{k}}(\mathbf{r})$ have been proposed, including LCAO [19] and real-space grid [32] methods.

5.2.2.1 Bypassing Periodicity: Cluster Models

While the above formalism is rather robust, there are problems for which a periodic description is not feasible. For example, in determining how a polar molecule such as water interacts with an extended π-electron system, one of the current authors had to employ a cluster model description for graphene in order to move beyond the DFT level. Here, systems of increasing size were "cut" from the graphene sheet, with dangling bonds capped with hydrogen. This allowed for the examination of the binding energy components to be studied with respect to increasing system size with the end goal of extrapolating to the periodic limit [33–35]. The trick with employing this type of approximation is balancing system size requirements versus accuracy and computational cost. It is highly desirable for the cluster to be representative of the periodic system, and should be sufficiently large enough to capture the relevant physics and chemistry. However, sometimes this is not feasible and a compromise has to be made. We will explore this approximation below in the context of electrochemical properties (Sect. 5.3.2) and binding energies (Sect. 5.3.3).

5.2.3 K-point Sampling

We now turn our attention to the selection of k-vectors. Regardless of our choice for $u_{\mathbf{k}}(\mathbf{r})$, we still frequently need to solve integrals of the form,

$$g = \int g(\mathbf{k})d\mathbf{k}, \tag{5.18}$$

which need to be integrated numerically. This necessitates the conversion of the integral in Eq. (5.18) in a summation over a grid, which is called k-space [22]. A common way to select this grid is the Monkhorst-Pack method. Here, the k-vectors are calculated based on the reciprocal space lattice vectors $\mathbf{b}_i$,

$$\mathbf{k} = \sum_{i=1}^{3} u_i \mathbf{b}_i \tag{5.19}$$

with u_i calculated as,

$$u_i = \frac{2r - q_r - 1}{2q_r}, r = 1, 2, ..., q_r, \tag{5.20}$$

where q_r is an integer specifying the upper bound of the k-vector. This provides a k-space that is evenly distributed [36]. It should be noted that as the k-vectors are defined in terms of reciprocal space, as the unit cell volume *decreases*, the larger your k-space needs to be; conversely, the larger your unit cell volume, the smaller your k-space needs to be. The determination the requisite size of your k-space is based upon the desired accuracy versus computational cost.

5.2.4 Density of States (DOS) and Analysis of Orbitals

For periodic systems, there are numerous molecular orbitals that lie relatively close to each other. While it is common to show molecular orbital diagrams for non-periodic systems, for periodic systems these can become rather complex. An alternative visualization method is through the density of states (DOS),

$$\mathrm{DOS}(\epsilon) = \sum_{n}\sum_{k} \delta(\epsilon - \epsilon_n(\mathbf{k})), \tag{5.21}$$

where δ is the well-known δ-function, ϵ is the energy on a grid, and $\epsilon_n(\mathbf{k})$ is the orbital energy for orbital (or band) n and k-point k. In practical terms, this means that we setup a grid of energies (represented by ϵ) and create a set of bins. We then go through the various eigenvalues ($\epsilon_n(\mathbf{k})$). If $\epsilon_n(\mathbf{k})$ lies within that bin (represented by the δ-function), then we increment that bin by 1. In this way, we gain a ready visualization of where the various electronic states lie in relation to each other! Additionally, one can perform a projection along the various angular momenta moments which results in a projected DOS (or PDOS). This representation is handy as it allows a ready visualization of which type of electrons are playing a role in chemical bonding. Trends in PDOS's have resulted in the formation of several important models, the most famous of which is the d-band model of Nørskov and Hammer [37]. This model has been instrumental in explaining several important trends in catalysis, such as effect of stress and strain [38–41] and the identification of novel catalysts [40, 42–45].

Another way to visualize the DOS is the crystal orbital overlap population (COOP) [46] method. This method weighs the density of states (DOS) with a Mulliken factor [47, 48],

$$\mathrm{COOP}(\epsilon) = 2\sum_n \sum_k c_i^* c_j S_{ij} \delta(\epsilon - \epsilon_n(\mathbf{k})), \tag{5.22}$$

where $c_{i/j}$ are the coefficients of atom i/j, S_{ij} is the overlap matrix between atoms i and j, n denotes the band, k denotes the relevant k-point, and $\delta(\epsilon - \epsilon_n(\mathbf{k}))$ is the delta function from Eq. (5.21). Peaks where COOP are negative denote anti-bonding orbitals, whilst positive peaks represent bonding orbitals; regions where the peak is zero (compared to the standard DOS) denote orbitals that are non-bonding [46]. This technique has been utilized by Jenness and coworkers to describe a variety of chemical phenomena, including metal-support interactions [49, 50], chemical reactivity [51], and Lewis acidity [11].

5.2.5 *Self-Interaction Errors*

Above, we briefly discussed the Hartree-Fock (HF) method in Sect. 5.2.1; the Hamiltonian of which is defined as,

$$h_i^{\mathrm{HF}} = -\frac{1}{2}\nabla_i^2 - \sum_{A=1}^{M} \frac{Z_A}{r_{iA}} + \sum_i^{N/2} (2J_i - K_i), \tag{5.23}$$

where J_i is the Coulomb operator $\left(J_i = \int d\mathbf{r}_2 \psi_i^*(2) r_{12}^{-1} \psi_i(2)\right)$ and K_i is the exchange operator $\left(K_i \psi_a(1) = \left[\int d\mathbf{r}_2 \psi_i^*(2) r_{12}^{-1} \psi_a(2)\right] \psi_a(1)\right)$. The operators in Eq. (5.6) are analogous to the energy expressions in Eqs. (5.9) and (5.10). While

J_i is a purely classical expression, K_i is quantum mechanical in nature and arises due to the indistinguisability of the electrons [17]. Comparison of Eqs. (5.23)–(5.11) reveals that while both HF and DFT methods include Coulomb interactions *explicitly*, only HF includes exchange *explicitly*; DFT includes exchange *implicitly* through the exchange correlation potential ($V_{XC}[\rho]$). While DFT offers a scaling of $\mathcal{O}(N^3)$ versus the $\mathcal{O}(N^4)$ of HF, unfortunately it suffers from two key drawbacks: (1) each electron interacts with itself through the Coulomb potential (termed self-interaction error, or SIE) and (2) it produces wavefunctions that are more localized than HF.

One solution to the problem is to employ hybrid functionals. These are density functionals that include a fraction of HF exchange. The two most common hybrid functionals are B3LYP [52] and PBE0 [53]. In B3LYP, the exchange-correlation energy is defined as,

$$E_{\mathrm{XC}}^{\mathrm{B3LYP}} = E_{\mathrm{XC}}^{\mathrm{LYP}} + a\left(E_{\mathrm{exchange}}^{\mathrm{HF}} - E_{\mathrm{exchange}}^{\mathrm{LYP}}\right) + b\Delta E_{\mathrm{exchange}}^{\mathrm{B88}} + c\Delta E_{\mathrm{correlation}}^{\mathrm{LYP}}, \tag{5.24}$$

where $a = 0.20$, $b = 0.72$, and $c = 0.81$. The terms $E_{\mathrm{exchange}}^{\mathrm{B88}}$ and $E_{\mathrm{XC}}^{\mathrm{LYP}}$ come from Becke [54] and Lee et al. [55]. In contrast, the PBE0 is based on the PBE functional [56] and defines the exchange-correlation energy in a simpler fashion [53],

$$E_{\mathrm{XC}}^{\mathrm{PBE0}} = E_{\mathrm{XC}}^{\mathrm{PBE}} + \frac{1}{4}\left(E_{\mathrm{exchange}}^{\mathrm{HF}} - E_{\mathrm{exchange}}^{\mathrm{PBE}}\right). \tag{5.25}$$

This formalism increases the overall accuracy of the underlying DFT method (see the book by Cramer [18] and the excellent review by Laurent et al. [57] for examples and comparisons), however it comes at the cost of computational resources as it now scales as $\mathcal{O}(N^4)$. Moreover while hybrid functionals are available for periodic systems, the computational costs are drastically much higher than what is seen for non-periodic systems, which limits their usage [22].

Another solution is the so-called DFT+U methods. Here we will follow the nomenclature of Stausholm-Møller et al. [58] (which in turn is based off of the work of Dudarev et al. [59]). In the DFT+U method, the total energy is represented as,

$$E_{\mathrm{DFT}+U} = E_{\mathrm{DFT}} + \sum_{a}^{\#\text{ of atoms}} E_{\mathrm{orb}}^{a}, \tag{5.26}$$

where E_{orb}^{a} is a modified orbital energy,

$$E_{\mathrm{orb}}^{a} = \frac{U^a}{2}\sum_{\sigma}\mathrm{Tr}\left(\hat{\rho}_{\sigma}^{a}\right) - \mathrm{Tr}\left(\left[\hat{\rho}_{\sigma}^{a}\right]^2\right), \tag{5.27}$$

and $\hat{\rho}_{\sigma}^{a}$ is atomic orbital occupation matrix [58]. The U^a term in Eq. (5.27) is what gives DFT+U its name, and represents an *energy penalty* for occupying certain orbitals. This in turn forces electrons to act more like HF electrons, reduces SIE, and still keeps the same DFT scaling of $\mathcal{O}(N^3)$. However, U^a is an empirical term whose value can vary from problem to problem. Methods to generate U^a parameters from ab initio using linear response have been proposed [60, 61]; unfortunately, these methods are still being tested and development is ongoing.

5.2.6 The Problem of Electron Correlation

In order to provide a tractable equation for solving electron-electron interactions, it was necessary to treat these interactions in an *average* fashion, which gives rise to the one electron operators that we have seen in Eqs. (5.6), (5.11), and (5.23). In other words, this means that we do not consider the effect of electron motions being coupled to every other electron. The effect of electron-electron motion is called electron correlation, and is a non-local effect [17, 18].

Within DFT, we have a term called the exchange-correlation potential (V_{XC}); in principal this term does account for electron correlation, albeit in an approximate fashion. In order to see why, we will expand out the exact electron density (ρ_{exact}) in terms of an initial, zeroth order electron density (ρ_0, which is calculated from the choice of basis set),

$$\rho_{\text{exact}} = \rho_0 + \nabla\rho_0 + \frac{1}{2}\nabla^2\rho_0 + \ldots \tag{5.28}$$

Functionals depending only on ρ_0 are referred to as being derived within the localized density approximation (LDA) [55]. Adding in gradient terms ($\nabla\rho_0$) results in generalized gradient approximations (GGA) [62]. A third class, meta-GGA's, utilized the Lagrangian of the density $\left(\nabla^2\rho_0\right)$, however, they will not be discussed here as they are not typically employed for surface chemistry. Obviously, the more terms we include in Eq. (5.28), the more electron correlation we capture; however practically we are limited to the gradient on average (GGA's).

Within the literature, correcting DFT for electron correlation is typically referred to as the dispersion energy. In order to see why, we consider the Möller-Plesset second order perturbation energy term for electron correlation for orbital i $\left(E_i^{(2)}\right)$ is,

$$E_i^{(2)} = \sum_{i \neq n} \frac{\langle i|\mathcal{V}|n\rangle \langle n|\mathcal{V}|i\rangle}{E_i^0 - E_n^0} \tag{5.29}$$

where n is the other molecular orbitals, E_i^0 is the HF energy for orbital i, $|i\rangle$ is the wavefunction for orbital i, and $\mathcal{V}$ is the electron-electron interaction operator

[17, 21]. Now, if we compare Eq. (5.29) to that for the dispersion energy between two atoms A and B $\left(E_{\text{disp}}^{AB}\right)$, [63]

$$E_{\text{disp}}^{AB} = -\sum_{m\neq 0}^{M_A}\sum_{n\neq 0}^{N_B} \frac{\langle 00|\mathcal{V}|mn\rangle \; \langle mn|\mathcal{V}|00\rangle}{E_m^A + E_n^B - E_0^A - E_0^B}. \tag{5.30}$$

We will not go into detail on every term in Eq. (5.30); however it is the similarities between Eqs. (5.29) and (5.30) has led to the assumption that DFT fully lacks dispersion. However, Jenness and coworkers have demonstrated that this is not the case, and in reality the missing electron correlation terms spans the whole region of intermolecular interactions [34, 35, 64].

Regardless, one of the more popular methods for correcting for electron correlation in DFT is through the adding a Lennard-Jones "dispersion" term, or the DFT+D method. Similar to DFT+U, we correct the total DFT energy by a pair-wise additive term,

$$E_{\text{DFT+D}} = E_{\text{DFT}} - s_6 \sum_{AB} \frac{C_6^{AB}}{R_{AB}}, \tag{5.31}$$

where the C_6^{AB} is a dispersion term that is either computed or fitted beforehand, and s_6 is a scaling term to match the reference data (typically CCSD(T), MP2, or experiment) [65–68]. Care needs to be taken with the s_6 term as you can end up overestimating your "dispersion" contribution, resulting in an overprediction of binding energies [34, 35]. However, these methods are quick and easy to implement, and consequently are widely available. It should be noted that this method does not modify the wavefunction like the DFT+U method; thus any follow on methods involving wavefunction analysis will have its accuracy limited by the underlying functional.

An alternative approach is the non-local density functional (also called van der Waals density functional or vdW-DF). Instead of adding on an empirical correction to the total energy, we modify the exchange-correlation energy as,

$$E_{\text{XC}} = E_{\text{exchange}} + E_{\text{correlation}}^{\text{local}} + E_{\text{correlation}}^{\text{non-local}}, \tag{5.32}$$

where $E_{\text{correlation}}^{\text{local}}$ is taken from the LDA approximation. $E_{\text{correlation}}^{\text{non-local}}$ is a new term, and is defined as,

$$E_{\text{correlation}}^{\text{non-local}} = \frac{1}{2}\int\int \rho(r)\phi(r,r')\rho(r')drdr' \tag{5.33}$$

where $(r,\ r')$ denote two separate points in space, $\rho(r)$ is our electron density, and $\phi(r,r')$ is an interaction kernal [68, 69]. A variety of different kernals exist with different density functionals (for a review see Berland et al. [69]); as a consequence there exist a variety of vdW-DF's to choose from such as vdW-DF2 [70–72] and optPBE-vdW [73, 74]. For a comparison of different vdW-DF's, we refer the reader to the work of Klimeš et al. [73, 74] These methods, while harder to implement, are more robust and accurate than DFT+D methods. Furthermore, if correlation effects are important in analyzing the wavefunction, vdW-DF's can capture that effect.

5.2.7 Forces, Hellmann-Feynman Theorem, and Geometry Optimization

Above, we discussed the methodology used in the calculation of the ground state energy for a molecular system. However, how can we minimize the positions of the individual atoms within a molecule so we can have the lowest energy structure? For this problem, we will need to minimize the energy with respect to position, which will require knowledge of the atomic forces.

From classical mechanics, we know that the force on a object (**F**) is calculated as,

$$\mathbf{F} = -\nabla U(r) = -\left(\frac{\partial}{\partial x} + \frac{\partial}{\partial y} + \frac{\partial}{\partial z}\right) U(r), \tag{5.34}$$

where $\{x, y, z\}$ is a set of Cartesian coordinates describing the position, r is the distance from origin ($r = \left(x^2 + y^2 + z^2\right)^{1/2}$), and $U(r)$ is a potential depending upon r. However, it is well known from the early days of quantum mechanics that classical descriptors break down in the limit of atoms and molecules. Thus, the question is now how can we calculate the forces of a quantum mechanical system?

The solution to the problem is the Hellmann-Feynman theorem; here we will summarize following the nomenclature of Feynman [75]. Here, our $U(r)$ is,

$$U = \int \psi^* \mathcal{H} \psi dv, \tag{5.35}$$

where ψ is our wavefunction, ψ^* it's complex conjugate, $\mathcal{H}$ is a Hamiltonian (e.g., Eq. (5.11). It should be noted here that this discussion is agnostic of the Hamiltonian used), and dv is a volume element. For a generic variable λ, we can write the derivative of Eq. (5.35) as,

$$\frac{\partial U}{\partial \lambda} = \int \psi^* \left(\frac{\partial \mathcal{H}}{\partial \lambda}\right) \psi dv + \int \left(\frac{\partial \psi^*}{\partial \lambda}\right) \mathcal{H} \psi dv + \int \psi^* \mathcal{H} \left(\frac{\partial \psi}{\partial \lambda}\right) dv. \tag{5.36}$$

One of the properties of Hamiltonians like Eq. (5.11) is the fact that they are Hermitian and as such, form their own self-adjoint [17, 76]. Consequently, we can write,

$$\int \left(\frac{\partial \psi^*}{\partial \lambda}\right) \mathcal{H}\psi dv = \int \left(\frac{\partial \psi}{\partial \lambda}\right) \mathcal{H}\psi^* dv, \tag{5.37}$$

and since $\mathcal{H}\psi = U\psi$ and $\mathcal{H}\psi^* = U\psi^*$ (i.e., Eq. (5.1) with U substituted for E), Eq. (5.36) becomes,

$$\frac{\partial U}{\partial \lambda} = \int \psi^* \left(\frac{\partial \mathcal{H}}{\partial \lambda}\right) \psi dv + U\left[\int \left(\frac{\partial \psi^*}{\partial \lambda}\right) \psi dv + \int \psi^* \left(\frac{\partial \psi}{\partial \lambda}\right) dv\right]. \tag{5.38}$$

Due to Eq. (5.37), the last two terms in Eq. (5.38) cancel. This gives the final expression for our forces in a quantum mechanical framework,

$$\frac{\partial U}{\partial \lambda} = \int \psi^* \left(\frac{\partial \mathcal{H}}{\partial \lambda}\right) \psi dv. \tag{5.39}$$

Or, in other words, the classical expression for the forces in Eq. (5.34) can be applied to quantum mechanical Hamiltonians! [75] The Hellmann-Feynman theorem is an extremely powerful tool, and as been applied not only to the consideration of atomic forces, but also problems in diffusion [77], calculating electronegativities [78], and in thermodynamic considerations [79].

Once forces are generated through the Hellmann-Feynman theorem, it is a simple matter of optimizing the geometry of the molecular system with respect to atomic positions. A number of optimizers exist and have been implemented in several software packages, including the LBFGS [80] and FIRE [81] methods (for a comparison of the efficiency for different optimizers, see Sheppard et al. [82]). Forces are converged to a predetermined criteria using the following inequality,

$$\max\left(\sum_{i=\{x,y,z\}} F_i^2\right) \leq F_{\max}^2, \tag{5.40}$$

where the sum is over the {x, y, z} components of the force on atom i, and $F_{\max}$ is a user defined criteria. For periodic DFT calculations, a $F_{\max}$ of 0.05 electron volt per Ångstrom is used (eV Å^{-1}).

5.3 Case Study: Adsorption of Munitions in Soils

In Sect. 5.2.1, we discussed the theoretical methodology used in current studies of surface interactions. In this Section, we explore how the above techniques can be used in a problem our group has been working on for several years: the effect of soil constituency on the fate and transport of munition compounds. Here, we will focus our discussion on our efforts involving components of arid soils interacting with munition compounds (Fig. 5.1). Additional techniques such as binding energies, charge density differences, electrochemical property prediction, Lewis acidity, and environmental transport will be discussed. Moreover, we will demonstrate how tools such as density of state analysis can provide critical insight into the nature of molecular interactions, as well as a comparison of cluster and periodic surface models.

5.3.1 *Binding Energies*

The first step in understanding the nature of adsorbate-surface interactions is the calculation of the binding energy. If we consider the adsorption process to be a chemical reaction of the type,

$$\text{Adsorbate} + \text{Surface} \longrightarrow (\text{Adsorbate-Surface})\,, \tag{5.41}$$

then we can write the formation energy ($E_{\text{formation}}$) as,

$$E_{\text{formation}} = \sum_{i}^{N_{\text{products}}} E_i^{\text{product}} - \sum_{i}^{N_{\text{reactants}}} E_i^{\text{reactant}}. \tag{5.42}$$

As the formation energy in Eq. (5.42) represents the energy contained when an adsorbate is bound to the surface, it is referred to as the binding energy, E_{BE}. While other terms are commonly used (e.g., complexation energy, interaction energy, etc.) for the purposes of this review we will refer to this quantity as the binding energy. Additionally, it is common in the physics field to reverse the sign on Eq. (5.42). For purposes of this review we will reverse the sign to be more in line with chemical convention.

The second most common components of arid and semi-arid soil are aluminas, or Al_2O_3 [83, 84]. Aluminas are typically found in the top layers of rocky soil [83], and are present due to the nature of sub-surface hydrothermal circulation [85]. Thus, it is instructive to discuss the binding of munition compounds onto this material. Alumina has several different phases, however the α phase (α-Al_2O_3 or corundum) is most energetically favored [86, 87], with the (0001) surface being the most preferred. This surface is shown in Fig. 5.2.

We begin with our 2013 study that utilized plane-wave DFT to study the interaction of TNT (Fig. 5.1a) on the α–Al_2O_3 (0001) surface [4]. We utilized an energy cutoff of 340 eV and the PBE exchange-correlation functional [56] and considered two configurations of TNT, a flat (called parallel in the original manuscript) and perpendicular. After minimization using the Hellmann-Feynman theorem (Sect. 5.2.7), a comparison of the gas-phase geometry to the adsorbed geometry reveals that while the perpendicular configuration maintains the various

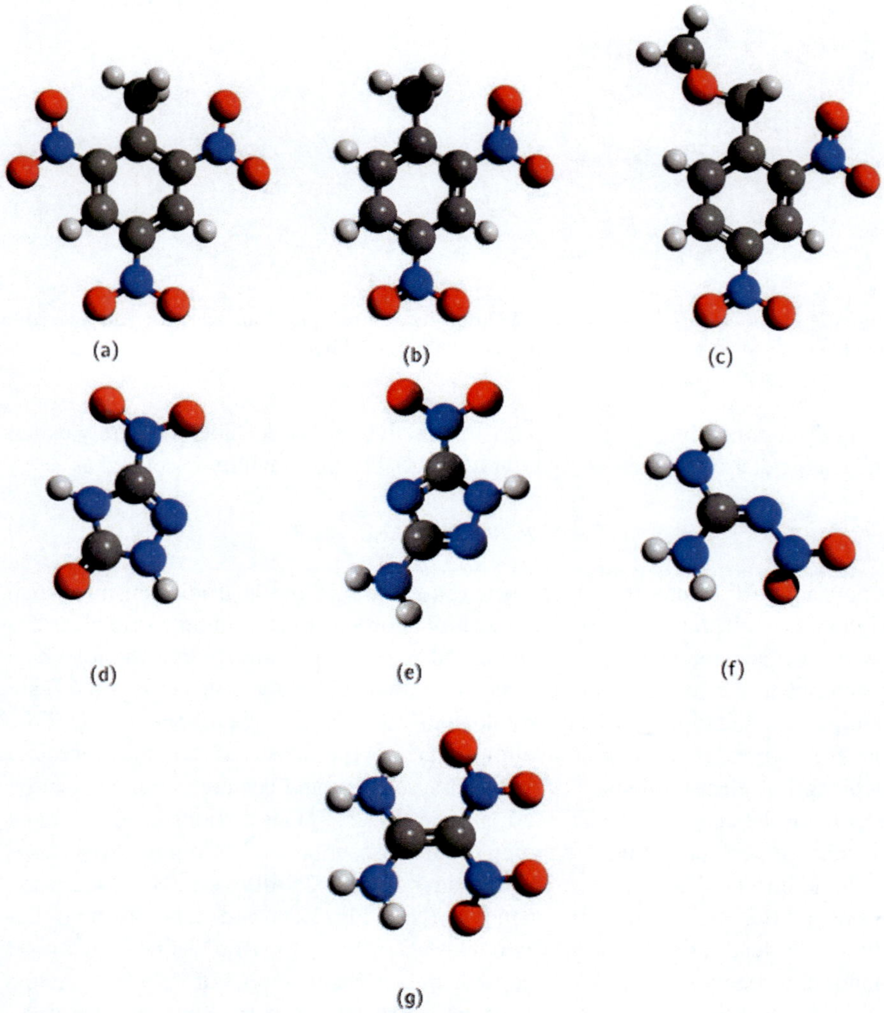

Fig. 5.1 Common munitions. (**a**) TNT. (**b**) DNT. (**c**) DNAN. (**d**) NTO. (**e**) ANTA. (**f**) NQ. (**g**) FOX7

dihedral angles, the flat features several deviations in the dihedrals of the nitro groups. This rotation allows for a greater interaction between the α–Al_2O_3 surface and the three nitro groups of TNT. In comparing the binding energies, the perpendicular configuration has an energy of -1.09 eV, and the flat has an energy of -2.13 eV, nearly double that of the perpendicular. Finally, they also show that relaxing the surface results in a contraction of Al–O bonds due to the under-coordination surface sites; in the presence of the TNT, these under-coordinated sites become coordinated, and the interlayer spacing becomes similar to what is found in bulk α–Al_2O_3 [4].

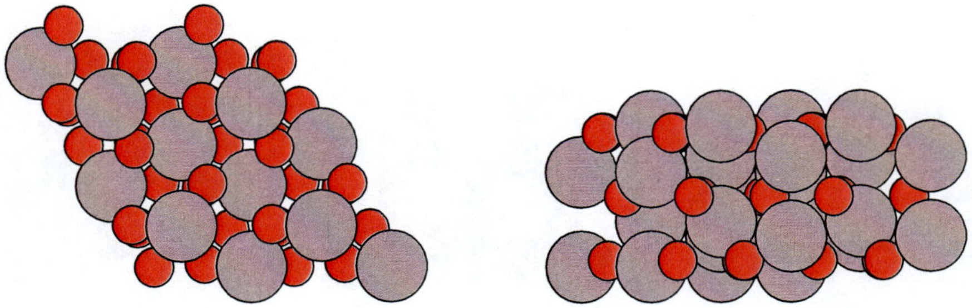

Fig. 5.2 The α–Al_2O_3 (0001) as viewed from the normal and from the side. Adapted with permission from [16]. Copyright (2020) American Chemical Society

In discussing binding energies, it is instructive to also consider a quantity called the charge density difference. Similar to Eq. (5.42), this quantity is defined as,

$$\rho_{\text{diff}} = \rho_{\text{ads-surf}} - \rho_{\text{ads}} - \rho_{\text{surf}}, \tag{5.43}$$

where ads/surf denotes the adsorbate/surface, and ρ is the electronic density (which is the wavefunction squared). This quantity allows us to determine where electrons have either accumulated or are depleted and gives an indication of how the adsorbate is binding to the surface. It is important to note that all three terms in Eq. (5.43) are computed at the same geometry. We demonstrate how this quantity is used for TNT on α–Al_2O_3 in Fig. 5.3. Examination of the charge density differences reveals a depletion of electron density at the binding centers and accumulation of electron density at the center of the TNT–Al^{+3} bond (Fig. 5.3. This demonstrates that there is a charge transfer component present in the adsorption of TNT onto α–Al_2O_3.

In addition to TNT, we can also consider 5-Nitro-2,4-dihydro-3H-1,2,4-triazol-3-one (NTO, Fig. 5.1d), nitroguanidine (NQ, Fig. 5.1f) and 1,1- Diamino-2,2-dinitroethylene (FOX7, Fig. 5.1g) on α–Al_2O_3 (0001) [3, 10]. NTO has two N–H bonds and functions as a weak acid with an experimental pK_a of 3.76 [88]. As the oxygens on the α–Al_2O_3 surface can act a Brønsted base (i.e., accepts a proton), NTO undergoes *dissociative* adsorption. In this process, instead of retaining its molecular identity, NTO donates a proton to the surface. This process is shown in Fig. 5.4. First, NTO is shown to adsorb in a flat configuration, with an oxygen from the nitro (–NO_2) group coordinated with the Al^{+3} site. As the adsorption process proceeds, the NTO assumes a perpendicular configuration, and the N–H bond dissociates with the proton (H^+) attaching to a surface oxygen and forming a hydroxyl group (–OH). The entire process is ~ -0.45 eV exothermic and is shown in Fig. 5.4. Additionally, NTO can adsorb in a perpendicular fashion through the nitro or carboxylic groups. The flat configuration has a binding energy of -2.83 eV (see the note above in regards to the sign convention). However, values are not given for the two perpendicular configurations, but they do note that they are 2-3$\times$ lower in binding strength when compared to the flat. Finally, in agreement with their study on TNT, they note that there is an accumulation/depletion of electron density at bond

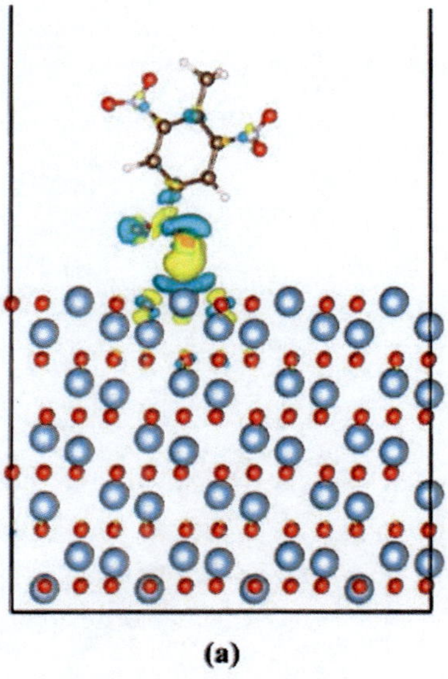

Fig. 5.3 Charge density differences for TNT on α–Al_2O_3 (0001): (**a**) perpendicular and (**b**) flat. Yellow regions show areas where electron density are accumulated and blue regions shows areas where electron density is depleted. Al atoms are blue, O red, C brown, and H white. Reprinted (adapted) with permission from [4]. Copyright (2013) American Chemical Society

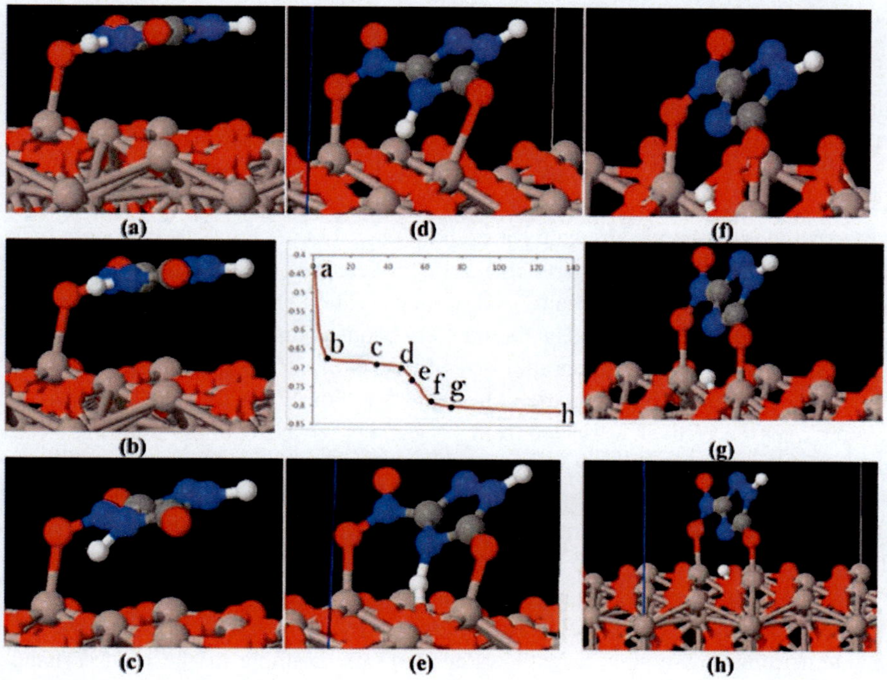

Fig. 5.4 Dissociation of NTO on the α–Al_2O_3 (0001) surface. Caption letters correspond to the energy points in the center graph. Reprinted (adapted) with permission from [3]. Copyright (2014) American Chemical Society

Table 5.1 Binding energies for NQ and FOX7 on clean α–Al_2O_3 (0001) and hydroxylated α–Al_2O_3 (0001) from Shukla et al. [10]. Units in eV. Table created with permission using data from Ref. [10], copyright (2017) American Chemical Society

	NQ			
Surface	Perpendicular	Flat I[a]	Flat II[b]	FOX7
α–Al_2O_3 (0001)	−1.78	−3.01	−3.40	−3.62
hydroxylated α–Al_2O_3 (0001)	−0.96	−0.96		−1.09

[a] Called parallel I in original manuscript

[b] Called parallel II in original manuscript

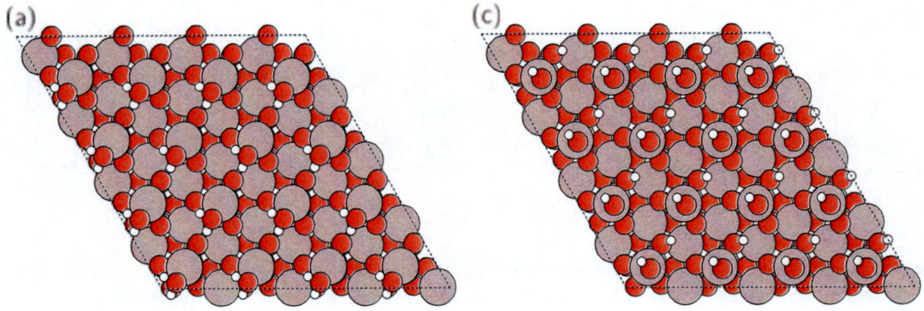

Fig. 5.5 Associative (**a**) and dissociative water (**b**) binding on α–Al_2O_3 (0001) surfaces. Black dotted lines denote the presence of the periodic boundaries used in the DFT calculations. Adapted with permission from [16]. Copyright (2020) American Chemical Society

and atomic centers, respectively. Sites near the NTO also assume a more bulk-like inter-layer distance as the under-coordinated sites become coordinated [3].

For NQ and FOX7, Shukla et al. noticed that upon adsorption to the clean α–Al_2O_3 surface the C–N bonds undergo a contraction relative to the gas-phase, whereas the N–O bonds elongate for both NQ and FOX7. For NQ, they note that the degree of these bond changes depends on the relative orientation of the NQ with respect to the surface, i.e., if both –NH_2 groups (called parallel II by Shukla et al. [10]) are coordinated, then the N–O and C–N bonds change more than if only one –NH_2 is coordinated (called parallel I by Shukla et al. [10]). The binding energies for these complexes are summarized in Table 5.1. For NQ, a flat binding motif is favored over the vertical, as is seen from TNT and NTO, as the binding strength (the absolute value of the binding energy) nearly doubles. Moreover, the structure where both –NH_2 groups are bound to the surface (Flat II) is more favorable, which can be attributed to the increased number of hydrogen bonds. FOX7 binds more strongly than NQ [10].

It is well known that alumina can become hydroxylated in the presence of water [89–92], thus in this study, they considered the effect of surface hydroxylation due to interactions of the surface with water [10]. An example of the hydroxylated α–Al_2O_3 surface is shown in Fig. 5.5b. In order to better understand the effect of surface hydroxylation, the binding energies for NQ and FOX7 on the hydroxylated

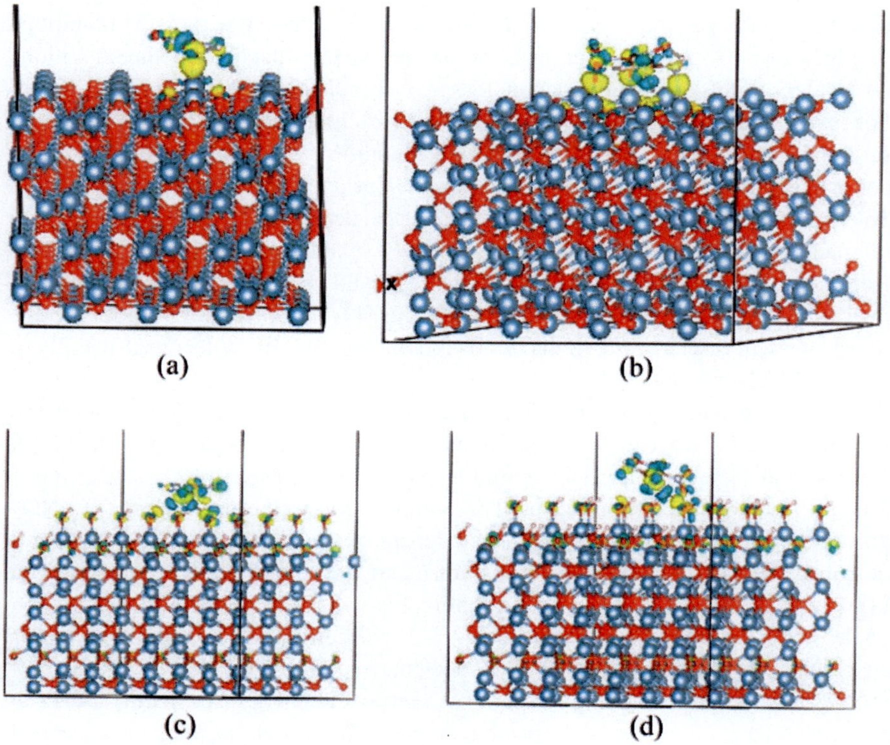

Fig. 5.6 Charge density differences for NQ and FOX7 on α–Al_2O_3 (0001): (**a**) NQ on clean α–Al_2O_3, (**b**) FOX7 on clean α–Al_2O_3, (**c**) NQ on hydroxylated α–Al_2O_3, and (**d**) FOX7 on hydroxylated α–Al_2O_3. Yellow regions show areas where electron density are accumulated and blue regions shows areas where electron density is depleted. Al atoms are blue, O red, C brown, and H white. Reprinted (adapted) with permission from [10]. Copyright (2017) American Chemical Society

α–Al_2O_3 are summarized in Table 5.1. Compared to the clean surface, the hydroxylated surface has greatly reduced binding strengths (a three-fold decrease is noted for the flat configurations of both molecules). In terms of geometry, the changes discussed above are not present with the hydroxylated α–Al_2O_3 surface; instead NQ assumes a geometry that is very mildly distorted from its gas-phase counterpart. These trends for NQ also hold for FOX7. We can gain a better understanding of why the clean α–Al_2O_3 (0001) surface changes binding strength/bond lengths more if we examine the charge density differences as shown in Fig. 5.6. Here we see the same pattern from Shukla and Hill [3, 4]; however, in the case of the hydroxylated surface (Fig. 5.6c–d), we see that the hydroxylated surface does still donated charge density but to a lesser extent [10].

At this point, we would like to note that the change in binding energy with respect to orientation as shown by Shukla et al. [10] ties into the nature of the interaction, i.e., Lewis acid-base interactions. Metiu and coworkers [93–95] demonstrated that as the orientation of the Lewis base (in the case of the current review the munitions in Fig. 5.1), so does the overlap between the surface bands and the munitions molecular orbitals. As the overlap increases, so does the strength of the interaction. For an in depth examination of surface band structure and molecular orbitals, we refer the reader to the seminal work by Hoffman [96–98]. A full discussion on the computational prediction of Lewis acidity and surfaces is presented in Sect. 5.3.4. To overcome the issue associated with geometrical dependence of surface–adsorbate interactions, the authors of the current review examined four munitions (TNT (Fig. 5.1a), DNAN (Fig. 5.1b), NTO (Fig. 5.1d), and NQ (Fig. 5.1f)) and how they interact with the α–Al_2O_3 and α–Fe_2O_3 surfaces [11]. However, for purposes of this review we will discuss TNT as the trends established for this molecule also apply to DNAN, NTO, and NQ [11].

In order to handle the geometry affect of Lewis acids [93–95], we used the minima hopping algorithm of Goedecker [99] with a Hookean constraint [100]. Jenness et al. [11] considered three binding motifs: two vertical (referenced as mode 1 and 2 in the original manuscript. These correspond to the position of the –NO_2 groups. Using the methyl as a marker, mode 1 corresponds to the –NO_2 at the para-position, and mode 2 at the ortho-position), and a flat. These three modes served as the starting point for the minima hopping. First, a short molecular dynamics run in the NVT ensemble is performed (for a discussion on various thermodynamic ensembles, we recommend the book by McQuarrie [101]) for 20–25 fs. Once time-propagated, the geometry is minimized with respect to energy (see Sect. 5.2.7). The new geometry is compared to the prior geometries. If it is unique, then it is appended onto the list of unique structures and if not unique, it is discarded. Examination of the unique minima allows for us to determine basic statistics (mean and standard deviation) about the nature of the binding energy. An example for TNT on α–Al_2O_3 and α–Fe_2O_3 is shown in Fig. 5.7b [11].

Overall the binding of TNT to α–Al_2O_3 (0001) is favorable, and is between ~ -1.8–1.9 eV. In Fig. 5.8, they report the changes in bond lengths; similar to what is reported by our group's earlier work [3, 4, 10], the C–N bonds experience a contraction while the N–O bonds elongate. Upon adsorption onto α–Fe_2O_3 (0001), TNT gains an extra $\sim$0.40–0.50 eV in stability, and the N–O/C–N bonds are more distorted [11]. These results indicate that the N–O bonds in soil rich in iron oxides would undergo a greater degree of change than those with a greater amount of aluminas. It needs to be stressed here that the effect of both on the kinetics of change have yet to be elucidated, and is an open area of research. However, environments rich in iron have been shown to transform –NO_2 groups [102–109], thus we can state that our study [11] provides a theoretical rationale behind the observed chemical reductions.

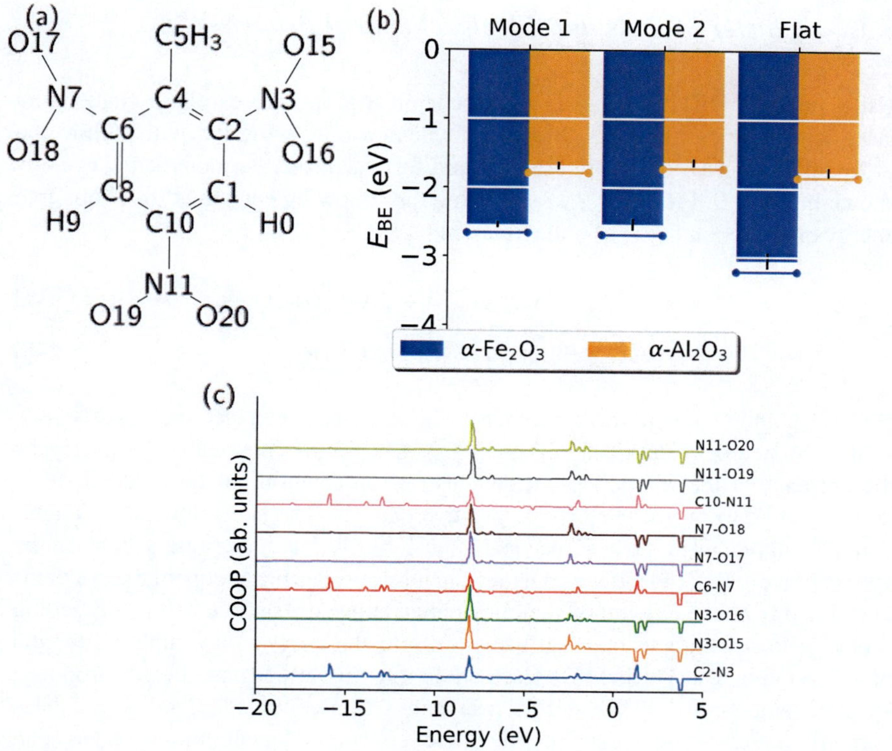

Fig. 5.7 (**a**) Schematic of 2,4,6-trinitrotoluene (TNT) with the atoms numbered, (**b**) mean binding energies for TNT on the α–Al_2O_3 (orange bars) and α–Fe_2O_3 (blue bars) surfaces, and (**c**) COOP curves for the gas-phase TNT. From Eq. (5.22), negative peaks represent anti-bonding interactions whilst positive peaks represent bonding interactions. Reproduced from Ref. [11] with permission from the PCCP Owner Societies

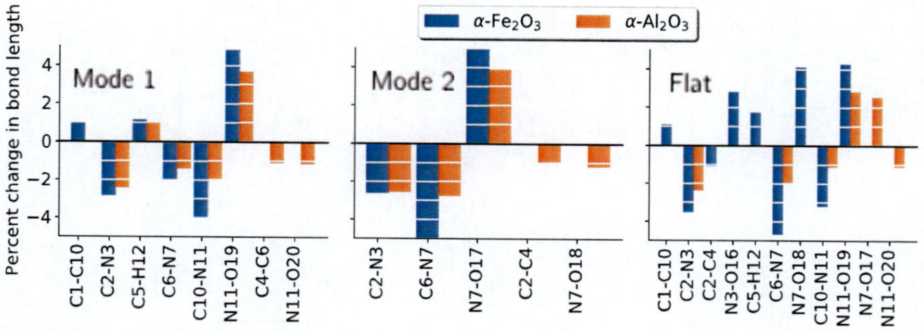

Fig. 5.8 Percent change in the bond lengths from the gas-phase to the adsorbed phase for TNT. Negative percent change denotes the bond contracting relative to the gas-phase, whilst a positive percent change denotes the bond elongating relative to the gas-phase. Numbering corresponds to Fig. 5.7a. Reproduced from Ref. [11] with permission from the PCCP Owner Societies

5.3.2 Cluster Models and Electrochemical Properties

Using periodic DFT to calculate electrochemical properties can be a challenging task, especially for surfaces. In order to demonstrate why, we will outline the procedure we have previously used [6] and for which the thermodynamic cycle is shown in Fig. 5.9. From here, we can write the following equations for Gibbs free energy of reduction $\left(\Delta G^0_{\text{red}}\right)$ and oxidation $\left(\Delta G^0_{\text{ox}}\right)$,

$$\Delta G^0_{\text{red}} = \Delta G^0\left(R^-_{\text{ads}}\right) - \Delta G^0\left(O_{\text{ads}}\right) \quad (5.44a)$$

$$\Delta G^0_{\text{ox}} = \Delta G^0\left(O^+_{\text{ads}}\right) - \Delta G^0\left(R_{\text{ads}}\right). \quad (5.44b)$$

Here, R_{ads} and O_{ads} denote the geometry with and without an electron, respectively. In the absence of a charge sign (e.g., R^-_{ads}), this denotes the neutral geometry. In the presence of the charge sign, it denotes that an electron has been added (for a negative charge) or removed (for a positive charge). The subscript "ads" denotes that the complex is in it's adsorbed state [5, 6]. For surfaces, a problem arises in the application of Eq. (5.44). If we consider a neutral surface, the energy of the system is uncoupled from the amount of vacuum applied above the surface. However, adding a charge to a 2-dimensional surface results in the energy becoming a function of the vacuum gap [110, 111]! Corrections for this effect have been introduced by Freysoldt et al. [110], and Komsa and Pasquarello [111]; however, these corrections are not widespread outside of the semiconductor community and as such their applicability to electrochemical predictions is non-existent. Consequently, the majority of work regarding electrochemical properties of munitions on soil surfaces have relied upon the cluster approximation, discussed in Sect. 5.2.2.1.

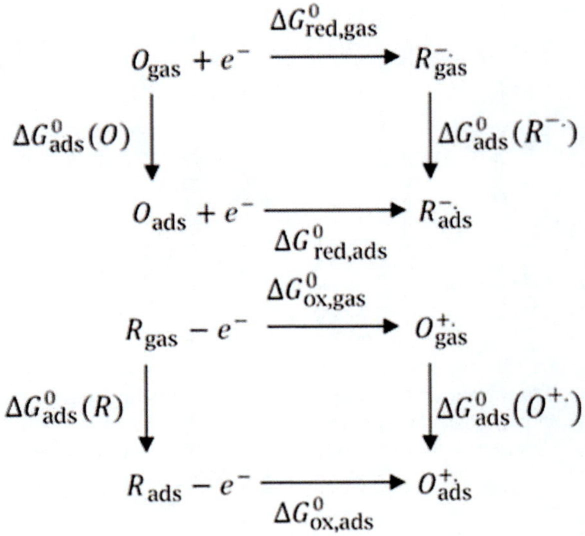

Fig. 5.9 Thermodynamic cycle used by Sviatenko et al. [6]. Reproduced from Ref. [6] with permission from the Wiley Publishing

Table 5.2 Electron affinities (E_A) and Gibbs free energy of reduction (ΔG^0_{red}) for ANTA, TNT, DNT, DNAN, and NTO in the gas-phase, adsorbed on silica, and in water from Sviatenko et al. [5, 6]. Units in electron volts (eV). ANTA data is reproduced with permission from Ref. [5], copyright (2015) American Chemical Society. TNT, DNT, DNAN, and NTO data is reproduced with permission from Ref. [6], copyright (2015) Wiley Publishing

Molecule	E_A			ΔG^0_{red}			Reference
	Gas-phase	Silica	Water	Gas-phase	Silica	Water	
ANTA(a) (–NO_2)	−0.53	−2.28	−3.50	−0.53	−2.16	−3.52	[5]
ANTA(b) (–NH_2)		−1.99			−1.95		[5]
TNT	−2.21	−3.08	−4.32	−2.17	−2.93	−4.24	[6]
DNT	−1.66	−1.90	−3.89	−1.68	−1.81	−3.92	[6]
DNAN	−1.46	−2.50	−3.82	−1.48	−2.37	−3.83	[6]
NTO	−1.47	−2.54	−4.05	−1.50	−2.40	−4.06	[6]

In 2015, our group published two papers in conjunction with Jackson State University examining the electron affinities and Gibbs free energy of reduction (ΔG^0_{red}) for TNT, DNT, DNAN, NTO, and ANTA in the gas-phase, adsorbed on silica, and in water [5, 6]. As these calculations involve the addition and subtraction of electrons from the system, a cluster model of the hydroxylated α-SiO_2 (100) surface was employed. Sviatenko et al. [6] used the M05-2x functional with the tzvp basis set. The electron affinities were calculated by attaching an electron to the munitions system (either gas-phase, adsorbed, or in water) and taking the energy difference with respect to the neutral. Their results are shown in Table 5.2. In terms of geometry, Sviatenko et al. observed the munition compounds moving closer to the surface upon addition of an electron, with an increase in the number of hydrogen bonds forming when the munition-silica system had an electron attached. Removing an electron resulted in the opposite trend [6].

In terms of the electrochemical properties in Table 5.2, it should be noted that the sign convention in Eq. (5.44) denotes that a more negative $E_A/\Delta G^0_{red}$ means a greater ability to attach an electron/undergo electrochemical reduction. Based on their results, it is clear that the munitions are able to undergo reduction more when in water versus being in the gas-phase (here, water is modeled using an implicit solvation model). In particular, TNT is the most likely to undergo reduction, with ANTA being the least reducible. Additionally, when adsorbed onto the silica surface, there is a greater propensity for electron attachment and reduction (when compared to the gas-phase)! While it has been demonstrated through experiments and theory that oxides can act as a reducing agent [49, 50, 92, 93, 112–125]; however, silicas such as α-SiO_2 have been demonstrated to be rather "neutral" in this regard [112].

5.3.3 Comparison of Cluster and Periodic Surface Models

At this point, we would like to discuss the differences between cluster and periodic surfaces. In 2015 our group in conjunction with Jackson State University [7] presented a comprehensive study of munition on kaolinite with both cluster model and periodic surface. While the differences between cluster and periodic were not the focus of this study, it is one of the few in which a careful comparison of both methods was performed. Kaolinite (shown in Fig. 5.10) is an aluminum-silicate clay comprised of alternating octahedral and tetrahedral sites. These sites are occupied with Al and Si atoms, respectively. Consequently, as the crystal is cleaved along the (100) plane (the lowest energy plane), either a Al- or Si-terminated plane will be presented. The binding energies of TNT (Fig. 5.1a), DNT (Fig. 5.1b), DNAN (Fig. 5.1c), and NTO (Fig. 5.1d) on kaolinite are summarized in Table 5.3.

Summarizing their results, Scott et al. [7] found that the Al-terminated surfaces bind the munitions more strongly than the Si-terminated surfaces; however, they

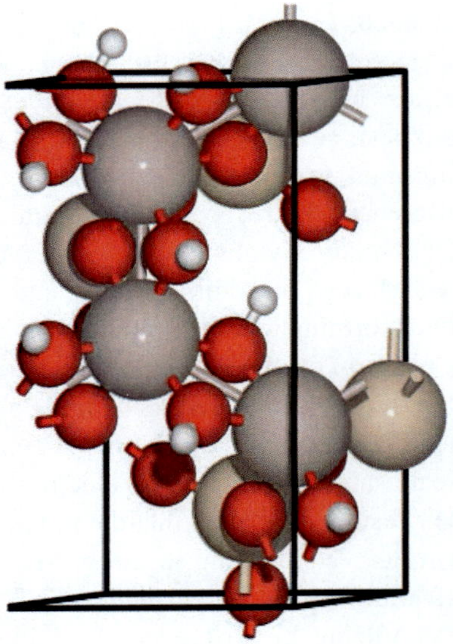

Fig. 5.10 Kaolinite 3D crystal structure. Red spheres are oxygen, gray aluminum, and beige silicon. Solid black line denotes the periodic boundary. Al atoms sit at octohedral sites, and Si atoms at tetrahedral sites

Table 5.3 TNT, DNT, DNAN, and NTO binding energies on kaolinite clay from Scott et al. [7] (units in eV). Table created from data published in Ref. [7]

Termination	TNT	DNT	DNAN	NTO
Cluster[a]				
Al	−1.33	−1.15	−1.22	−1.20
Si	−0.82	−0.92	−1.06	−0.59
Periodic[b]				
Al	−1.19	−1.05	−1.09	−0.93
Si	−0.60	−0.61	−0.59	−0.53

[a] Method: BLYP+D2/6-31+G(d,p)
[b] Method: BLYP+D2/Planewave

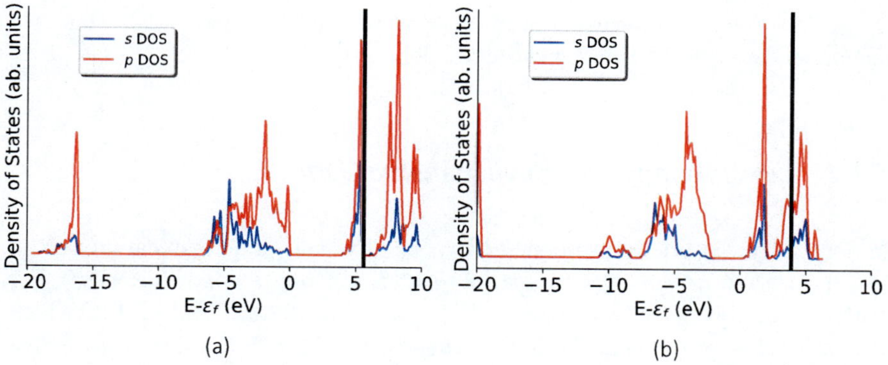

Fig. 5.11 Projected density of states for M^{+3} binding site on the (**a**) α–Al_2O_3 (0001) and (**b**) α–Fe_2O_3 (0001) surfaces. Energy is zeroed to the Fermi level (taken to be the energy of the highest occupied band), with the vertical black lines denoting the position of the E_s^* energy. Reproduced from Ref. [92] with permission from the PCCP Owner Societies

also note that for both terminations the surface–munition distances are roughly equivalent. These results indicate that under conditions in which the Al-termination is favored, DNT would bind twice as a strong as for those under which Si-termination flourishes. Comparison of the gas-phase to adsorbed DNT geometry shows that the N–O bonds are relatively unaffected while the C–N bonds undergo a mild contraction. However, this contraction only holds for the Al-terminated surface as the Si-terminated the geometry is roughly equivalent to the gas-phase [7]. Using the arguments outlined in Sect. 5.3.4, we can justify these trends by looking at the band gap. The theoretical band gap has been calculated to be between 6.2 and 8.2 eV [126], which is significantly higher than the band gaps for α–Al_2O_3 and α–Fe_2O_3 shown in Fig. 5.11a,b, respectively. From this we can conclude, based on the Lewis acid arguments from Jenness and coworkers [11, 92] that kaolinite is a relatively weak Lewis acid, and as such would not transform/degrade the –NO_2 groups significantly.

In terms of comparing with the cluster model, the periodic surface shows similar bond lengths and trends as the cluster model. However, the use of a cluster model overestimates the binding of TNT, DNT, DNAN, and NTO by ~0.1–0.2 eV. This demonstrates that while geometrical trends can be established with a high level of accuracy with a cluster model (and potentially allowing one to do electrochemical analysis similar to Sect. 5.3.2), the binding energies need to use a periodic surface or else they would be more strongly bound. This would have the consequence of making the error in understanding transport phenomena hard to assess.

On a final note regarding cluster models, it is important to note that dangling bonds represent defects in the surface. It has been previously demonstrated that the band-gap picture changes radically as defects are introduced as "gap-states" between the valence and conduction bands [127–132] form. This would partially explain the differences in binding energies reported in Table 5.3 and would hinder molecular orbital type analyses such as those presented in Sect. 5.2.4. Thus, we recommend caution when using cluster models.

5.3.4 Lewis Acidity and Environmental Fate

In Sect. 5.3.1, we briefly mentioned Lewis acidity. This is a chemical process in which a system *accepts* electron density (the acid) from a donor (the base). For example, aluminas have been shown to be Lewis acidic [89, 92, 93, 133–136] which in turn drives several chemical processes [89, 90, 92, 124, 125, 134, 136–148]. In this section, we discuss the nature of Lewis acidity, and how it can create a conceptual picture for understanding the fate of munitions in soil.

As mentioned above, a solid Lewis acid *accepts* electron density from the highest occupied molecular orbital (HOMO) of the adsorbate. By definition the HOMO lies near the Fermi level of the surface of the Lewis acid; thus, the relative position of the conduction band with respect to the Fermi level would be a key identifier for Lewis acidity. Consequently Jenness et al. introduced the mean of the s-conduction band,

$$E_s^* = \frac{\int_{\epsilon_F}^{\infty} \rho_s(\epsilon)\epsilon d\epsilon}{\int_{\epsilon_F}^{\infty} \rho_s(\epsilon) d\epsilon}, \tag{5.45}$$

where ϵ_F is the Fermi level (taken to be the energy of the highest occupied band of the surface), $\rho_s(\epsilon)$ is the energy-dependent s-orbital contribution from the projected density of states (PDOS), and ϵ is the energy as a measure of Lewis acidity [92]. It should be noted at this point that the E_s^* descriptor is of primary interest in a comparative sense. For example, the Lewis acid properties of the various heterogeneous Al^{+3} sites on the γ-Al_2O_3 can be correlated with the reactivity of this material (as demonstrated for dehydration and etherification of ethanol) [92].

In 2018, we used this argument to provide an origin for the bond changes in TNT when adsorbed onto α–Al_2O_3 and α–Fe_2O_3 (shown in Fig. 5.8). We found that α–Fe_2O_3 results in a greater contraction of C–N bonds and a greater elongation of N–O bonds when compared to α–Al_2O_3. Using E_s^* as a descriptor for Lewis acidity, we find that α–Fe_2O_3 is more Lewis acidic than α–Al_2O_3 (Fig. 5.11b vs. a) as the E_s^* of α–Fe_2O_3 (0001) is much closer to the Fermi level than that of α–Al_2O_3 (0001). What this means is that we can expect to see stronger binding of our munitions onto the α–Fe_2O_3 surface when compared to the α–Al_2O_3. Moreover, we can also expect to see a greater elongation/contraction effect for the N–O/C–N bonds. Both of these predictions are correct (Figs. 5.7 and 5.8).

We next examined the COOP curve (Eq. (5.22)) for the isolated TNT, shown in Fig. 5.7c. On the energy scaled employed, the HOMO energy lies ~1 eV. For the N–O bonds, the COOP shows a strong bonding character at the HOMO; thus, as the Lewis acid Al^{+3} centers depopulate this region the bond order *decreases* and the bonds elongate. Similarly, for the C–N bonds the nature is of the HOMO is anti-bonding and as this region is depopulated the bond order *increases* and the bonds contract. This is supported by examination of the charge density differences as the N–O bonds are depleted of charge and the C–N bonds gain charge [11]. This is in agreement with prior studies on α–Al_2O_3 [3, 4, 10] and provides an explanation for the reported charge density differences shown here in Figs. 5.3 and 5.6. A similar analysis was performed for DNAN, NTO, and NQ. We concluded this study by stating it is these elongations that would allow for the –NO_2 groups to be reduced.

5.3.5 Environmental Transport

While these prior studies are key in understanding the *fate* of munitions in soil, so far little work has gone into understanding *transport* through these soils. A key problem here is that when thermodynamic contributions are computed for molecules, an ideal gas is assumed; however, in reality these molecules do not behave as an ideal gas, nor are they in the gas-phase! To face these issues, we have introduced a correction to the free energy of the munition (G_{mun}). We start by assuming an equilibrium between the aqueous and vapor phases [101]

$$G_{\text{mun}}^{\text{aq}} = G_{\text{mun}}^{\text{vapor}}, \tag{5.46a}$$

$$= G_{\text{mun}}^{\circ}(T) + k_{\text{B}}T\ln\left(\frac{P_{\text{mun}}}{P^{\circ}}\right) \tag{5.46b}$$

where P_{mun} is the partial pressure of adsorbate mun, G_{mun}° is the Gibbs free energy at a standard pressure of P° (here taken to be 1 atmosphere), and the $k_{\text{B}}T\ln\left(\frac{P_{\text{mun}}}{P^{\circ}}\right)$ is the deviation from standard. In this fashion, they are able to write G_{mun} as a function of the partial pressure of an adsorbate, and relate the partial pressure to aqueous concentration by Henry's Law,

$$H = \frac{C}{P_{\text{mun}}}, \tag{5.47}$$

where H is Henry's Law constant, C is the aqueous phase concentration, and P_{mun} is the partial pressure. Thus,

$$G_{\text{mun}}^{\text{aq}} = G_{\text{mun}}^{\circ}(T) + k_{\text{B}}T\ln\left(\frac{C}{H \times P^{\circ}}\right). \tag{5.48}$$

In Eq. (5.48), $G_{\text{mun}}^{\circ}(T)$ is calculated in the usual fashion (i.e. with the ideal-gas approximation) and $k_{\text{B}}T\ln\left(\frac{C}{H\times P^{\circ}}\right)$ represents a modification that accounts for adsorbate concentration. In addition to this modification, they considered several thermodynamic states:

1. Munition
2. Water monolayer (Fig. 5.5a)
3. Dissociatively adsorbed water monolayer (fully hydroxylated surface, Fig. 5.5b)
4. Munition with water
5. Munition with dissociatively adsorbed water (partially hydroxylated surface)

and calculated the probability of each state forming with a Boltzmann distribution,

$$\text{Probability} = \frac{e^{-\Delta E_i^f / k_B T}}{\sum_i^{\text{states}} e^{-\Delta E_i^f / k_B T}} \times 100. \tag{5.49}$$

ΔE_i^f is the formation energy of thermodynamic state i from the above list. We refer the reader to Ref. [16] for further details on the methodology. The salient point here is that while prior studies have considered the effect of a fully hydroxylated surface or a clean surface, this is the first study to consider intermediate surface states, such as hydroxyls and the munition bound to the surface, as opposed to the munition being hydrogen bonded to the hydroxyl network [16].

The application of Eqs. (5.46)–(5.49) is shown in Figs. 5.12 and 5.13 for TNT and DNAN, respectively. It should be noted that while both the water monolayer (Fig. 5.5a) and hydroxylated (Fig. 5.5b) states were considered, it was found that the hydroxylated state was preferred 100%. Thus, the rest of the discussion was centered around the hydroxylated α–Al_2O_3 (0001) surface. From Figs. 5.12 and 5.13, the first thing of note is that while DNAN largely prefers the flat configuration, TNT favors the two vertical modes as well as the flat (albeit the flat moreso). For DNAN, increasing the concentration results in more of the vertical modes being preferred, however the majority of the surface coverage would still be flat. For TNT increasing the concentration results in an increase in the vertical binding modes, which would be expected based off of prior studies of aromatics on metal

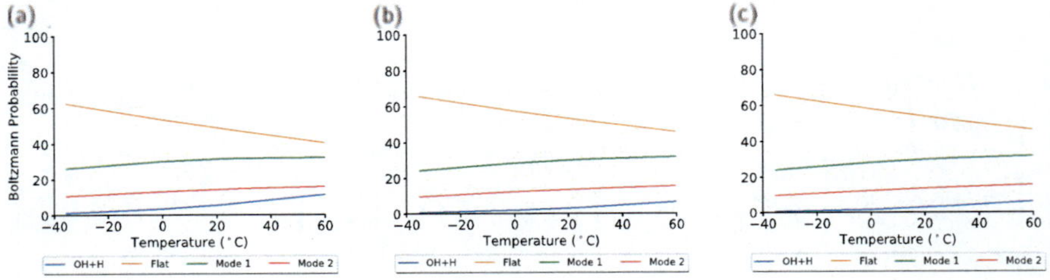

Fig. 5.12 TNT Boltzmann probabilities (Eq. 5.49) with respect to concentration and temperature on α–Al_2O_3 (0001). Binding modes of the munitions are in the presence of dissociatively adsorbed water (i.e., hydroxylated surface). The concentrations are (**a**) 0.01 mg L^{-1}, (**b**) 30 mg L^{-1}, and (**c**) 60 mg L^{-1}. Reproduced with permission from [16]. Copyright (2020) American Chemical Society

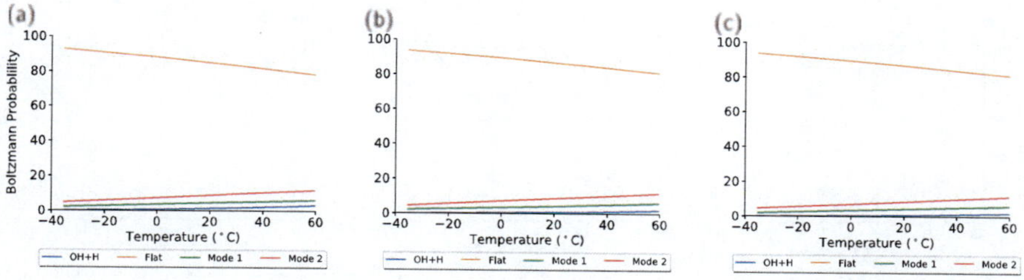

Fig. 5.13 DNAN Boltzmann probabilities (Eq. (5.49)) with respect to concentration and temperature on α–Al_2O_3 (0001). Binding modes of the munitions are in the presence of dissociatively adsorbed water (i.e., hydroxylated surface). The concentrations are (**a**) 0.01 mg L^{-1}, (**b**) 30 mg L^{-1}, and (**c**) 60 mg L^{-1}. Reproduced with permission from [16]. Copyright (2020) American Chemical Society

surfaces [149, 150]. This discrepancy for DNAN might be explained by the model employed by Jenness et al. not accounting for adsorbate-adsorbate interactions; thus the energetic shift to vertical binding is missing [16]. However, accounting for these terms is a computationally demanding process with no easy solution. Despite this drawback, these results do demonstrate that DNAN and TNT are soil bound for soils rich in α–Al_2O_3.

We have also applied this formalism to the various thermodynamic binding states for α–Fe_2O_3 (0001). Similar to α–Al_2O_3, the preferred water state was to dissociatively adsorb, resulting in surface hydroxyls (similar to what is shown in Fig. 5.5b for α–Al_2O_3). However, unlike the case for α–Al_2O_3 (0001), we found for TNT 100% surface coverage across all three concentration considered in our study (Fig. 5.14). Considering the stronger Lewis acid properties of α–Fe_2O_3 over α–Al_2O_3 [11], this is to be expected. However, what is curious is that for DNAN, there is a 0% surface coverage, and the α–Fe_2O_3 surface remains in it's hydroxylated state! This means that while TNT is a soil bound contaminant for soils rich in α–Fe_2O_3, DNAN is a aqueous phase contaminant under those conditions [16] (Fig. 5.15).

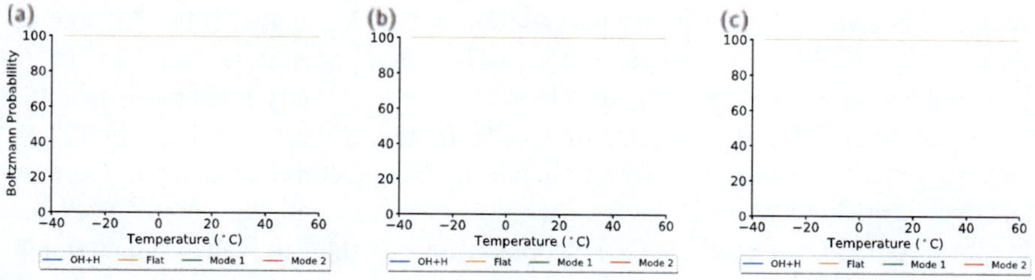

Fig. 5.14 TNT Boltzmann probabilities (Eq. (5.49)) with respect to concentration and temperature on α–Fe_2O_3 (0001). Binding modes of the munitions are in the presence of dissociatively adsorbed water (i.e., hydroxylated surface). The concentrations are (**a**) 0.01 mg L^{-1}, (**b**) 30 mg L^{-1}, and (**c**) 60 mg L^{-1}. Reproduced with permission from [16]. Copyright (2020) American Chemical Society

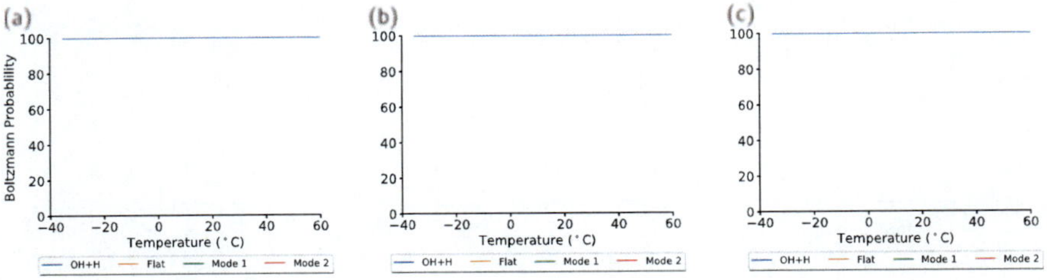

Fig. 5.15 DNAN Boltzmann probabilities (Eq. 5.49) with respect to concentration and temperature on α–Fe_2O_3 (0001). Binding modes of the munitions are in the presence of dissociatively adsorbed water (i.e., hydroxylated surface). The concentrations are (**a**) 0.01 mg L^{-1}, (**b**) 30 mg L^{-1}, and (**c**) 60 mg L^{-1}. Reproduced with permission from [16]. Copyright (2020) American Chemical Society.

5.4 Looking to the Future

5.4.1 Breathing New Life into an Old Method: Density Functional Tight Binding

In the above discussions, rigorous first principals (or ab initio) DFT calculations were performed (this methodology is outlined above in Sect. 5.2.1). These calculations, while efficient and accurate, still limit our system sizes. Advancements such as projected augmented wavefunctions (PAWs) [32, 151, 152] and highly parallizable codes such as `GPAW` [26] allow us to study systems containing upwards of 1000 atoms. However, we wish to explore the effect of the natural environment on the fate and transport of key materials and chemicals; unfortunately, the natural environment is an exceedingly complex system, and it is necessary to find more approximate quantum chemistry methods that can deliver similar accuracy at a fraction of the cost.

The semi-empirical density functional tight-binding method (DFTB) is one such technique. This method has been applied to a wide variety of problems ranging from intermolecular interactions [65, 153–158], organic molecules [159–167], metal-organic frameworks [168], molecular vibrations [158], metals [169–175], metal oxides [162, 166, 173, 176], solvents [166, 177–181], optical properties [175, 182], semiconductor defects [170, 183], and chemical reactions [184]. For a more in depth description of DFTB and its applications, we refer the reader to the literature [185–191]. This method is also described in detail in "Introductory Roadmap to Current Reactive Force-field Methodologies", which is a chapter in the current volume. Here, we would like to highlight the key points of the method, as well as our attempts to bring this method closer to its DFT roots.

DFTB is an approximate Hamiltonian method, in which we write the zeroth-order molecular Hamiltonian as,

$$H^0_{\mu\nu} = \begin{cases} \epsilon_{\mu}^{\text{free-atom}} & \text{if } \mu = \nu \\ \langle \phi_{\mu}^{A} | \hat{T} + V_0^A + V_0^B | \phi_{\nu}^{B} \rangle & \text{if A} \neq \text{B} \\ 0 & \text{Otherwise} \end{cases} \tag{5.50}$$

Following diagonalization and summation, a repulsive potential is then added to the solution of the above Hamiltonian. While the terms in Eq. (5.50) can largely be derived from DFT, the repulsive potential is still a largely empirical term and is defined as the difference between the DFT energy of the molecule and the DFTB Hamiltonian [159, 185, 186, 190–192]. Splines are typically used for reproducing the repulsive energy curve. We can also form a second-order Hamiltonian with the addition of atomic charges [159],

$$E^{\text{SCC-DFTB}} = \sum_{i}^{\text{occupied}} \langle \phi_i | \hat{H}_0 | \phi_i \rangle + \frac{1}{2} \sum_{\alpha,\beta}^{N} \gamma_{\alpha\beta} \Delta q_{\alpha} \Delta q_{\beta} + E_{\text{repulsion}} \tag{5.51}$$

This method is referred to as the self-consistent charge (SCC) method due to the charges (Δq_{α}) being determined by a Mulliken charge analysis in a self-consistent manner. This allows for the fluctuations in the atomic environment arising from the molecular environment to be accounted for.

The Hamiltonian terms are pre-computed and stored into Slater-Koster tables for each atomic pair as a function of atom-atom distance. These tables are read in at the start of the DFTB calculation and specific distance values are interpolated. While several orders of magnitude faster than DFT, the interactions for each atom pair needs to be calculated and stored before hand which limits applicability as the majority of DFTB codes do not allow for the calculation of the relevant matrix elements on the fly. Moreover, as mentioned above, the repulsion terms are also dependent on both atom-atom pairs and having a ready data-base of DFT repulsive curves. This makes for a rather arduous parameterization process, and there has been a lot of work into automatic parameter generation [161, 167, 193–195].

Recently [196], we have introduced an orbital dependence to the confinement potential (we called this method the orbital dependent confinement (ODC). Where the confinement plays a role in to DFTB is discussed in "Introductory Roadmap to Current Reactive Force-field Methodologies" in this volume). Moreover, examination of the radial wavefunction allows for the necessary r_0 parameters to be generated automatically. This method, called AODC, was able to replicate the geometries of a variety of organic molecules that have strong lone-pair electron effects. In Table 5.4 we present scaling factors from both ODC (which were hand fitted) and AODC.

Table 5.4 Confinement scaling factors for ODC and AODC. Reproduced with permission from [196]. Copyright (2020) American Chemical Society

	ODC			AODC		
Atom	1*s*	2*s*	2*p*	1*s*	2*s*	2*p*
Hydrogen	4.00			4.62		
Carbon	2.00	1.85	1.75	0.37	2.14	2.44
Oxygen	2.00	2.50	1.50	0.27	1.57	1.77
Nitrogen	1.75	1.25	2.00	0.32	1.81	2.04

Table 5.5 Molecular topologies for H_2, H_2O, and NH_3. Reproduced with permission from [196]. Copyright (2020) American Chemical Society

Parameterization	3rd Order	H Correction	H_2	H_2O	NH_3
Orbital Independent Confinement (OIC)^a	No	No	0.74 Å	0.97 Å 101.0°	1.03 100.2°
Orbital Dependent Confinement (ODC)^b	No	No	0.73 Å	0.96 Å 107.0°	1.03 105.6°
Auto-Orbital Dependent Confinement (AODC)^b	No	No	0.75 Å	0.96 Å 106.2°	1.02 Å 105.8°
mio-1-1[159]	No	No	0.74 Å	0.97 Å 107.2°	1.02 Å 110.0°
3ob-3-1 [164]	Yes	Yes	0.74 Å	0.96 Å 108.2°	1.01 Å 109.3°
DFT			0.75 Å	0.97 Å 103.9°	1.02 Å 106.2°
Experiment [197]			0.75 Å	0.96 Å 104.6°	1.02 Å 106.8°

^{a} r_0 is set to $1.85r_{\text{covalent}}$ for all atoms
^{b} r_{covalent} scaling factors given in Table 5.4

In Table 5.5 we present the results of our new DFTB confinement model with respect to previous DFTB parameterizations (mio-1-1 [159] and 3ob-3-1 [164]), DFT, and experiment. From here, we can readily see that introduction of orbital based confinement improves the description of lone-pair electrons as the ODC and AODC bond angles are much closer to the experimental and DFT values. Mio-1-1 and 3ob-3-1 employ an orbital independent confinement (OIC), and as such their bonds deviate from the optimal value. We have also introduced the use of combination rules for exponential functions that will allow for only homogeneous curves to be fitted for the repulsive interactions. While these results are as good as traditionally fitted DFTB repulsion, parameterization is an order of magnitude faster and requires a smaller training suite [196]. We believe these advances will breathe new life into the DFTB method, and will enable a further expansion of the systems being studied.

5.4.2 Artificial Intelligence and Machine Learning (AI/ML)

In addition to method refinement and new method development, other opportunities for boosting capabilities can be found by leveraging expertise from other subject areas. In particular, through the combination of increasingly available and powerful computing processing power, large sets of data, and easily accessible software tools, data science approaches have proven to have much to offer traditional computational chemistry. Artificial intelligence (AI) and machine learning (ML) have been adopted numerous times to a variety of chemical problems. The aim of this section is only to provide awareness for some of the considerations for AI/ML methods in computational chemistry and to highlight some relevant examples that illustrate the utility of these increasingly applied techniques.

In some instances, ML models have been used as a substitute for quantum or molecular mechanics. While the data acquisition and model building for such applications is time consuming, the model, once constructed, may be orders of magnitude faster than the pure computational chemistry method. In other cases, AI/ML is used to develop supporting information, acting as a subcomponent to the overarching computational approach. For example, where many "flavors" of methods exist (e.g., density function theory with its alphabet soup of available functionals), ML may be used to generate a tastier flavor or potentially match functionals up with chemical systems or properties in order to achieve higher accuracy.

Further diversity in the application of AI/ML approaches in computational chemistry can be explained in terms of the underlying data or features that are supplied to the ML models. For chemistry, the structure of a compound or system of interest contains detailed data in a compact form. A trained chemist is able to make predictions and rationalizations from these representations alone. To enable an ML model to make use of such "intuitive" information, new representations (i.e. a machine readable form of the structure) are needed.

Starting from a three dimensional structure, some approaches utilize the Coulomb matrix, which is essentially the nuclear-nuclear repulsion terms for each atom pair in a molecule. There are several considerations that may be taken into account when using the Coulomb matrix as a molecular representation for ML—e.g., ensuring same size for different molecules via padding with zeros, storing only the upper triangle to take advantage of symmetry, etc. The benefit of the Coulomb matrix is that it captures the three-dimensional structure of the system; but it comes at the requirement of having or generating a "good" (i.e., reasonable and representative) three-dimensional structure of the system. Chemical fingerprint methods, which can include any number of properties or attributes, are often connectivity based, and can avoid the need for a three-dimensional structure. As such, the fingerprint features can be derived starting from the SMILES (Simple Molecular Input Line Entry System)[198–200] string for a compound. Some fingerprints may provide information on what structural groups are present and how many of each, relying on chemist-defined moieties, while others follow a graph-based approach.

Ultimately, ML models are not a "one and done approach": feature optimization (including novel, custom approaches), method selection, and method hyperparamterization need to be retooled for every different problem, dataset, or target property. For energetic materials, the challenges are "how best" to apply ML, and how to maximize data quantity and quality. In the following sections, we discuss: (1) some successful applications of ML for energetic material property prediction, and (2) efforts where ML has been applied to support the underlying theory and input to DFT and computational methods rather than being the predictive model itself.

5.4.2.1 Machine Learning and Energetics

Energetic materials exist as a highly specialized subset of organic molecules. Despite their distinct molecular properties, machine learning has successfully been applied to these materials. Applications fall under many different umbrellas: property prediction, [201–203] candidate screening, [204] synthetic route assessment, [205] and microstructure generation [206]. We will focus our discussion on the more straightforward examples of property prediction.

High quality, quantitative property predictions for energetic materials are obtained via costly quantum mechanical calculations. While such in silico approaches have their advantages relative to experiment (in terms of cost and safety, for example), such predictions are limited due to the computational resources required. To address this issue, several efforts have worked to utilize ML models to predict the performance of energetic materials [201–204].

In their proof-of-concept work, Elton et al. [201] tested numerous featurizations and ML models for the prediction of energetic performance properties (e.g., detonation pressure and velocity, explosive energy, etc.). Models included kernel ridge regression, ridge regression, support vector regression, random forest, and k-nearest neighbors. All method saw at least some level of success, but the kernel ridge and ridge regressions typically performed best. The optimal featurizations for their small test set (109 molecules total) was the sum over bonds approach. They achieved errors ranging from 4 to 11% for the density and detonation volatility, detonation pressure and explosive energy, respectively.

Barnes et al. [202] built on the previously described proof of concept work by utilizing a larger data set and expanding the types of ML models. In particular, they apply a neural network model (multilayer perceptron) to predict energetic performance properties. Another key development in their work is the use of feature sensitivity to explore the importance of different atoms/atom groups in the model development. Doing so allowed for better interpretation of the ML results, along with comparison to chemical intuition.

These works illustrate that ML models can be successfully applied to energetic materials, but they also highlight that challenges for improved developments lay in model optimization (feature and model selection) and most importantly, the underlying data. Larger data sets will enable for advanced ML approaches (such as neural networks) to detect patterns and features that do not manifest on a smaller scale. Some efforts utilize quantum chemical and other predicted values to create a substantial dataset of approximately 18,000 molecules [203]. This approach is feasible for sufficiently small molecules, but will be limited quickly with increasing system size.

This is an opportunity where DFTB may work in tandem with ML, providing the underling dataset for the ML model at a lower cost (in terms of time and resources) compared to traditional quantum chemical methods.

5.4.2.2 AI/ML and DFT

In this section, we discuss in more detail a particular machine learning model, neural networks, along with applications for ML in DFT method development.

A powerful model that has seen effective use in a range of subject areas is the deep neural network (DNN). The machinery behind DNNs are inspired by our nervous system, where our nerve endings take in input data that is passed to our brain, which then interprets the data and elucidates a response. Typically, the more neurons between the signal input and signal processing center, the more reliable the data is interpreted. In terms of DNN, these neurons are represented as "nodes", or "layers". Thus, the more nodes a DNN has, the "deeper" the neural network is, and the more accurate it may become; but the size of the input layer, number of nodes or hidden layers, and other hyperparameters must be optimized for each research problem.

Translating our nervous system into an artificial neural network (ANN) was first achieved in 1958 by psychologist Frank Rosenblatt to understand visual data and object recognition [207]. However, it would take an additional 50 years before applications to chemistry were developed [208] as there were significant challenges that had to be overcome to allow DNNs to predict molecular structures. The first challenge is the generation of training and testing databases. DNNs are designed to mimic our neural anatomy, and as a result it must learn from its "surroundings". This is achieved by providing a set of training data that is subsequently used to teach/train the DNN to predict the properties of interest. Despite the wide applicability of the methods discussed in Sect. 5.2.1, the required data for a robust training set has been elusive due to the cost of computing a wide variety of chemical systems. In fact, until recently, each group interested in generating training data had to either generate it in-house, or farm the literature. However, databases such as those from Smith et al. [209], and Truhlar and coworkers [210] have recently become available.

The second challenge for DNN is how to interpret a 3-dimensional chemical structure. Behler and Parrinello achieved the encoding by using 2- and 3-body symmetry functions [208]. Following this seminal work, the field of DNN in predicting geometries split into two main groups: methods based on 2- (radial) and 3-body (angular) symmetry functions, [208, 211, 212] and message-passing techniques that use non-invariant radial functions [213–215]. The former are more computationally efficient as there are not feed back loops which add to the computational cost. However, message-passing based methods enable properties to be predicted that are not associated with the geometry, such as charge states [216, 217]. These type of DNN models open the door to a wide range of chemical systems due to their transferability and extensibility, and will enable the computation of properties that were previously too expensive.

The first application of DNN to chemical problems was achieved in 2007 by Behler and Parrinello [208]. In this landmark study, they were successful in predicting the energies and forces of a set of chemical structures and were able to optimize large systems not included in the training set. In this initial work, they showed errors per atom between DFT and DNN to be between 0.1–0.35 kcal mol^{-1} (with 1 eV $\approx$ 23.06 kcal mol^{-1}) [208]. These errors are reasonably low as the target error is less than 2 kcal mol^{-1} [218]. The ANI-1 model of Smith et al. [212], which uses a similar architecture to Behler and Parrinello [208], was able to achieve a RMSE of 1.8 kcal mol^{-1}, which was the lowest RMSE when compared to DFTB (10.2 kcal mol^{-1}), PM6 (22.0 kcal mol^{-1}) and AM1 (16.2 kcal mol^{-1}) [212]. Moreover the ANI-1 model was shown to be highly extensible, and has since been used in problems related to the expedition of drug discovery [219].

In regards to message-passing techniques, the Hierarchically Interacting Particle Neural network (HIP-NN) predicted the ground state energy of organic molecules to within a 0.26 kcal mol^{-1} mean absolute error. This method employs a hierarchical approach due to several energy predictions being required. At the zeroth order level, the energy is based purely on the atom types, which is then followed by higher energy predictions based on the neighboring atoms [215]. This type of message-passing allows HIP-NN to predict not only the energy and atomic forces, but also the charge of the systems [216, 217]. Here, the charge prediction can be based on either Hirshfeld [220], NBO [221, 222], MSK [223], or CM5 [224] charge schemes with a mean absolute error ~0.01 electrons [217]. HIP-NN prediction of charges enables the calculation of IR spectra, which has been shown to be comparable to reference IR spectra. Thus, the HIP-NN can be used to compute the IR spectra for a wide range of chemicals, which reduces the computational cost [216, 217].

In addition to these groundbreaking studies, Burke and coworkers [225–227] utilized ML techniques within the framework of DFT to reduce the computational cost of generating electron densities. They initially examined fermions in a 1-dimensional (1D) box for a proof-of-concept. If this work can lead to an ML-developed kinetic energy functional ($T_s[\rho]$) for the Kohn-Sham (KS) equation, then this would allow for orbital-free DFT to be used, which would greatly decrease the computational cost [225–227]. This proof-of-concept illustrated that $T_s[\rho]$ is not only achievable, but also accurate. Unfortunately, the corresponding functional derivatives exhibited numerical instabilities which limits this implementation [225, 227]. Better results were obtained when then ML model learned the Hohenberg-Kohn (HK) mapping of the potential to density ($V(r) \rightarrow \rho(r)$). This model allowed for the molecular dynamics of malondialdehyde based on the mapping of the potential to density with errors of 4–14 kcal mol^{-1}; these errors are much larger compared to the errors obtained from the DNN PES models (ANI-1, HIP-NN) but are a promising proof-of-concept [227]. A recent study by Meyer et al. [228] took this concept of learning the $T_s[\rho]$ to feed into orbital-free DFT with success. They found that if the ML model is trained on both $T_s[\rho]$ and its derivative, it leads to decreased errors and increased in extensibility for compounds outside the training data [228].

5.5 Conclusions

In the current review, we examined several methodologies for examining surface interactions. We have presented an overview of periodic DFT, its shortcomings, and how to overcome these shortcomings. From there, we have demonstrated how these techniques can be applied in calculating electrochemical, adsorption, charge transfer, Lewis acidity, and transport properties. Our group has extensively applied these techniques, and have used them to explain a variety of phenomena involving munitions in soil. We finally presented our recent work on reducing empiricism in DFTB, which we believe will be an invaluable methodology for studying the natural environment from a quantum chemical perspective.

Acknowledgments We are grateful to Drs. Caitlin Bresnahan, Timothy Schutt, and Ping Gong for their comments and edits on the current manuscript. The use of trade, product, or firm names in this report is for descriptive purposes only and does not imply endorsement by the U.S. Government. The tests described and the resulting data presented herein, unless otherwise noted, were obtained from research funded under the Installations and Operational Environments Office of the Technical Director of the United States Army Corps of Engineers and the Environmental Security Technology Certification Program of the Department of Defense by the USAERDC. Permission was granted by the Chief of Engineers to publish this information. The findings of this report are not to be construed as an official Department of the Army position unless so designated by other authorized documents. This work was supported by a grant of computer time from the DOD High Performance Computing Modernization Program at ERDC, Vicksburg. This document has been approved for public release (Distribution Statement A).

References

1. Van Loon GW, Duffy SJ (2000) Environmental chemistry: a global perspetive. Oxford University, Oxford
2. Hill FC, Sviatenko LK, Gorb L, Okovytyy SI, Blaustein GS, (2012) Leszczynski. J Chemosphere 88:635–643
3. Shukla MK, Hill F (2014) J Phys Chem C 118:310–319
4. Shukla MK, Hill FJ (2013) Phys Chem C 117:13136–13142
5. Sviatenko LK, Gorb L, Hill FC, Leszczynska D, Leszczynski, J (2015) J Phys Chem A 119:8139–8145
6. Sviatenko LK, Isayev O, Gorb L, Hill FC, Leszczynska D, Leszczynski J (2015) J Comput Chem 36:1029–1035
7. Scott AM, Burns EA, Lafferty BJ, Hill FC (2015) J Mol Model 21:21–37

8. Sviatenko L, Gorb L, Hill F, Leszczynska D, Shukla M, Okovytyy S, Hovorun D, Leszczynski J (2016) Environ Sci Technol 50:10039–10046
9. Gorb L, Shukla MK (2017) J Mol Model 23:81
10. Shukla MK, Wang J, Seiter J (2017) J Phys Chem C 121:11560–11567
11. Jenness GR, Seiter J, Shukla MK (2018) Phys Chem Chem Phys 20:18850–18861
12. Pittman KM, McAlexander HR, Tschumper GS, Shukla MKJ (2019) Phys Chem A 123:3504–3509
13. Schutt TC, Shukla MK (2019) J Phys Chem A 123:4973–4979
14. McAlexander HR, Giles SA, Crouch RA, Peel HR, Jones S, Bednar AJ, Shukla MK (2020) Struct Chem 31:975–982
15. Moores LC, Jones SJ, George GW, Henderson DL, Schutt TCJ (2020) Photo Chem Photobiol A Chem 386:112094
16. Jenness GR, Giles SA, Shukla MK (2020) J Phys Chem C 124:13837–13844
17. Szabo A, Ostlund NS (1996) Modern quantum chemistry: introduction to advan ed electronic structure theory. Dover Publications Inc, New York
18. Cramer C (2008) Essentials of computational chemistry: theories and models, vol 344, 2nd ed. Wiley, United Kingdom, pp 215–221
19. Larsen AH, Vanin M, Mortensen JJ, Thygesen KS, Jacobsen KW (2009) Phys Rev B 80:195112-1–195112-10
20. Raghavachari K, Trucks GW, Pople JA, Head-Gordon M (1989) Chem Phys Lett 157:479–483
21. Möller C, Plesset MS (1934) Phys Rev 46:618–622
22. Sholl DS, Steckel JA (2009) Density functional theory: a practical introduction. Wiley, Hoboken
23. Sherrill CDJ (2010) Chem Phys 132:110902
24. Becke ADJ (2014) Chem Phys 140:18A301
25. Parrish RM et al. (2017) J Chem Theory Comput 13:3185–3197
26. Enkovaara J et al (2010) J Phys Condens Matter 22:253202–253226
27. Frisch MJ et al (2009) Gaussian 09 Revision D.01. Gaussian Inc. Wallingford CT
28. Kresse G (1993) Hafner. J Phys Rev B 47:558–561
29. Kresse G (1994) Hafner. J Phys Rev B 49:14251–14269
30. Kresse G (1996) Furthmüller. J Comput Mater Sci 6:15–50
31. Kresse G (1996) Furthmüller. J Phys Rev B 54:11169–11186
32. Mortensen J, Hansen L, Jacobsen K (2005) Phys Rev B 71:035109
33. Jenness GR, Jordan KD (2009) J Phys Chem C 113:10242–10248
34. Jenness GR, Karalti O, Jordan KD (2010) Phys Chem Chem Phys 12:6375–6381
35. Jenness GR, Karalti O, Al-Saidi WA, Jordan KDJ (2011) Phys Chem A 115:5955–64
36. Monkhorst HJ, Pack JD (1976) Phys Rev B 13:5188–5192
37. Hammer B, Nørskov J (2000) Impact of surface science on catalysis, vol 45, pp 71–129
38. Kitchin J, Nørskov J, Barteau M (2004) Chen. J Phys Rev Lett 93:4–7
39. Tang W, Henkelman GJ (2009) Chem Phys 130:194504
40. Skoplyak O, Menning CA, Barteau MA, Chen JGJ (2007) Chem Phys 127:114707-1–114707-11
41. Miller SD, Kitchin JR (2009) Surf Sci 603:794–801
42. Ínoğlu N, Kitchin JR (2011) ACS Catal 1:399–407
43. Calle-Vallejo F, Martínez JI, Garía-Lastra JM, Sautet P, Loffreda D (2014) Angew Chemie - Int Ed 53:8316–8319
44. Xu Y, Greeley J, Mavrikakis M (2005) J Am Chem Soc 127:12823–12827
45. Acerbi N, Tsang SCE, Jones G, Golunski S, Collier P (2013) Angew Chem Int Ed Engl 52:7737–7741
46. Hughbanks T, Hoffmann R (1983) J Am Chem Soc 105:3528–3537
47. Mulliken RS (1955) J Chem Phys 23:1833
48. Mulliken RS (1950) J Am Chem Soc 72:4493–4503
49. Jenness GR, Schmidt JR (2013) ACS Catal 3:2881–2890
50. Hermes ED, Jenness GR, (2014) Schmidt J Mol Simul 41:123–133

51. Jenness GR, Wan W, Chen JG, Vlachos DG (2016) ACS Catal 6:7002–7009
52. Becke AD (1993) J Chem Phys 98:5648–5652
53. Adamo C, Barone V (1998) Chem Phys Lett 298:113–119
54. Becke AD (1988) Phys Rev A 38:3098–3100
55. Lee C, Yang W, Parr RG (1988) Phys Rev B 37:785–789
56. Perdew JP, Burke K, Ernzerhof M (1996) Phys Rev Lett. 77:3865–3868
57. Laurent AD, Jacquemin D (2013) Int J Quantum Chem 113:2019–2039
58. Stausholm-Møller J, Kristoffersen HH, Hinnemann B, Madsen GKH, Hammer B (2010) J Chem Phys 133:144708
59. Dudarev SL, Savrasov SY, Humphreys CJ, Sutton AP (1998) Phys Rev B 57:1505–1509
60. Cococcioni M, de Gironcoli S (2005) Phys Rev B 71:35105
61. Kulik HJ, Marzari N (2011) J Chem Phys 135:194105
62. Perdew JP (1991) In: Ziesche P, Eschrig H (eds) Electron struct solids. Akademie, Berlin
63. Stone AJ. (1996) The theory of intermolecular forces, 2nd ed. Oxford University, Oxford
64. Wang F-F, Jenness G, Al-Saidi Wa, Jordan KD (2010) J Chem Phys 132:134303
65. Grimme S (2004) J Comput Chem 25:1463–1473
66. Grimme S (2006) J Comput Chem 27:1787–1799
67. Grimme S, Antony J, Ehrlich S, Krieg H (2010) J Chem Phys 132:154104
68. Grimme S (2011) Wiley Interdiscip Rev Comput Mol Sci 1:211–228
69. Berland K, Cooper VR, Lee K, Schröder E, Thonhauser T, Hyldgaard P, Lundqvist BI (2015) Reports Prog Phys 78:066501
70. Langreth DC et al (2009) J Phys Condens Matter 21:84203
71. Román-Pérez G, Soler JM (2009) Phys Rev Lett 103:96102
72. Lee K, Murray ÉD, Kong L, Lundqvist BI, Langreth DC (2010) Phys Rev B 82:81101
73. Klimeš J, Bowler DR, Michaelides A (2010) J Phys Condens Matter 22:022201
74. Klimes J, Bowler DR, Michaelides A (2011) Phys Rev B - Condens Matter Mater Phys 83:1–13
75. Feynman RP (1939) Phys Rev 56:340–343
76. Fleisch DA (2020) A student's guide to the Schröodinger equation. Cambridge University, New York
77. Wyrick J, Einstein TL, Bartels L (2015) J Chem Phys 142:101907
78. Putz MV, Russo N, Sicilia E (2005) Theor Chem Acc 114:38–45
79. Wu Q, Van Voorhis T (2005) Phys Rev A 72:024502
80. Liu DC, Nocedal J (1989) Math Prog 45:503–528
81. Bitzek E, Koskinen P, Gähler F, Moseler M, Gumbsch P (2006) Phys Rev Lett 97:1–4
82. Sheppard D, Terrell R, Henkelman G (2008) J Chem Phys 128:134106
83. Heidari A, Raheb A (2020) J Mt Sci 17:1652–1669
84. Curtiss B, Adams JB, Ghiorso MS (1985) Geochim Cosmochim Acta 49:49–56
85. Naumov MV (2005) Geofluids 5:165–184
86. McHale JM (1997) Science (80-) 277:788–791
87. Lee C-k, Cho E, Lee H-s, Seol KS, Han S (2007) Phys Rev B 76:1–7
88. Nandi AK, Singh SK, Kunjir GM, Singh J, Mandal AK, Pandey RK (2013) Cent Eur J Energ Mater 10:113–122
89. Hirunsit P, Faungnawakij K, Luadthong C (2015) RSC Adv 5:11188–11197
90. Digne M, Sautet P, Raybaud P, Toulhoat H, Artacho E (2002) J Phys Chem B 106:5155–5162
91. Kwak JH, Mei D, Peden CHF, Rousseau R, Szanyi J (2011) Catal Lett 141:649–655
92. Jenness GR, Christiansen MA, Caratzoulas S, Vlachos DG, Gorte RJ (2014) J Phys Chem C 118:12899–12907
93. Metiu H, Chretien S, Hu Z, Li B, Sun X, Chrétien S, Hu Z, Li B, Sun X (2012) J Phys Chem C 116:10439–10450
94. Chretien S, Buratto S, Metiu H (2007) Curr Opin Solid State Mater Sci 11:62–75
95. Chrétien S, Metiu H (2007) J Chem Phys 126:104701
96. Hoffmann R (1988) Rev Mod Phys 60:601
97. Hoffmann R (1987) Angew Chemie Int Ed English 26:846–878

98. Hoffmann R (1988) Solids and surfaces: a Chemist's view of bonding in extended strutures. VCH Publishers, New York
99. Goedecker S (2004) J Chem Phys 120:9911–9917
100. Peterson Aa (2013) Top Catal 57:40–53
101. McQuarrie DA (1975) Statistical mechanics. In: Harper's chemistry series. Harper and Row, New York
102. Abadin H, Smith C, Ingerman L, Llados FT, Barber LE, Plewak D, Diamond GL (2012) Toxicological Profile for RDX. Tech rep January; US Dept of Health and Human Services, pp 1–215
103. Suliman RJ (2016) A Free Radiial Metabolite of Munition RDX (Hexahydro-1,3,5-trinitro-1,3,5-triazine) and 1-nitorso Degradation Product MNX Determined with Electron Paramagneti Resonance (EPR) and Study the Toxic Effects of Deepwater Horizon Crude Oil in Rats. Doctor of Pharmacy, University of Louisiana at Monroe.
104. Wani AH, Davis JL 2003 Influence of Ubiquitous electron acceptors on in situ anaerobi biotransformation of RDX in groundwater. Tech rep September. ERDC, New York
105. Naja G, Halasz A, Thiboutot S, Ampleman GUY (2008) Environ Sci Technol 42:4364–4370
106. Zhao J-s, Halasz A, Paquet L, Beaulieu C (2002) Hawari J 68:5336–5341
107. Ariyarathna T, Vlahos P, Smith RW, Fallis S, Groshens T, Tobias C (2017) Environ Toxicol Chem 36:1170–1180
108. Fuller ME, McClay K, Hawari J, Paquet L, Malone TE, Fox BG, Steffan RJ (2009) Appl Microbiol Biotechnol 84:535–544
109. Hawthorne SB, Lagade AJM, Kalderis D, Lilke AV, Miller DJ (2000) Environ Sci Technol. 34:3224–3228
110. Freysoldt C, Neugebauer J, de Walle CG, Van de Walle CG (2011) Phys Status Solidi 248:1067–1076
111. Komsa H-P, Pasquarello A (2013) Phys Rev Lett 110:095505
112. Stakheev A, Kustov L (1999) Appl Catal A Gen 188:3–35
113. Diebold U (2003) Surf Sci Rep 48:53–229
114. Deshlahra P, Iglesia E (2016) J Phys Chem C. acs.jpcc.6b04604
115. Hayek K, Fuchs M, Klotzer B, Reichl W, Rupprechter G (2000) Top Catal 13:55–66
116. Hong J, Marceau E, Khodakov AY, Gaberová L, Griboval-Constant A, Gi-rardon J-S, Fontaine CL, Briois V (2015) ACS Catal: 1273–1282
117. Stevenson SA, Lisitsyn A, Knoezinger H (1990) J Phys Chem 94:1576–1581
118. Kočí K, Obalová L, Lacný Z (2008) Chem Pap 62:1–9
119. Inoue H, Brankovi SR, Wang JX, Adži RR (2002) Electrochim Acta 47:3777–3785
120. Habisreutinger SN, Schmidt-Mende L, Stolarczyk JK (2013) Angew Chemie—Int Ed 52:7372–7408
121. Yan Z, Bukur DB, Goodman DW (2011) Catal Today 160:39–43
122. Goulas KA, Mironenko AV, Jenness GR, Mazal T, Vlachos DG (2019) Nat Catal 2:269–276
123. Jenness GR, Vlachos DG (2015) J Phys Chem C 119:5938–5945
124. O'Brien CP, Jenness GR, Dong H, Vlachos DG, Lee IC, O'Brien CP, Jenness GR, Dong H, Vlachos DG, Lee IC, (2016) J Catal 337:122–132
125. Salavati-fard T, Vasiliadou ES, Jenness GR, Lobo RF, Caratzoulas, S, Doren DJ (2019) ACS Catal 9:701–715
126. Nisar J, Århammar C, Jämstorp E, Ahuja R (2011) Phys Rev B 84:75120
127. Mooney PM, Watkins KP, Jiang Z, Basile aF, Lewis RB, Bahrami-Yekta V, Masnadi-Shirazi M, Beaton Da, Tiedje T (2013) J Appl Phys 113:133708
128. Lany S, Zunger A (2008) Phys Rev B 78:235104
129. Comes RB, Spurgeon SR, Heald SM, Kepaptsoglou DM, Jones L, Ong PV, Bowden ME, Ramasse QM, Sushko PV, Chambers SA (2016) Adv Mater Interfaces 3:n/a–n/a
130. Jacobs R, Booske J, Morgan D (2012) Phys Rev B 86:054106
131. Luo G, Yang S, Jenness GR, Song Z, Kuech TF, Morgan D (2017) NPG Asia Mater 9:e345
132. Mayeshiba T, Wu H, Angsten T, Kaczmarowski A, Song Z, Jenness G, Xie W, Morgan D (2017) Comput Mater Sci 126:90–102

133. Christiansen MA, Mpourmpakis G, Vlachos DG (2013) ACS Catal 3:1965–1975
134. Busca G (2014) Adv Catal, 1st ed, vol 57. Elsevier, Amsterdam, pp 319–404. Chapter 3
135. Wischert R, Copéeret C, Delbecq F, Sautet P (2011) Chem Commun (CAMB) 47:4890–4892
136. Digne M, Sautet P, Raybaud P, Euzen P, Toulhoat H (2004) J Catal 226:54–68
137. Phung TK, Lagazzo A, Rivero Crespo MÁ, Sánchez Escribano V, Busca G (2014) J Catal 311:102–113
138. Liu X, Truitt RE (1997) J Am Chem Soc 119:9856–9860
139. Larmier K, Chizallet C, Cadran N, Maury S, Abboud J, Lamic-Humblot, A.-F, Mar eau E, Lauron-Pernot H (2015) ACS Catal 5:4423–4437
140. Digne M, Sautet P, Raybaud P, Euzen P, Toulhoat H (2002) J Catal 211:1–5
141. Wischert R, Laurent P, Cop eret C, Delbe q F, Sautet P (2012) J Am Chem Soc 134:14430–14449
142. Krokidis X, Raybaud P, Gobichon AE, Rebours B, Euzen P, Toulhoat H (2001) J Phys Chem B 105:5121–5130
143. Phung TK, Proietti Hern andez L, Lagazzo A, Bus a G (2015) Appl Catal A Gen 493:77–89
144. Nortier P, Fourre P, Saad ABM, Saur O, Lavalley JC, Mohammed AB, Lavalley JC, Saad ABM, Saur O, Lavalley JC (1990) Appl Catal 61:141–160
145. Shabaker J (2003) J Catal 215:344–352
146. Morterra C, Magnacca G (1996) Catal Today 27:497–532
147. DeWilde JF, Chiang H, Hi kman DA, Ho CR, Bhan A (2013) ACS Catal 3:798–807
148. Roy S, Mpourmpakis G, Hong, D-Y, Vlachos DG, Bhan A, Gorte RJ (2012) ACS Catal 2:1846–1853
149. Wang S, Vorotnikov V, Vlachos DG (2015) ACS Catal 5:104–112
150. Nakagawa Y, Tamura M, Tomishige K (2013) ACS Catal 3:2655–2668
151. Blöhl PE (1994) Phys Rev B 50:17953–17979
152. Kresse G (1999) G Phys Rev B 59:1758–1775
153. Elster M, Hobza P, Frauenheim T, Suhai S, Kaxiras, E (2001) J Chem Phys 114:5149
154. Zhechkov L, Heine T, Patchkovskii S, Seifert G, Duarte HA (2005) J Chem Theory Comput 1:841–847
155. McNamara JP, Hillier IH (2007) Phys Chem Chem Phys 9:2362–2370
156. Grimme S (2012) Chem A Eur J 18:9955–9964
157. Brandenburg JG, Grimme S (2014) J Phys Chem Lett 5:1785–1789
158. Grimme S, Bannwarth C, Shushkov P (2017) J Chem Theory Comput 13:1989–2009
159. Elstner M, Porezag D, Jungni kel G, Elsner J, Haugk M, Frauenheim T, Suhai S, Seifert G (1998) G Phys Rev B 58:7260–7268
160. Elstner M, Jalkanen KJ, Knapp-Mohammady M, Frauenheim T, Suhai S (2001) Chem Phys 263:203–219
161. Gaus M, Chou CP, Witek H, Elstner M (2009) J Phys Chem A 113:11866–11881
162. Luschtinetz R, Frenzel J, Milek T, Seifert G (2009) J Phys Chem C 113:5730–5740
163. Gaus M, Cui Q, Elstner M (2011) J Chem Theory Comput 7:931–948
164. Gaus M, Goez A, Elstner M (2012) J Chem Theory Comput 9:338–354
165. Gaus M, Lu X, Elstner M, Cui Q (2014) J Chem Theory Comput 10:1518–1537
166. Heckel W, Elsner BAM, Schulz C, Müller S (2014) J Phys Chem C 118:10771–10779
167. Krishnapriyan A, Yang P, Niklasson AM, Cawkwell MJ (2017) J Chem Theory Comput 13:6191–6200
168. Lukose B, Supronowicz B, Petkov PS, Frenzel J, Ku AB, Seifert G, Vayssilov G, Heine T (2011) Phys Status Solidi B 249:335
169. Zheng G, Witek HA, Bobadova-Parvanova P, Irle S, Musaev DG, Prabhakar R, Morokuma K, Lundberg M, Elstner M, Kohler C, Frauenheim T (2007) J Chem Theory Comput 3:1349–1367
170. Hourahine B, Aradi B, Frauenheim T (2010) J Phys Conf Ser 242:012005(1–6)
171. Rincon L, Hasmy A, Marquez M, Gonzalez C (2011) Chem Phys Lett 503:171–175
172. Mäkinen V, Koskinen P, Häkkinen H (2013) Eur Phys J D 67:1–6
173. Vilhelmsen LB, Hammer B (2014) J Chem Phys 141:044711
174. Aktürk A, Sebet i A (2016) AIP Adv 6:055103

175. Alkan F, Aikens CM (2018) J Phys Chem C 122:23639–23650
176. Hellström M, Jorner K, Bryngelsson M, Huber SE, Kullgren J, Frauenheim T, Broqvist P (2013) J Phys Chem C 117:17004–17015
177. Cui Q, Elstner M, Kaxiras E, Frauenheim T, Karplus M (2001) J Phys Chem B 105:569–585
178. Pu J, Gao J, Truhlar DG (2004) J Phys Chem A 108:5454–5463
179. Geerke DP, Thiel S, Thiel W, van Gunsteren WF (2008) Phys Chem Chem Phys 10:297–302
180. Hou G, Zhu X, Cui QJ, Chem (2010) Theory Comput 6:2303–2314
181. Goyal P, Elstner M, Cui Q (2011) J Phys Chem B 115:6790–6805
182. Niehaus TA (2009) J Mol Struct THEOCHEM 914:38–49
183. Zobelli A, Ivanovskaya V, Wagner P, Suarez-Martinez I, Yaya A, Ewels CP (2012) Phys Status Solidi Basi Res 249:276–282
184. Vagánek A, Rimarčík J, Ilčin M, Škora P, Lukeš V, Klein E (2013) Comput Theor Chem 1014:60–67
185. Elstner M (2006) Theor Chem Acc 116:316–325
186. Aradi B, Hourahine B, Frauenheim T (2007) J Phys Chem A 111:5678–5684
187. Seifert G (2007) H Phys Chem A 111:5609–5613
188. Koskinen P, Mäkinen V (2009) Comput Mater Sci 47:237–253
189. Paxton AT (2009) Multiscale Simul Methods Mol Sci, vol 42, pp 145–176
190. Seifert G, Joswig, J-O (2012) WIREs Comput Mol Sci 2:456–465
191. Hourahine B et al (2020) J Chem Phys 152:124101
192. Porezag D, Frauenheim T, Köhler T, Seifert G, Kaschner R (1995) Phys Rev B 51:947–957
193. Kranz JJ, Kubillus M, Ramakrishnan R, Von Lilienfeld OA, Elstner M (2018) J Chem Theory Comput 14:2341–2352
194. Goldman N, Fried LE, Koziol LJ (2015) Chem Theory Comput, 11:4530–4535
195. Doemer M, Liberatore E, Knaup JM, Tavernelli I, Rothlisberger U (2013) Mol Phys 111:3595–3607
196. Jenness GR, Bresnahan CG, Shukla MK (2020) J Chem Theory Comp 16:6894–6903
197. Clough SA, Beers Y, Klein GP, Rothman LS (1973) J Chem Phys 59:2254–2259
198. Weininger D (1988) J Chem Inf Comput Sci 28:31–36
199. Weininger D, Weininger A, Weininger JL (1989) J Chem Inf Comput Sci 29:97–101
200. Weininger D (1990) J Chem Inf Comput Sci 30:237–243
201. Elton DC, Boukouvalas Z, Butrico MS, Fuge MD, Chung PW (2018) Sci Rep 8:1–12
202. Barnes BC, Elton DC, Boukouvalas Z, Taylor DCE, Mattson WD, Fuge MD, Chung PW (2018) ArXiv e-prints
203. Barnes BC (2020) AIP Conf Proc, 2272. DOI: 10.1063/12.0001089
204. Kang P, Liu Z, Abou-Rachid H, Guo H (2020) J Phys Chem A 124:5341–5351
205. Fortunato M, Coley C, Barnes B, Jensen K (2020) AIP Conf Proc 2272:070014
206. Chun S, Roy S, Nguyen YT, Choi JB, Udaykumar HS, Baek SS (2020) Sci Rep 10:1–15
207. Rosenblatt F (1958) Psychol Rev 65:386–408
208. Behler J, Parrinello M (2007) Phys Rev Lett 98:146401
209. Smith JS, Zubatyuk R, Nebgen B, Lubbers N, Barros K, Roitberg AE, Isayev O, Tretiak S (2020) Sci Data 7:1–10
210. Verma P, Wang Y, Ghosh S, He X, Truhlar DG (2019) J Phys Chem A 123:2966–2990
211. Yao K, Herr JE, Toth DW, Kintyre M, Parkhill R, (2018) J Chem Sci 9:2261–2269
212. Smith JS, Isayev O, Roitberg AE (2017) Chem Sci 8:3192–3203
213. Schütt KT, Arbabzadah F, Chmiela S, Müller KR, Tkatchenko A (2017) Nat Commun 8:6–13
214. Unke OT, Meuwly M (2019) J Chem Theory Comput 15:3678–3693
215. Lubbers N, Smith JS, Barros K (2018) J Chem Phys 148:241715
216. Sifain AE, Lubbers N, Nebgen BT, Smith JS, Lokhov AY, Isayev O, Roitberg AE, Barros K, Tretiak S (2018) J Phys Chem Lett 9:4495–4501
217. Nebgen B, Lubbers N, Smith JS, Sifain AE, Lokhov A, Isayev O, Roitberg AE, Barros K, Tretiak S (2018) J Chem Theory Comput 14:4687–4698
218. Bogojeski M, Vogt-Maranto L, Tuckerman ME, Müller KR, Burke K (2020) Nat Commun 11:5223

219. Smith JS, Roitberg AE, Isayev O (2018) ACS Med Chem Lett 9:1065–1069
220. Hirshfeld FL (1977) Theo Chim Acta 44:129–138
221. Reed AE, Weinhold F (1985) J Chem Phys 83:1736–1740
222. Reed AE, Weinstock RB, Weinhold F (1985) J Chem Phys 83:735–746
223. Singh UC, Kollman PAJ (1984) Comput Chem 5:129–145
224. Marenich AV, Jerome SV, Cramer CJ, Truhlar DG (2012) J Chem Theory Comput 8:527–541
225. Li L, Snyder JC, Pelaschier IM, Huang J, Niranjan UN, Duncan P, Rupp M, Muller KR, Burke K (2016) Int J Quantum Chem 116:819–833
226. Snyder JC, Rupp M, Hansen K, Muller, K-R, Burke K (2012) Phys Rev Lett 108:253002
227. Brockherde F, Vogt L, Li L, Tuckerman ME, Burke K, Müller K-R (2016) Nat Commun 8:872
228. Meyer R, Weichselbaum M, Hauser AW (2020) J Chem Theory Comput 16:5685–5694

Chapter 6
Application of Computational Approaches to Analysis of Multistep Chemical Reactions of Energetic Materials: Hydrolysis of Hexahydro-1,3,5-Trinitro-1,3,5-Triazine (RDX) and Octahydro-1,3,5,7-Tetranitro-1,3,5,7-Tetrazocine (HMX)

Liudmyla K. Sviatenko, Leonid Gorb, Danuta Leszczynska, Sergiy I. Okovytyy, Manoj K. Shukla, and Jerzy Leszczynski

Abstract Cyclic nitramines RDX (hexahydro-1,3,5-trinitro-1,3,5-triazine) and HMX (octahydro-1,3,5,7-tetranitro-1,3,5,7-tetrazocine) are energetic materials used in military application. They may contaminate soil and natural water by release to the environment during manufacturing, transportation, storage, and disposal. Among different methods for the cyclic nitramines remediation, an alkaline hydrolysis is one of the most promised techniques. Knowledge of detailed mechanism and kinetics of cyclic nitramines decomposition under alkaline conditions is helpful to improve the technique of safe and effective removal of RDX

L. K. Sviatenko
Department of General and Biological Chemistry N2, Donetsk National Medical University, Kropyvnytskyi, Ukraine

L. Gorb
Department of Molecular and Quantum Biophysics, Institute of Molecular Biology and Genetics, National Academy of Sciences of Ukraine, Kyiv, Ukraine

D. Leszczynska
Interdisciplinary Center for Nanotoxicity, Department of Civil and Environmental Engineering, Jackson State University, Jackson, MS, USA

S. I. Okovytyy
Department of Organic Chemistry, Oles Honchar Dnipro National University, Dnipro, Ukraine

M. K. Shukla
US Army Engineer Research and Development Center, Vicksburg, MS, USA

J. Leszczynski (✉)
Interdisciplinary Center for Nanotoxicity, Department of Chemistry and Biochemistry, Jackson State University, Jackson, MS, USA
e-mail: jerzy@icnanotox.org

J. Leszczynski, M. K. Shukla (eds.), *Practical Aspects of Computational Chemistry V*, https://doi.org/10.1007/978-3-030-83244-5_6

and HMX from contaminated groundwater and soil. Experimental studies confirm that the mechanism of hydrolysis is quite complex, multistep and influenced by pH, concentration, and temperature. Computational studies are necessary to understand the complete reaction mechanism of the hydrolysis and to predict its kinetics under different conditions. In this chapter, we discuss insights gained from more than a decade of computational investigations of energetic materials. We focus here on studies of the reaction mechanism for hydrolytic decomposition of RDX and HMX leading to stable products. In the context of these molecules, we also report on recent studies of the reaction kinetics.

6.1 Introduction

Cyclic nitramines RDX (hexahydro-1,3,5-trinitro-1,3,5-triazine) and HMX (octahydro-1,3,5,7-tetranitro-1,3,5,7-tetrazocine) are energetic materials with different applications. They may be released to the environment during manufacturing, transportation, storage, training, and disposal. Wastes from explosive manufacturing processes are classified as hazardous wastes by the US Environmental Protection Agency (EPA) [1]. The EPA has classified RDX as a possible human carcinogen, weight-of-evidence carcinogenic classification of C [2]. Therefore, there is a strong need for fundamental knowledge and technological implementation of the chemical processes that lead to chemical degradation of energetic nitrocompounds. The alkaline hydrolysis is one of the most promised methods for the cyclic nitramines remediation. The mechanism of hydrolysis is typically quite complex, multistep and influenced by relative stabilities of various conformers, pH, concentration, and temperature. Computational studies are indispensable to understand the complete reaction mechanism of the hydrolysis. In this chapter, we discuss the reaction mechanism for hydrolytic decomposition of RDX and HMX involving initial deprotonation and nitrite elimination, cycle cleavage, and further transformation of cycle-opened intermediate to the end products caused by a series of C-N bonds ruptures, hydroxide attachments, and proton transfers. We focus on studies of the reaction mechanism and kinetics using computational approach to the analysis of multistep chemical reactions [3–5]. This procedure was successfully applied to the prediction of kinetics of alkaline hydrolysis of such energetic materials as 2,4,6-trinitrotoluene, 2,4-dinitrotoluene and 2,4-dinitroanisole [6]. The applied protocol includes generation of a multistep Gibbs free-energy reaction profile appropriate for the transformations of the reagents to products using quantum-chemical approximation, followed by evaluation of the rate constants, construction and solution of the corresponding kinetic equations. Such a procedure allows one to significantly extend the number of steps through computational prediction.

6.1.1 Short Survey of Experimental Data on RDX Hydrolysis

RDX can migrate to groundwater easily since its water solubility is 0.04 g/L [7]. This compound is stable against hydrolysis in an aqueous solution at normal range of pH in surface water and groundwater systems [8], however, hydrolysis can occur under alkaline conditions [9–22]. Degradation of RDX in natural coastal seawaters, away from light, may proceed by alkaline hydrolysis [9, 10]. Alkaline hydrolysis was found to be one of the most effective treatment approaches to remediate RDX-contaminated soil, sediment, and water [11–13]. It was extensively studied at different conditions (temperature and pH level) [13–22]. Hydrated lime was effectively used for RDX transformation on military training ranges [14]. Kinetics of the alkaline hydrolysis of RDX was studied in aqueous solutions, in aqueous acetone, and soil slurries [13, 15–19]. The reaction of interaction between RDX and OH^- was shown to be a second-order reaction [13, 15–17]. Kinetics may be evaluated by pseudo first order rate constant and half-life time for RDX (Table 6.1). An increase in reaction temperature and pH of the contaminated water resulted in enhancing efficacy of the hydrolysis. The free energy of activation for RDX alkaline hydrolysis was determined to be 20.7 kcal/mol at 25 °C (Table 6.2).

The products of RDX hydrolysis at pH > 10 were nitrite, ammonia, nitrous oxide, formaldehyde, and formate [13, 15, 16, 22]. The production of these products depended on temperature, pH, and reaction time during the experiment.

Table 6.1 Pseudo first order rate constant (k) and half-life ($t_{1/2}$) for alkaline hydrolysis of RDX and HMX

pH	T (°C)	k × 10^{-3} (min^{-1})	$t_{1/2}$	Ref.
RDX				
11	25	0.8	17.9 h	[17]
11.5	25	1.7	7.5 h	[17]
12	25	2.3	5.3 h	[17]
12	25	2.7	4.4 h	[12]
12.2	25	7.9	1.5 h	[17]
12.5	25	8.3	1.4 h	[12]
12.6	25	22.3	0.6 h	[17]
13	25	27.7	0.5 h	[17]
13	25	26.8	0.4 h	[12]
13.3	25	52.3	0.2 h	[12]
11.18	50	9.3	75 min	[13]
11.32	50	13	53 min	[13]
12	50	58.2	12 min	[13]
12.3	50	127.2	5.5 min	[13]
HMX				
10	30	0.0017	407,647 min	[15]
10.34	50	0.09	7788 min	[13]
11.32	50	0.99	700 min	[13]
12.36	50	1.1	641 min	[13]

Table 6.2 Energy (E_a), changes in enthalpy (ΔH^*), entropy (ΔS^*), and Gibbs free energy (ΔG^*) of activation for alkaline hydrolysis of RDX and HMX

Compound	E_a, kcal/mol	ΔH^*, kcal/mol	ΔS^*, cal/mol × K	ΔG^*, kcal/mol
RDX	14.0 [20], 23.9 [13], 17.3 [21]	22.6 [16]	8.0 [16]	20.7 [16]
HMX	25.0 [20], 26.7 [13], 26.6 [21]	24.5 [22]	7.5 [22]	22.3 [22]

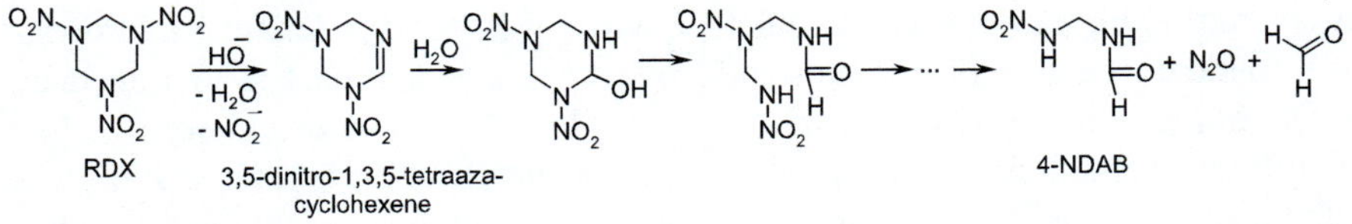

Scheme 6.1 Proposed degradation pathway for alkaline hydrolysis of RDX

At temperature in the range of 25–30 °C and pH 10 RDX transformation leads to formation of 0.98 equivalent of nitrite ion, 0.95 equivalent of 4-nitro-2,4-diazabutanal (4-NDAB), 0.94 equivalent of formaldehyde, and 0.77 equivalent of nitrous oxide [15, 22]. At higher pH and/or temperature 4-NDAB completely degrades producing formate and ammonia [13, 15]. Based upon these experimental investigations, the degradation mechanism begins from the initial deprotonation of RDX and nitrite ion elimination resulting in the formation of intermediate 3,5-dinitro-1,3,5-triazacyclohexene (Scheme 6.1). This intermediate may attach a water molecule to double C=N bond. A proton transfer between oxygen and nitrogen atoms leads to a ring-opened intermediate. Further pathways for its decomposition may include cleavage of C-N bonds, attachment of water molecule or hydroxide ion, that resulted in products such as nitrous oxide, formaldehyde, and 4-NDAB. Different intermediates on the degradation pathway were proposed [15]. Despite the detailed experimental data, the theoretical modeling was necessary to uncover all reaction steps that could have been missed during the experimental analysis and to predict and evaluate step-by-step chemical mechanism for alkaline hydrolysis of RDX.

6.1.2 *Short Survey of Experimental Data of HMX Hydrolysis*

Hydrolysis of HMX under alkaline conditions is exothermic. The heat of HMX base hydrolysis in aqueous sodium hydroxide, measured in calorimetry experiment, was found to be 1.5 kJ/g of HMX [23]. HMX alkaline hydrolysis required 22.3 kcal/mol of Gibbs free energy of activation (Table 6.2) [22]. The kinetics for base hydrolysis of HMX was extensively investigated at different temperatures (Table 6.1) [13, 20, 22–25]. Kinetics of HMX decomposition follows a second-order rate equation [13, 20, 22, 26]. The rate-limiting step for HMX hydrolysis is probably the initial E2 elimination with the formation of 3,5,7-trinitro-1,3,5,7-tetraazacyclooctene (Scheme 6.2), however this intermediate was not experimentally detected [22].

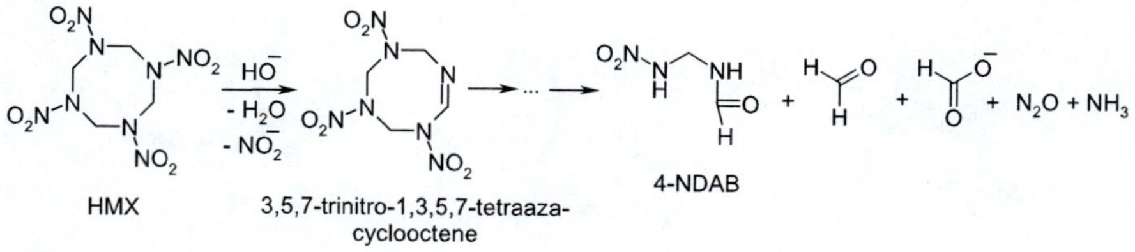

Scheme 6.2 Proposed degradation pathway for alkaline hydrolysis of HMX

The products of HMX hydrolysis at 30 °C and pH 10 were 1.82 equivalent of formaldehyde, 1.48 equivalent of nitrous oxide, 1.15 equivalent of nitrite ion, and 0.86 equivalent of 4-NDAB [15]. The presence of ammonia, and formate was not confirmed during the first 5% of the alkaline hydrolysis at pH 10. It should be noted, these compounds were formed during hydrolysis of HMX at pH 12 (Scheme 6.2) [15]. Similar products (formaldehyde, nitrite, 4-NDAB) were obtained during degradation of HMX in coastal waters [10]. The observed reaction products at 1.5 M sodium hydroxide and high temperature (105–155 °C) were nitrite, formate, nitrate, acetate, nitrogen, nitrous oxide, and ammonia [25]. The obtained results suggest that the mechanism of HMX alkaline hydrolysis consists of the initial nitrite elimination, followed by a ring cleavage and spontaneous formation of the final products (Scheme 6.2) [15]. The structure of some reactive species is difficult to determine experimentally and, therefore, computational study was necessary to understand the complete mechanism of HMX alkaline hydrolysis.

6.2 Computational Modeling of Hydrolysis of RDX

6.2.1 Conformational Analysis of RDX Structure

It is well known that RDX may exist in several conformational forms in gas phase [27–30]. These forms include AAA, AAE, EEA, twist, and boat forms. All these structures were adopted as the starting point to minimize the energy of the conformers in water solution using PCM(Pauling)/M06-2X/6-311++G(d,p) approach [31]. After the energy minimization was completed, the AAE conformer relaxed to the AAA form, and boat conformer converged into twist one. The calculated relative Gibbs free energies for the stable conformations of RDX are shown in Fig. 6.1. Based on the obtained results the most stable structure in water solution is AAA, which was also suggested to be the most likely conformer in gas phase. It is also the global minimum structure in β-solid RDX and in acetone, dimethyl sulfoxide, acetonitrile solutions [28, 29]. AAA and AAE conformers were also reported to contribute to the RR spectra of RDX in acetonitrile solution and in gas phase [27]. The calculated equilibrium constants for transformations AAA→twist and AAA→EEA are in the range of 10^{-3}–10^{-4}, that predicts negligible

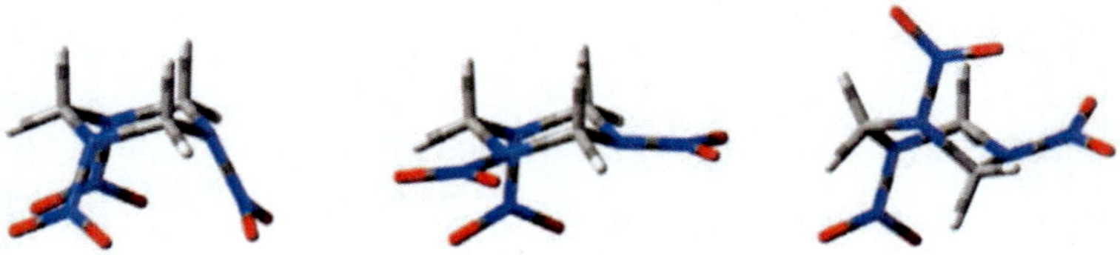

Fig. 6.1 Relative Gibbs free energies for stable conformers of RDX predicted using the PCM(Pauling)/M06-2X/6-311++G(d,p) level of theory (in kcal/mol). Reprinted from Ref. [31]. Copyright (2015), with permission from Elsevier

populations of twist and EEA conformers. For modeling the mechanism of alkaline hydrolysis the most stable conformer of RDX (AAA) was chosen.

6.2.2 Mechanism of RDX Alkaline Hydrolysis

The mechanism of hydrolytic decomposition of RDX under alkaline condition was theoretically investigated at PCM(Pauling)/M06-2X/6-311++G(d,p) level of theory using isolated hydroxide ion (OH^-) as well as hydrated OH^- with three water molecules ($OH^-(H_2O)_3$) to account for specific solvation [31, 32]. Significant difference between the results of both models was not found, except the increase of activation barrier for steps involving hydrated hydroxide ion. The scheme of reaction mechanism and corresponding Gibbs free energy diagram are shown in Fig. 6.2. The alkaline hydrolysis of RDX begins with nucleophilic attack of hydroxide ion onto methylene hydrogen leading to deprotonation and sequential nitrite elimination resulted in formation of C=N bond. Release of large amount of energy during this process provides evidence of its irreversibility and transformation of RDX to stable unsaturated intermediate INT1. Experimental studies tentatively determine the presence of INT1 in decomposition pathway of RDX during alkaline hydrolysis [15, 16]. Theoretically obtained free energy of activation for INT1 formation in case of model with hydrated hydroxide ion (21.2 kcal/mol) was found to be close to reported experimental one (20.7 kcal/mol) than for model with nonhydrated hydroxide (10.7 kcal/mol) [16]. This observation evidences requirement of taking into account of specific hydration of hydroxide ion to get more realistic activation energies and, correspondently, to create a more accurate kinetic model for the investigated process of alkaline hydrolysis.

Nucleophilic attack of hydroxide ion onto C2 carbon atom of double C=N bond of INT1 leads to formation of unstable negatively charged intermediate INT2. Further transformation of INT2 may occur by two possible pathways with small activation barriers (Fig. 6.2). The direct C2-N3 bond breaking and sequential iso-

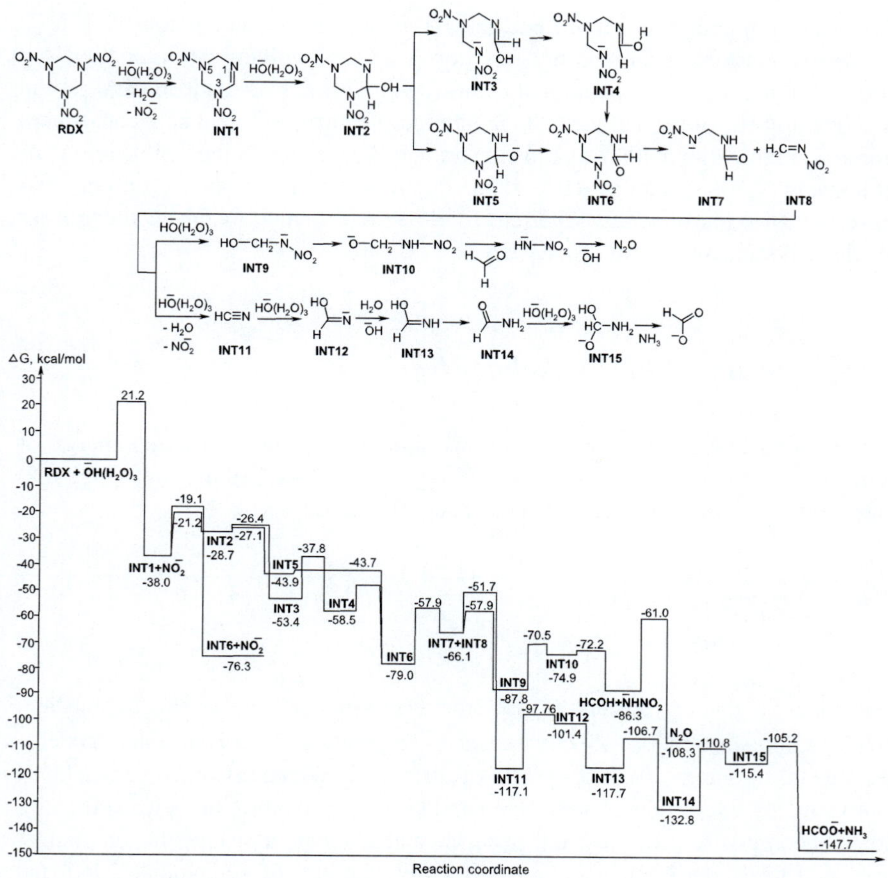

Fig. 6.2 The pathways of RDX alkaline hydrolysis, based on corresponding Gibbs free energy modeled using PCM(Pauling)/M06-2X/6-311++G(d,p) level of theory. Reproduced from Ref. [32] with permission from the Royal Society of Chemistry

merization leads to stable ring-opened intermediate INT4. Further tautomerization of INT4 leads to INT6. Another pathway of INT2 transformation begins with a proton transfer from oxygen atom to nitrogen N1 atom, facilitated by water molecule as a catalyst. Easy C2-N3 bond breaking in formed INT5 results in formation of stable ring-opened intermediate INT6. Further transformation of INT6 leads to negatively charged anion of 4-NDAB (INT7) and uncharged methylenenitroamine (INT8). 4-NDAB was experimentally observed as a stable intermediate during alkaline hydrolysis of nitramines [15]. Further transformation of INT8 occurs by hydroxide ion attachment to carbon atom of C=N double bond, a proton transfer between oxygen and nitrogen atoms and C-N bond rupture with release of formaldehyde. An elimination of hydroxide from nitramide anion leads to formation of nitrous oxide. The details of experimentally reported product distribution for alkaline hydrolysis of RDX corresponds well to the computationally predicted results [15, 32]. Additional formation of ammonium and formate may be considered as a side pathway of INT8 decomposition that occurs through reaction pathway involving deprotonation and nitrite elimination with formation of cyanic acid. Further hydroxide ion and proton attachment leads to formamide, which undergoes hydroxide ion attack, consequent ammonia elimination with formation of formate. Absorption in the visible region of simulated UV-Vis spectra of RDX, stable intermediates and products is absent that correspond to experimentally reported colorless reaction mixture during the process of RDX alkaline hydrolysis [16, 31, 33].

6.2.3 *Kinetics of RDX Alkaline Hydrolysis*

Kinetics of RDX alkaline hydrolysis was simulated using Gibbs free energy of activation calculated at PCM(Pauling)/M06-2X/6-311++G(d,p) level (Fig. 6.2) [32]. The rate constants were calculated according to Eq. (6.1):

$$k_{uni} = \frac{k \cdot T}{h} \cdot e^{-\frac{\Delta G_T^{\neq}}{RT}} \left(\mathrm{s}^{-1}\right) \quad k_{bi} = \frac{k \cdot T}{h} \cdot e^{-\frac{\Delta G_T^{\neq}}{RT}} \cdot \left(\frac{1}{c}\right) \left(\mathrm{L} \cdot \mathrm{mol}^{-1} \cdot \mathrm{s}^{-1}\right) \tag{6.1}$$

where h is the Planck constant, k—Boltzmann constant, R—universal gas constant, $\Delta G_T^{\neq}$—Gibbs free energy of activation, T—temperature, c—transformation coefficient equal to one mol/L. The system of differential equations (6.2) for reactants, intermediates, and products was solved and kinetics plots were modeled (Fig. 6.3). To fit theoretically predicted and experimental kinetics, a scaling factor of 0.94 was used to decrease values of all Gibbs free energies of activation. Calculated fitted value of rate constant (5.78×10^{-3} h^{-1}) was close to experimental one (7.21×10^{-3} h^{-1}) [15].

$$\sum_i \frac{dn_i}{dt} = \sum_{j(j \neq i)} k_{ji} n_j - n_i \sum_{l(l \neq i)} k_{il} \quad \sum_i \frac{dn_i}{dt} = \sum_{j(j,m \neq i)} k_{ji} n_j n_m - n_i \sum_{l(l,o \neq i)} k_{il} n_o \tag{6.2}$$

Comparative analysis of kinetic plots shows good agreement between theoretically modeled and experimental data. The deviation for nitrous oxide arises probably because of unaccounted side reactions. Concentrations of intermediates appearing during the reaction process are negligible because of their rapid decomposition. Absence of formate and ammonia on simulated kinetic curves confirms that the major pathway of INT8 transformation leads to nitrous oxide. Experimentally observed minor products (0.075 equivalent of ammonium and 0.07 equivalent of formate) may arise during degradation of 4-NDAB under base conditions [15]. Theoretically predicted period for 99% RDX decomposition by alkaline hydrolysis at 298 K and pH 10 is approximately 32 days. The 10 K increase in temperature

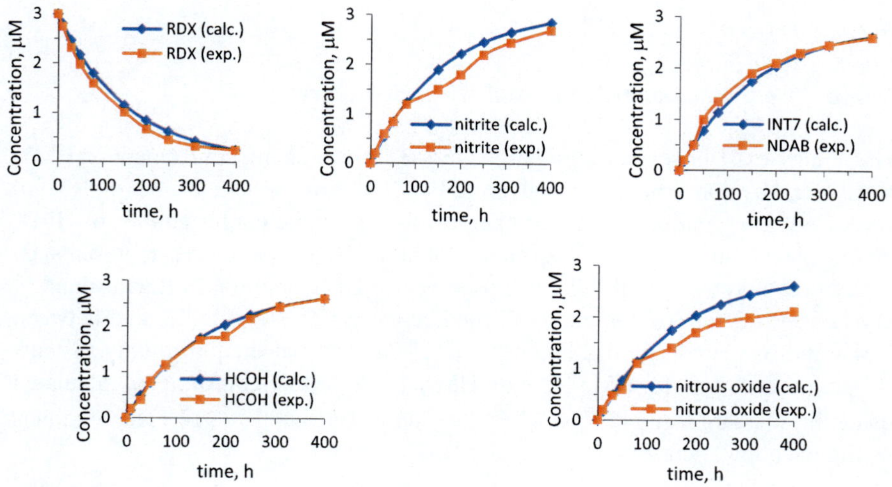

Fig. 6.3 Plots of concentration vs. time for RDX and products of alkaline hydrolysis calculated at PCM(Pauling)/M06-2X/6-311++G(d,p) level and experimental data. Reproduced from Ref. [32] with permission from the Royal Society of Chemistry

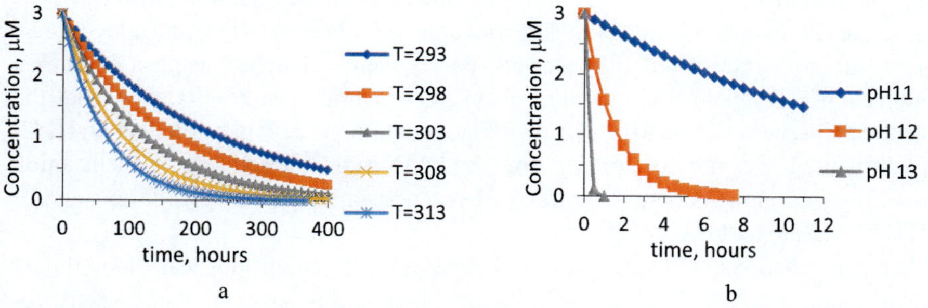

Fig. 6.4 Plots of concentration vs. time for RDX degradation during alkaline hydrolysis calculated at PCM(Pauling)/M06-2X/6-311++G(d,p) level at pH 10 and different temperatures (**a**) and at 298 K and different pH (**b**). Reproduced from Ref. [32] with permission from the Royal Society of Chemistry

leads to increase in reaction rate by 1.9-fold in temperature range of 293–313 K (Fig. 6.4a). Rate of RDX hydrolysis is significantly influenced by pH (Fig. 6.4b). Theoretically predicted half-life times for RDX are about 11 h, 54 min, and 6 min at pH 11, pH 12, and pH 13, respectively.

6.2.4 *Hydrolysis of HMX*

6.2.4.1 Conformational Analysis of HMX Structure

HMX may exist in several polymorphs [34–37]. Five stable conformers of HMX were found by conformational analysis at PCM(Pauling)/M06-2X/6-311++G(d,p) level [38]. The relative Gibbs free energies for the stable conformations of HMX are displayed in Fig. 6.5. The most stable HMX(I) conformer represents α-HMX polymorph structure [34, 39], observed in DMF solution [40] and found in crystalline structure of complexes formed between HMX and DMF, and between HMX and N-methyl-2-pyrrolidinone [35, 41]. The estimated equilibrium constants for transformations of HMX(I) into HMX(II)-HMX(V) conformers, calculated using the standard formula $K = e^{-\frac{\Delta G}{RT}}$, are in the range of 10^{-2}–10^{-3}, that predicts negligible populations of conformers HMX(II)-HMX(V).

6.2.4.2 Mechanism of HMX Alkaline Hydrolysis

Theoretical investigation of base and acid effects on the degradation of HMX in gas phase and in water solution was performed at B3LYP/6-311++G(d, p) level [42]. The solvation energies of the reaction species were calculated using a combined molecular force field and quantum chemistry method. The results show that the base can increase the rate of HMX degradation while the acid has no effect on HMX hydrolysis. The removing a proton from the HMX molecule and further nitrite anion release is the most energetically favored reaction direction for the initial stage of HMX alkaline hydrolysis.

The mechanism of HMX hydrolysis under alkaline conditions was theoretically investigated at PCM(Pauling)/M06-2X/6-311++G(d,p) level of theory and provided on Fig. 6.6 [38]. Only the most stable α-HMX conformer was considered for the study. The initial deprotonation on the first step of HMX alkaline hydrolysis occurs under action of hydroxide ion and leads to unstable carbanion INT1a (Fig. 6.6). The elimination of nitrite from INT1a results in formation of stable 3,5,7-trinitro-1,3,5,7-tetraazacyclooctene (INT1). These two sequential processes occur

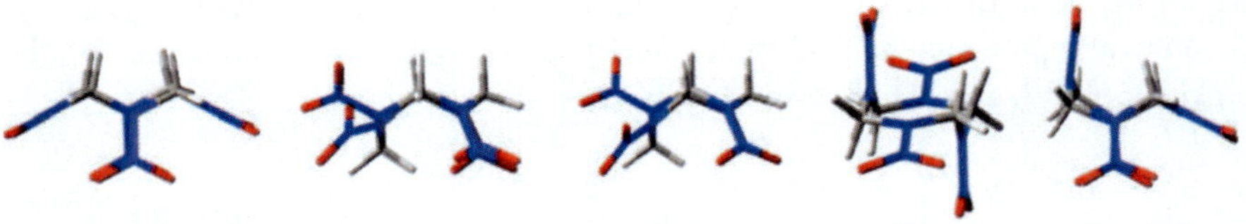

HMX(I)	HMX(II)	HMX(III)	HMX(IV)	HMX(V)
0.00	2.62	3.04	3.21	3.75

Fig. 6.5 Stable conformers of HMX and their relative Gibbs free energies predicted using PCM(Pauling)/M06-2X/6-311++G(d,p) level of theory (in kcal/mol). Reprinted with permission from Ref. [38]. Copyright (2016) American Chemical Society

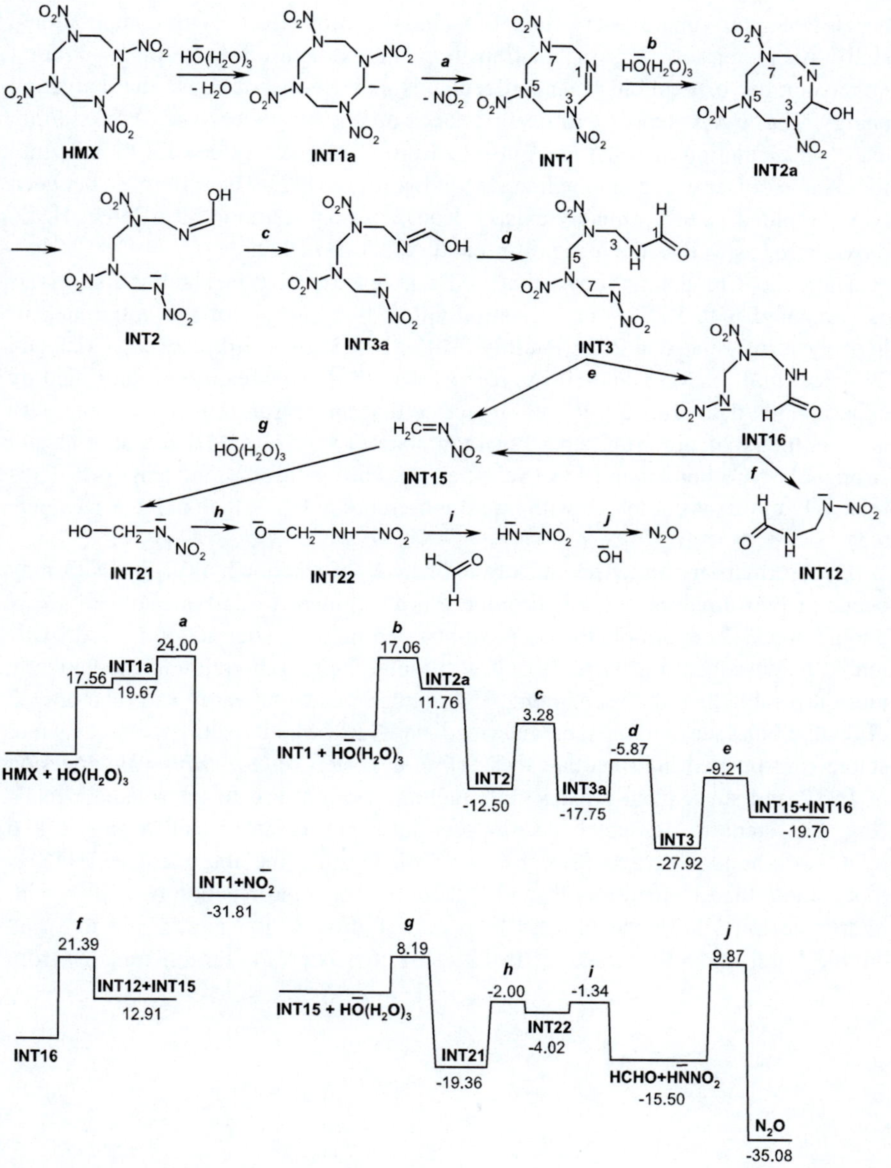

Fig. 6.6 The pathways of HMX alkaline hydrolysis, based on corresponding Gibbs free energy modeled using PCM(Pauling)/M06-2X/6-311++G(d,p) level of theory. Reprinted with permission from Ref. [38]. Copyright (2016) American Chemical Society

in one step with an activation energy of 24.0 kcal/mol (Fig. 6.6, step *a*). It should be noted that experimental Gibbs free energy of activation for the first step of HMX alkaline hydrolysis was determined to be 22.3 kcal/mol, that is 1.7 kcal/mol smaller than theoretically predicted one [22, 38]. The difference (~7%) presumably arises because of approximate description of specific solvation effects in the computational study. An increase of stability of transition state structure in case of involving a larger number of explicit water molecules is expected to decrease the activation energy. The initial step of hydrolysis is accompanied by release of 31.8 kcal/mol energy. Calculation at B3LYP/6-311++G(d,p) level gives value of 49.9 kcal/mol for change of free energy in the first step of hydrolysis [42]. The difference between two computed values probably exists because of consideration of different HMX polymorphs as well as the using different theoretical methods.

Theoretical modeling shows that HMX is less reactive in alkaline hydrolysis as compared with RDX. The activation Gibbs free energy for the initial step of hydrolysis, calculated at PCM(Pauling)/M06-2X/6-311++G(d,p) level, is 21.2 and 24.0 kcal/mol for RDX and HMX, respectively [38]. This feature is confirmed by experimental data [15, 22]. It was suggested that the reason of different reactivity is a geometry of nitramine and transition state (TS) [32]. HMX has shorter and stronger N-NO_2 bond than RDX (Scheme 6.3). The change of bond lengths in TS is relatively more (as compared with initial nitramine) in HMX than in RDX (Scheme 6.3).

Transformation of intermediate of the first step of alkaline hydrolysis INT1 may occur in two directions: (1) hydroxide ion attachment to carbon atom of C=N double bond, (2) deprotonation in positions 3, 5, and 7 under action of hydroxide ion with subsequent nitrite release. It was found the first direction is energetically more favorable than the second one. Hydroxide ion attachment to carbon atom C2 of double bond leads to unstable intermediate INT2a which readily transforms into stable ring-opened intermediate INT2 (Fig. 6.6, step *b*). *Cis-trans* isomerization of INT2 and subsequent amide-imide tautomerization lead to intermediate INT3 (Fig. 6.6, steps *c*, *d*), which decomposes into intermediates INT15 and INT16 by C7-N6 bond cleavage (Fig. 6.6, step *e*). Despite the intermediate INT3 is more stable than formed INT15 and INT16, further rapid reaction of INT15 with hydroxide ion allows the process to occur. Hydroxide ion attachment to C=N double bond of INT15 leads to INT21 (Fig. 6.6, step *g*). Proton transfer from

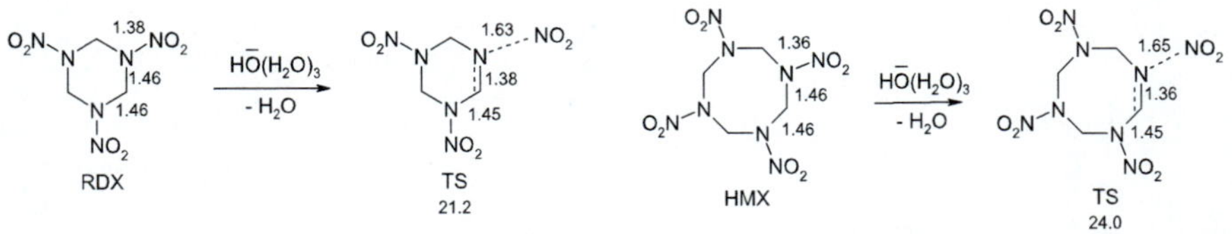

Analysis of N-N bond in RDX and HMX

Compound	Electron density, $e \cdot Å^{-3}$	Laplacian of electron density, $e \cdot Å^{-5}$	Energy density, au	Energy of homolytic bond cleavage, kcal/mol
RDX	0.3525	0.1581	-0.3370	55.81
HMX	0.3661	0.1708	-0.3607	57.08

Scheme 6.3 Comparison of RDX and HMX. Reproduced from Ref. [32] with permission from the Royal Society of Chemistry

oxygen atom to nitrogen atom and subsequent rapid C-N bond rupture resulted in release of formaldehyde and formation of nitramide ion, which degrades to nitrous oxide and hydroxide (Fig. 6.6, steps *h–j*). The most favorable pathway for INT16 decomposition is a C-N bond rupture with formation of intermediates INT12 and INT15 (Fig. 6.6, step *f*). INT12 represents experimentally observed 4-NDAB [15]. Calculated results suggest that alkaline hydrolysis of HMX represents a highly exothermic multistep process leading to products such as 4-NDAB, nitrite, formaldehyde, and nitrous oxide [38]. All these alkaline hydrolysis products are experimentally observed species [15]. The slowest step in the reaction channel is the release of nitrite.

6.2.4.3 Kinetics of HMX Alkaline Hydrolysis

Kinetics of HMX alkaline hydrolysis was simulated using Gibbs free energy of activation calculated at PCM(Pauling)/M06-2X/6-311++G(d,p) level (Fig. 6.6) and is shown in Fig. 6.7 along with experimental data [15, 38]. To fit experimentally observed and computationally predicted kinetics, a scaling factor of 0.93 was applied to decrease all activation energies.

Comparative analysis of kinetic plots shows good agreement between experimental and theoretically modeled data (Fig. 6.7). The biggest deviations are observed for formaldehyde and nitrous oxide. One might explain these differences due to unaccounted side processes. Good agreement between theoretical and experimental kinetics allows to predict the influence of temperature and pH on rate of HMX

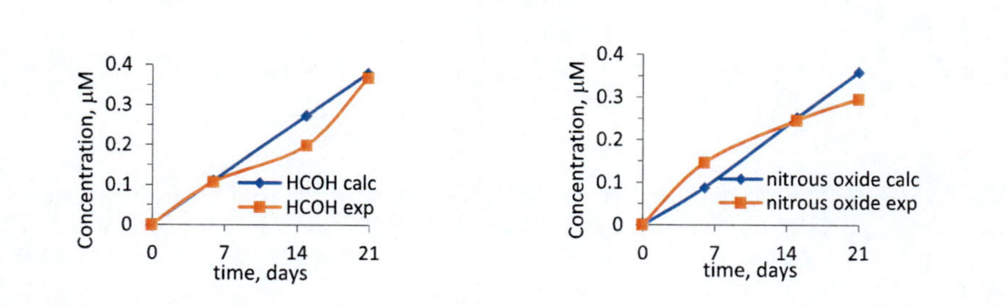

Fig. 6.7 Plots of concentration vs. time for HMX and products of alkaline hydrolysis calculated at PCM(Pauling)/M06-2X/6-311++G(d,p) level and experimental data. Reprinted with permission from Ref. [38]. Copyright (2016) American Chemical Society

degradation. Such relationships have not been measured experimentally and computational study provides important insight into these processes. The 10 K increase in temperature leads to increase in reaction rate by 2.1-fold in temperature range of 293–313 K (Fig. 6.8a). It should be noted that these data are obtained within the accuracy provided by computational methods included in the Gaussian 09 suit of programs. The PCM solvent model included in the Gaussian 09 does not incorporate prediction of a temperature dependence of Gibbs free energy of solvation. Half-life times for HMX transformation decreased from 3.8 years at pH 9.5 to 7.3 h at pH 13, respectively (Fig. 6.8b). A period for 99% HMX degradation is expected to be 30 years at pH 9.5 while 2 days at pH 13. Simulated kinetics of HMX decomposition along with products formation under alkaline conditions is presented in Fig. 6.9. Transformation of 1 mol of HMX leads to formation of ~1 mol of 4-NDAB and nitrite, and ~2 mol of formaldehyde and nitrous oxide at pH 10 and 12 (Fig. 6.9).

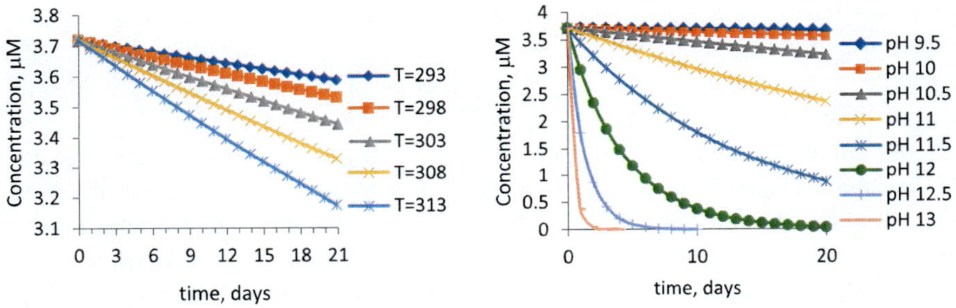

a b

Fig. 6.8 Plots of concentration vs. time for HMX degradation during alkaline hydrolysis calculated at PCM(Pauling)/M06-2X/6-311++G(d,p) level at pH 10 and different temperatures (**a**) and at 298 K and different pH (**b**). Reprinted with permission from Ref. [38]. Copyright (2016) American Chemical Society

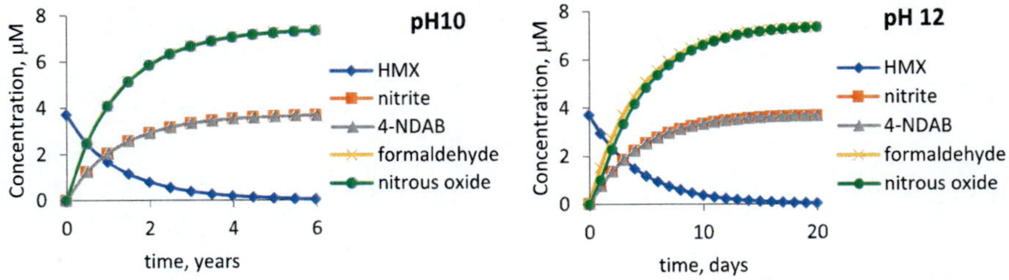

Fig. 6.9 Plots of concentration vs. time for HMX decomposition and products formation under alkaline hydrolysis of HMX calculated at PCM(Pauling)/M06-2X/6-311++G(d,p) level at 298 K and pH 10 (**a**) and pH 12 (**b**). Reprinted with permission from Ref. [38]. Copyright (2016) American Chemical Society

6.2.5 *Mechanism of 4-NDAB Decomposition Under Alkaline Conditions*

4-NDAB, as a stable intermediate product of base hydrolysis of HMX and RDX [15], is also formed during biotic degradation of nitramines [43, 44]. 4-NDAB has good solubility in water and stability in aqueous solutions at pH range close to neutral. Highly alkaline conditions may facilitate its transformation [15]. The mechanism of 4-NDAB decomposition under alkaline conditions was computationally studied at PCM(Pauling)/M06-2X/6-311++G(d,p) level and is shown in Fig. 6.10 [32].

There are two main directions for 4-NDAB transformation (Fig. 6.10). The first one starts from hydroxide ion attachment to carbon atom of carbonyl group and leads to unstable intermediate INT1. The last one easy transforms to stable INT2 by release of formate. Proton transfer in INT2 between nitrogen atoms facilitates rupture of C-N bond, which results in formation of INT3 and nitramide anion ($NHNO_2^-$). Further transformation of $NHNO_2^-$ leads to nitrous oxide. Hydroxide

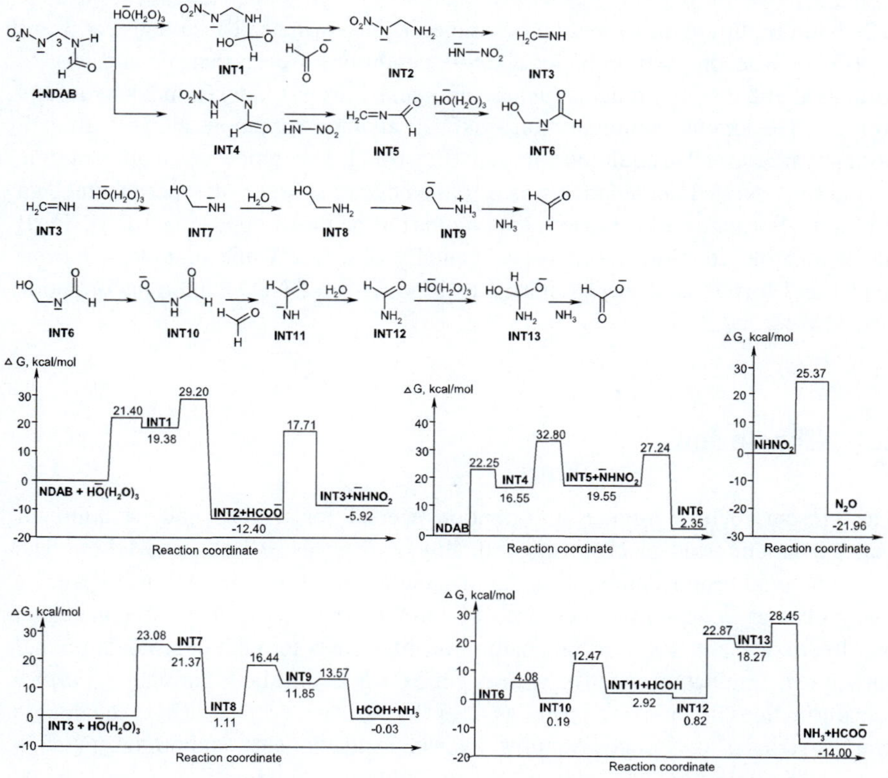

Fig. 6.10 Computer generated pathways of 4-NDAB alkaline hydrolysis, corresponding Gibbs free energy diagrams. Reproduced from Ref. [32] with permission from the Royal Society of Chemistry

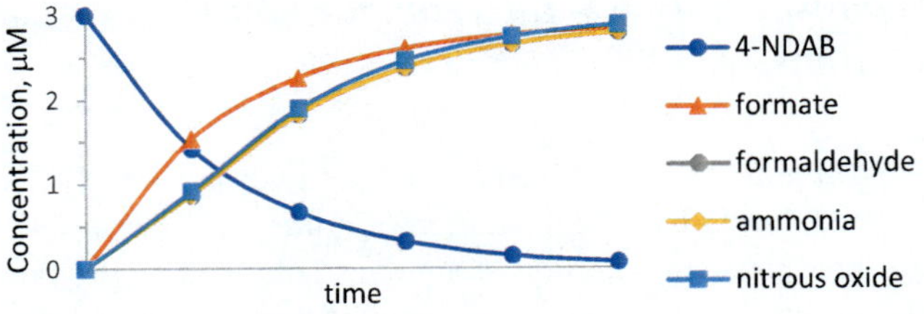

Fig. 6.11 Plots of concentration vs. time for alkaline hydrolysis of 4-NDAB calculated at PCM(Pauling)/M06-2X/6-311++G(d,p) level. Reproduced from Ref. [32] with permission from the Royal Society of Chemistry

attachment to INT3 and subsequent protonation leads to INT8. Proton transfer from oxygen atom to nitrogen atom promotes C-N bond rupture with formation of ammonia and formaldehyde. The second route for 4-NDAB transformation begins with proton transfer between nitrogen atoms and subsequent elimination of $NHNO_2^-$. Hydroxide attachment transforms intermediate INT5 into INT6. Proton transfer from oxygen atom to nitrogen atom leads to INT10, which decomposes with formaldehyde elimination. Further protonation and hydroxide attachment promote C-N bond rupture with formation of ammonia and formate. These both pathways of 4-NDAB decomposition under alkaline conditions are exothermic and lead to formation of the same products such as ammonia, nitrous oxide, formaldehyde, and formate. The kinetics simulation for 4-NDAB alkaline conditions suggests that all products are formed in equimolar amount (Fig. 6.11). Theoretical calculation of time period for 4-NDAB degradation was refrained due to absence of experimental data for fitting. Formation of formate and ammonia during the disappearance of 4-NDAB under alkaline conditions was experimentally observed while ammonia, nitrous oxide, and formic acid were reported as products of 4-NDAB biotransformation [15, 43, 44].

6.3 Conclusion

The present review provides information needed for a deep understanding of reaction kinetics and mechanism of alkaline hydrolysis of RDX and HMX. The computational results demonstrate that the studied reactions are multistep exothermic processes in which the presence of hydroxide ion is crucial due to conclusion that hydroxide initiates a decomposition of nitramines through deprotonation and subsequent elimination of nitrite with a cyclic C=N double bond formation. Further transformations include ring opening caused by hydroxide attachment to the double bond, a series of C-N bonds ruptures, hydroxide attachments, and proton transfers leading to stable products such as 4-nitro-2,4-diazabutanal, nitrite, nitrous oxide, and formaldehyde. Transformation of 4-nitro-2,4-diazabutanal in highly alkaline conditions leads to formation of ammonia, nitrous oxide, formaldehyde, and formate. The reaction mechanisms proposed based on the results of computational studies are confirmed by a good agreement of predicted stoichiometry of the products of the multistep process of RDX and HMX hydrolysis with experimental data. Simulated kinetics using computational approach to analysis of multistep chemical reactions allowed to predict the time for RDX and HMX hydrolytic decomposition at different pH and temperature. The estimated half-life for HMX and RDX degradation is approximately 1 year and 3.5 days, respectively, at pH 10 and temperature 298 K. The lower reactivity of HMX is due to a stronger N-NO_2 bond as compared with RDX. Hydrolysis rate increases by the orders of magnitude in case of pH 11–13. Reaction rate of alkaline hydrolysis increases by approximately twofold with increase in temperature by 10 K over temperature range of 293–313 K. Modeling of mechanism and kinetics provides an efficient way to study decomposition pathways of various energetic materials at different conditions.

Acknowledgments The use of trade, product, or firm names in this report is for descriptive purposes only and does not imply endorsement by the U.S. Government. The tests described and the resulting data presented herein, unless otherwise noted, were obtained from research conducted under the Environmental Quality Technology Program of the United States Army Corps of Engineers and the Environmental Security Technology Certification Program of the Department of Defense by the USAERDC. Permission was granted by the Chief of Engineers to publish this information. The findings of this report are not to be construed as an official Department of the Army position unless so designated by other authorized documents. This document has been approved for public release (Distribution Statement A). One of the coauthors (JL) acknowledge support from ARO grant # W911NF-20-1-0116, computation time by the Extreme Science and Engineering Discovery Environment (XSEDE) by National Science Foundation Grant Number OCI-1053575 and XSEDE award allocation Number DMR110088 and by the Mississippi Center for Supercomputer Research.

References

1. McLellan WL, Hartly WR, Brower ME (1988) PB90-273525. US Environmental Protection Agency, Office of Drinking Water, Washington, DC
2. United States Environmental Protection Agency (USEPA) (2014) Office of Solid Waste and Emergency Response (5106P). EPA 505-F-14-008
3. Sviatenko LK, Gorb L, Hovorun D, Leszczynski J (2012) J Phys Chem A 116:2333
4. Sviatenko LK, Gorb L, Hovorun D, Leszczynski J (2014) Biopolym Cell 30:239
5. Sviatenko LK, Gorb L, Hovorun D, Leszczynski J (2014) Chem Res Toxicol 27:981
6. Sviatenko L, Kinney C, Gorb L, Hill FC, Bednar AJ, Okovytyy S, Leszczynski J (2014) Environ Sci Technol 48:10465
7. Lynch JC, Myers KF, Brannon JM, Delfino JJ (2001) J Chem Eng Data 46:1549
8. Sikka HC, Bannerjee S, Pack EJ, Appleton HT (1980) Report TR 81-538. US Army Medical Research and Development Command, Frederick, MD
9. Hoffsommer JC, Rosen JM (1973) Bull Environ Contam Toxicol 10:78
10. Monteil-Rivera F, Paquet L, Giroux R, Hawari J (2008) J Environ Qual 37:858
11. Davis JL, Nestler CC, Felt DR, Larson SL (2007) Final report ERDC/EL TR-07-4. Environmental Laboratory US Army Engineer Research and Development Center, Vicksburg, MS

12. Gent DB, Johnson JL, Felt DR, O'Connor G, Larson SL (2010) Final report ERDC/EL TR-10-4. Environmental Laboratory US Army Engineer Research and Development Center, Vicksburg, MS
13. Heilmann HM, Wiesmann U, Stenstrom MK (1996) Environ Sci Technol 30:1468
14. Larson SL, Davis JL, Martin WA, Felt DR, Nestler CC (2008) ESTCP cost and performance report (ER-0216). Environmental Laboratory US Army Engineer Research and Development Center, Vicksburg, MS
15. Balakrishnan VK, Halasz A, Hawari J (2003) Environ Sci Technol 37:1838
16. Hoffsommer JC, Kubose DA, Glover DJ (1977) J Phys Chem 81:380
17. Hwang S, Felt DR, Bouwer EJ, Brooks MC, Larson SL, Davis JL (2006) J Environ Eng 132:256
18. Epstein S, Winkler CA (1952) Can J Chem 30:734
19. Brooks MC, Davis JL, Larson SL (2003) Final report ERDC/ELTR-03-19. Environmental Laboratory US Army Engineer Research and Development Center, Vicksburg, MS
20. Epstein S, Winkler CA (1951) Can J Chem 29:731
21. Makarov A, LoBrutto R, Christodoulatos C, Jerkovich A (2009) J Hazard Mater 162:1034
22. Croce M, Okamoto Y (1979) J Org Chem 44:2100
23. Bishop RL, Harradine DM, Flesner RLT, Larson SA, Bell DA (2000) Ind Eng Chem Res 39:1215
24. Bishop RL, Flesner RL, Dell'Orco PC, Spontarelli T, Larson SA, Bell DA (1998) Ind Eng Chem Res 37:4551
25. Bishop RL, Flesner RL, Dell'Orco PC, Spontarelli T, Larson SA, Bell DA (1999) Ind Eng Chem Res 38:2254
26. Heilmann HM, Stenstrom MK, Hesselmann RPX, Wiesmann U (1994) Water Sci Technol 30:53
27. Al-Saidi WA, Asher SA, Norman P (2012) J Phys Chem A 116:7862
28. Karpowicz RJ, Brill TB (1984) J Phys Chem 88:348
29. Rice BM, Chabalowski CF (1997) J Phys Chem A 101:8720
30. Molt Jr RW, Watson Jr T, Lotrich VF, Bartlett RJ (2011) J Phys Chem A 115:884
31. Sviatenko LK, Gorb L, Hill FC, Leszczynska D, Okovytyy SI, Leszczynski J (2015) Chemosphere 134:31
32. Sviatenko LK, Gorb L, Leszczynska D, Okovytyy SI, Shukla MK, Leszczynski J (2017) Environ Sci Processes Impacts 19:388
33. Hwang S, Min K, Davis JL (2004) Comparative assessment of nucleophilic alkaline hydrolysis for remediation of high explosives-contaminated groundwater. Environ Eng Res 9:13–22
34. Cady HH, Larson AC, Cromer DT (1963) The crystal structure of α-HMX and a refinement of the structure of β-HMX. Acta Cryst *16*:617–623
35. Cobbledick RE, Small RWH (1975) Acta Cryst B31:2805
36. Choi CS, Boutin HP (1970) Acta Cryst B26:1235
37. Konek CT, Mason BP, Hooper JP, Stoltz CA, Wilkinson J (2010) Chem Phys Lett 489:48
38. Sviatenko LK, Gorb L, Hill FK, Leszczynska D, Shukla MK, Okovytyy SI, Hovorun D, Leszczynski J (2017) Environ Sci Technol 50:10039
39. Main P, Cobbledick RE, Small RWH (1985) Acta Cryst C41:1351
40. Lenchitz C, Velicky R (1964) Tech memo 1343. Picatinny Arsenal, Dover, NJ
41. Haller TM, Rheingold AL, Brill TB (1985) Acta Cryst C41:963
42. Zhang C, Li Y, Xiong Y, Wang X, Zhou M (2011) J Phys Chem A 115:11971
43. Fournier D, Halasz A, Spain J, Spanggord RJ, Bottaro JC, Hawari J (2004) Appl Environ Microbiol 70:1123
44. Bhushan B, Paquet L, Halasz A, Spain JC, Hawari J (2003) Biochem Biophys Res Commun 306:509

Chapter 7
Dataset Modelability by QSAR: Continuous Response Variable

Alexander Golbraikh, Rong Wang, Vinicius M. Alves, Inta Liepina, Eugene Muratov, and Alexander Tropsha

Abstract Recently, we have introduced fast-calculated and reliable statistical criteria to estimate whether a predictive QSAR model can be built for a given chemical dataset. These modelability criteria were successfully applied to more than 100 datasets with a binary response variable. In this study, we have extended the modelability approach to datasets with a continuous response variable. The QSARome datasets with built *k* Nearest Neighbors (kNN) and Random Forest (RF) QSAR models for different IC_{50} values were used as the training set. Two modeling criteria were proposed based on our preliminary results: MODI_ssR^2 and MODI_q^2. Both criteria were based on the kNN approach without variable selection and optimization of the number of nearest neighbors. MODI_ssR^2 was based on leave-group-out (20%-out or fivefold) external cross-validation. MODI_q^2 values were calculated as q^2 values for the entire dataset (using leave-one-out cross-validation). Modeling criteria were calculated for 1, 3 and 5 nearest neighbors. QSAR models were built using 34 datasets with different methodologies and molecular descriptors. The highest coefficients of determination were found between QSAR_R^2 and the MODI_ssR^2 criterion calculated for $k = 5$ nearest neighbors: 0.91, 0.90, and 0.90, respectively, for training, test, and both training and test sets together (entire dataset). Four additional external datasets were employed to validate our approach. Our results demonstrate that the modelability criteria can be used as a good estimate of the predictive power of QSAR models even prior to model building. They establish the lower bound of the predictive power of QSAR models in terms of QSAR_R^2. Regression equations between QSAR_R^2 and modelability

A. Golbraikh · V. M. Alves · E. Muratov · A. Tropsha (✉)
UNC Eshelman School of Pharmacy, University of North Carolina at Chapel Hill, Chapel Hill, NC, USA
e-mail: alex_tropsha@unc.edu

R. Wang
School of Pharmacy, Guangdong Pharmaceutical University, Guangzhou, People's Republic of China

I. Liepina
Latvian Institute of Organic Synthesis, Rīga, Latvia

J. Leszczynski, M. K. Shukla (eds.), *Practical Aspects of Computational Chemistry V*, https://doi.org/10.1007/978-3-030-83244-5_7

criteria can also be used for estimation of the predictive power of QSAR models for datasets not modeled yet. The use of modelability criteria can eliminate the futile attempts to build QSAR models for "tough" datasets.

Abbreviations

BCF	Bioaccumulation factor
CCR	Correct classification rate (mean of sensitivity and specificity for a QSAR model built for a dataset with binary response variable)
CDK	Chemical development kit
CV	Cross-validation
GPCR	G protein-coupled receptors
HC	Hierarchical clustering (one of the machine learning methods used in QSAR)
HDAC	Histone deacetylase inhibitors
HiT QSAR	Hierarchical technology QSAR (software)
IC_{50}	Inhibitor concentration reducing activity by 50%
IGC_{50}	50% inhibition growth concentration
K_i	Inhibition constant
kNN	k nearest neighbors (one of the machine learning methods used in QSAR studies)
kNN_R^2	Coefficient of determination (square of the correlation coefficient) between predicted and observed activities of a QSAR model built using kNN
LGO-CV	Leave-group-out cross-validation
LOO-CV	Leave-one-out cross-validation
MODI	Prefix for modelability criteria
MODI_CCR	Modelability criterion based on CCR for the entire dataset
MODI_q^2	Modelability criterion based on q^2
MODI_ssCCR	Modelability criterion based on similarity search, leave 20%-out external cross-validation and CCR for union of external test sets
MODI_ssR^2	Modelability criterion based on similarity search, leave 20%-out external cross-validation and R^2 for union of external test sets
MRTD dataset	Maximum recommended therapeutic dose dataset
QSAR	Quantitative structure—activity relationships analysis
QSAR_R^2	Coefficient of determination (square of correlation coefficient) between predicted and observed activities of a QSAR model
RF	Random forest (one of the machine learning methods used in QSAR studies)
RF_R^2	Coefficient of determination (square of correlation coefficient) between predicted and observed activities of a QSAR model built using RF
RI	Regulatory information
SiRMS descriptors	Simplex representation of molecular structure descriptors
ss	Similarity search
SVM	Support vector machines (one of the machine learning methods used in QSAR studies)
USEPA	US Environmental Protection Agency

7.1 Introduction

Recently, we have introduced various modelability criteria for QSAR studies of datasets with binary response variables [1, 2]. The modelability index (MODI) showed to be a reliable, fast, and easy to use approach to determine if a particular dataset is modelable or not. This approach was validated and compared to QSAR models developed using *k* Nearest Neighbor (kNN) algorithm. We have shown a high correlation between the predictive power of QSAR models calculated as CCR and two modelability criteria. MODI has proved to be helpful for the dataset assessment before the actual modeling. Usually, unnecessary time and resources are spent to develop "sophisticated tricks" to build predictive QSAR models for a challenging dataset. Using MODI, one can understand that additional QSAR modeling might not be necessary. Finally, it was also demonstrated that the results of these calculations only moderately depend on the type of chemical descriptors used for the calculation of modelability criteria [1].

In this study, we extended our approach to datasets with continuous response variables. Recently, we published encouraging preliminary results [2]. We considered various modelability criteria and selected the MODI_q^2 (kNN_q^2 value in the entire descriptor space, i.e., without variable selection) and MODI_ssR^2 (kNN_R^2 with 20% leave out cross-validation in the entire descriptor space) as the best ones. We made calculations for 1, 3, and 5 nearest neighbors. In the preliminary studies [2], we considered 14 datasets as the training set and just four as the test set. QSAR models for all these datasets were built prior to that study, and some of these results were published [2–4]. Both coefficients of determination for 14 datasets between kNN_R^2 and MODI_q^2 for 3 and 5 nearest neighbors were as high as 0.90, and between kNN_R^2 and MODI_ssR^2 for 3 and 5 nearest neighbors they were 0.85 and 0.88, respectively. In this report, we used 34 datasets [5] with continuous responses as the training set and 14 datasets from the Regulatory Information (RI) [2] as the test set. For the training set, QSAR models were built using RF and kNN methodologies.

The results demonstrated that the above modelability criteria could be successfully used for the estimation of the predictive power of QSAR models prior to modeling. The highest predictive power was obtained for $k = 5$ nearest neighbors. The highest coefficients of determination R^2 between QSAR_R^2 values and the MODI_ssR^2 criterion were 0.91, 0.90, and 0.90 for training, test, and both training and test sets (i.e., the entire dataset), respectively. This MODI_ssR^2 modelability criterion was calculated for five nearest neighbors, QSAR models for the training sets were built with kNN using Dragon 5.5 descriptors [6], and for the test sets using different methods and descriptors (see below). Additionally, we validated our modelability criteria using four datasets from [2]. We demonstrated that the modelability criteria for a different number of nearest neighbors should be used for different datasets. Modelability criteria can be used as the estimation of the lower bound of the predictive power of QSAR models in terms of QSAR_R^2. Regression equations between kNN_R^2 and modelability criteria seemed to be slightly better than those between RF_R^2 and modelability criteria, especially for 1 nearest neighbor.

7.2 Methods

7.2.1 Datasets

7.2.1.1 Training Set Data

We collected data for 34 well-studied G protein-coupled receptor (GPCR) targets [5]. The data were collected from ChEMBL [7, 8] and the Psychoactive Drug Screening Program database [9]. The datasets were curated following a protocol previously described by our group [10]. Here the following targets were selected: ten serotonin receptors (5-HT_{1A}, 5-HT_{1B}, 5-HT_{1D}, 5-HT_{1E}, 5-HT_{2A}, 5-HT_{2C}, 5-HT_3, 5-HT_5, 5-HT_6, and 5-HT_7), seven adrenoceptors (α_{1A}, α_{1B}, α_{2A}, α_{2B}, α_{2C}, β_1, and β_2), five dopamine receptors (D_1, D_2, D_3, D_4, and D_5), five muscarinic receptors (M_1, M_2, M_3, M_4, and M_5), four histamine receptors (H_1, H_2, H_3, and H_4), and three neurotransmitter transporters [serotonin (SERT), norepinephrine (NET), and dopamine (DAT)]. The datasets included compounds with k_i values (in mol/L) tested on the human tissues, which were transformed to logarithmic value before modeling. QSAR models were built by RF and kNN methods along with Dragon 5.5 [6] descriptors (see Tables 7.1 and 7.2).

7.2.1.2 Test Set Data (14 Datasets)

Regulatory Information Datasets

Ten Regulatory Information datasets with activity data were retrieved from multiple sources (see Table 7.3). For the estimation of both methods of modelability index, Dragon 5.5 [6] descriptors were employed. Same descriptors were used for the development of QSAR models along with RF [11] algorithm.

P. promelas LC_{50} Dataset

This dataset containing 809 compounds was compiled from the ECOTOX database [12, 13] where the toxicity is expressed as the concentration (in mol/L), which kills 50% of fathead minnow individuals after 96 h. Dragon 5.5 descriptors were used in calculations of modelability criteria. Multiple QSAR models were built using HC [14], Dragon 5.5 [6], SiRMS [15], and CDK [16] descriptors with HC [14], kNN [17], and RF [18] methods. Two RF implementations were used: available in R [19] and developed by Polishchuk et al. [11]. The best model was built using the RF method and RF (as implemented in R statistical package [20]) with Dragon 5.5 descriptors ($R^2 = 0.69$ for the external validation set) and was used in this study.

Table 7.1 Random forest (RF) models for 34 QSARome datasets

Dataset	NComp	R^2	k-O-P	k-P-O	R_0^2-O-P	R_0^2-P-O	$R_{45}^2(q^2$-O-P)	$R_{45}^2(q^2$-P-O)	STDEV sqrt[$(P-O)^2$/Ncomp]	MAE |P-O|/Ncomp
5-HT1A	1274	0.67	0.99	1	0.41	0.67	0.4	0.67	0.76	0.58
5-HT1B	508	0.68	0.99	1	0.46	0.67	0.45	0.67	0.82	0.62
5-HT1D	504	0.71	0.99	1	0.52	0.71	0.52	0.71	0.83	0.62
5-HT1E	108	0.34	0.99	1	−0.4	0.33	−0.41	0.32	0.57	0.37
5-HT2A	1035	0.72	0.99	1	0.51	0.71	0.5	0.71	0.75	0.56
5-HT2C	892	0.66	0.99	1	0.38	0.65	0.37	0.65	0.72	0.55
5-HT3	153	0.48	0.98	1	−0.03	0.48	−0.05	0.48	0.83	0.5
5-HT5	181	0.66	0.99	1	0.46	0.66	0.45	0.65	0.67	0.47
5-HT6	1107	0.71	1	1	0.54	0.71	0.54	0.71	0.68	0.52
5-HT7	464	0.6	0.99	0.99	0.3	0.6	0.3	0.6	0.85	0.63
alpha1A	495	0.64	0.98	1	0.46	0.64	0.45	0.64	0.9	0.65
alpha1B	420	0.68	0.99	1	0.54	0.68	0.53	0.68	0.74	0.51
alpha2A	412	0.69	0.99	1	0.46	0.68	0.45	0.68	0.73	0.53
alpha2B	283	0.51	0.99	1	−0.12	0.51	−0.13	0.51	0.75	0.54
alpha2C	332	0.68	0.99	1	0.52	0.68	0.51	0.68	0.76	0.58
beta1	194	0.61	0.99	1	0.2	0.6	0.2	0.6	0.82	0.63
beta2	239	0.61	0.99	0.99	0.18	0.6	0.17	0.6	0.99	0.79
D1	485	0.68	1	0.99	0.46	0.67	0.46	0.67	0.7	0.5
D2	2060	0.60	0.99	1	0.02	0.58	0	0.58	0.72	0.53
D3	1496	0.64	0.99	1	0.31	0.63	0.31	0.63	0.79	0.61
D4	749	0.61	0.99	1	0.29	0.61	0.28	0.61	0.78	0.59
D5	250	0.57	0.98	1	0.16	0.57	0.15	0.57	0.85	0.61
DAT	876	0.76	0.99	1.00	0.62	0.75	0.62	0.75	0.62	0.45
H1	396	0.61	0.98	1.01	0.26	0.61	0.24	0.61	0.84	0.61

(continued)

Table 7.1 (continued)

Dataset	NComp	R^2	k-O-P	k-P-O	R_0^2-O-P	R_0^2-P-O	$R_{45}^2(q^2$-O-P)	$R_{45}^2(q^2$-P-O)	STDEV sqrt[$(P\text{-}O)^2$/Ncomp]	MAE |P-O|/Ncomp
H2	93	0.07	0.99	0.99	−1.88	−0.05	−1.89	−0.06	0.71	0.39
H3	1385	0.65	1.00	1.00	0.33	0.65	0.33	0.65	0.67	0.49
H4	444	0.56	0.99	0.99	0.06	0.55	0.06	0.55	0.73	0.55
M1	384	0.52	0.97	1.00	0.02	0.52	−0.01	0.52	1.00	0.71
M2	431	0.62	0.98	1.00	0.30	0.62	0.29	0.62	0.91	0.67
M3	424	0.66	0.99	1.00	0.43	0.66	0.42	0.66	0.89	0.65
M4	234	0.53	0.97	1.01	0.10	0.53	0.07	0.53	0.96	0.66
M5	253	0.50	0.98	1.00	−0.08	0.50	−0.11	0.50	0.86	0.57
NET	970	0.74	0.99	1.00	0.61	0.74	0.60	0.74	0.67	0.50
SRT	1262	0.80	0.99	1.00	0.70	0.79	0.69	0.79	0.73	0.54

NComp are counts of compounds in the datasets
R^2 are the coefficients of determination for the corresponding models
k-O-P are the slopes for trend lines through the origin for observed vs. predicted activities
k-P-O are the slopes for trend line through the origin for predicted vs. observed activities
R_0^2 (O-P) are the coefficients of determination for trend lines through the origin for observed vs. predicted activities
R_0^2 (P-O) are the coefficients of determination for trend lines through the origin for predicted vs. observed activities
R_{45}^2 (q^2-O-P) are coefficients of determination for bisector lines for observed vs. predicted activities
R_{45}^2 (q^2-P-O) are coefficients of determination for bisector lines for predicted vs. observed activities
STDEV are standard deviations between predicted and observed activities
MAE are mean absolute errors between predicted and observed activities

Table 7.2 kNN models for 34 QSARome datasets

Dataset	NComp	R^2	k-O-P	k-P-O	R_0^2-O-P	R_0^2-P-O	$R_{45}^2(q^2$-O-P)	$R_{45}^2(q^2$-P-O)	STDEV sqrt[$(P-O)^2$/Ncomp]	MAE |P-O|/Ncomp
5-HT1A	1274	0.69	1	0.99	0.49	0.69	0.68	0.49	0.73	0.55
5-HT1B	508	0.65	0.99	0.99	0.45	0.65	0.65	0.44	0.84	0.64
5-HT1D	504	0.72	1	0.99	0.53	0.71	0.71	0.53	0.82	0.62
5-HT1E	108	0.39	0.99	1	−0.25	0.38	0.38	−0.27	0.5	0.33
5-HT2A	1035	0.71	1	0.99	0.54	0.71	0.7	0.54	0.75	0.56
5-HT2C	892	0.65	1	0.99	0.44	0.65	0.64	0.44	0.72	0.55
5-HT3	153	0.42	0.99	0.99	0.07	0.4	0.4	0.06	0.86	0.51
5-HT5	181	0.69	1	0.99	0.55	0.69	0.69	0.55	0.62	0.44
5-HT6	1107	0.71	1	0.99	0.5	0.71	0.7	0.5	0.69	0.52
5-HT7	464	0.68	1	0.99	0.5	0.68	0.68	0.45	0.7	0.58
alpha1A	495	0.67	0.99	1	0.48	0.67	0.67	0.48	0.86	0.63
alpha1B	420	0.72	1	0.99	0.6	0.72	0.72	0.6	0.69	0.48
alpha2A	412	0.68	1	0.99	0.5	0.68	0.68	0.5	0.73	0.53
alpha2B	283	0.56	0.99	0.99	0.15	0.56	0.55	0.15	0.69	0.49
alpha2C	332	0.7	1	0.99	0.54	0.7	0.7	0.54	0.73	0.54
beta1	194	0.65	1	0.99	0.46	0.65	0.65	0.46	0.75	0.58
beta2	239	0.69	1	0.99	0.51	0.69	0.68	0.51	0.87	0.66
D1	485	0.68	1	0.99	0.45	0.68	0.68	0.45	0.7	0.51
D2	2060	0.60	1	0.99	0.23	0.59	0.59	0.22	0.71	0.52
D3	1496	0.64	0.99	1	0.38	0.64	0.64	0.38	0.79	0.6
D4	749	0.57	1	0.99	0.26	0.57	0.57	0.26	0.81	0.6
D5	250	0.51	1	0.98	0.26	0.51	0.5	0.26	0.9	0.62
DAT	876	0.78	1	1	0.67	0.77	0.77	0.67	0.58	0.43
H1	396	0.63	0.99	1	0.33	0.62	0.62	0.32	0.82	0.59

(continued)

Table 7.2 (continued)

Dataset	NComp	R^2	k-O-P	k-P-O	R_0^2-O-P	R_0^2-P-O	$R_{45}^2(q^2$-O-P)	$R_{45}^2(q^2$-P-O)	STDEV sqrt[$(P-O)^2$/Ncomp]	MAE |P-O|/Ncomp
H2	93	0.17	1	0.99	−0.96	0.1	0.096	−0.97	0.55	0.34
H3	1385	0.66	1	0.99	0.38	0.65	0.65	0.38	0.66	0.47
H4	444	0.55	1	0.99	0.011	0.54	0.53	0.011	0.74	0.57
M1	384	0.6	0.98	1	0.31	0.6	0.6	0.29	0.9	0.64
M2	431	0.69	0.99	1	0.49	0.68	0.68	0.48	0.81	0.62
M3	424	0.69	1	0.99	0.52	0.69	0.68	0.52	0.86	0.62
M4	234	0.55	0.98	1	0.3	0.55	0.55	0.29	0.92	0.65
M5	253	0.6	0.98	1	0.31	0.6	0.6	0.3	0.75	0.52
NET	970	0.77	1	0.99	0.67	0.77	0.77	0.67	0.63	0.48
SRT	1262	0.81	1	0.99	0.75	0.81	0.81	0.75	0.69	0.51

NComp are counts of compounds in the datasets
R^2 are the coefficients of determination for the corresponding models
k-O-P are the slopes for trend lines through the origin for observed vs. predicted activities
k-P-O are the slopes for trend line through the origin for predicted vs. observed activities
R_0^2 (O-P) are the coefficients of determination for trend lines through the origin for observed vs. predicted activities
R_0^2 (P-O) are the coefficients of determination for trend lines through the origin for predicted vs. observed activities
R_{45}^2 (q^2-O-P) are coefficients of determination for bisector lines for observed vs. predicted activities
R_{45}^2 (q^2-P-O) are coefficients of determination for bisector lines for predicted vs. observed activities
STDEV are standard deviations between predicted and observed activities
MAE are mean absolute errors between predicted and observed activities

Table 7.3 Regulatory Information (RI) datasets

No	Activity	Compounds after curation	Activity range	Ref
1	Cancer Potency Value_CalEPA	214	2.48–9.44	[1]
2	Oral Slope Factor Regional Screening Level - Superfund	189	2.12–7.97	[2]
3	Reference Concentration_Regional Screening Level – Superfund	125	3.11–9.71	[2]
4	Reference Dose Point of Departure EPA IRIS	269	1.81–7.88	[18]
5	Reference Dose Dog Office of Pesticide Programs	139	4.64–9.95	[24]
6	Reference Dose EPA IRIS	269	4.48–10.59	[18]
7	Reference Dose Point of Departure Dog Office of Pesticide Programs	139	2.64–7.87	[24]
8	Reference Dose Point of Departure Rat Office of Pesticide Programs	190	2.79–7.36	[24]
9	Reference Dose Rat Office of Pesticide Programs	190	4.79–9.86	[24]
10	Reference Dose Regional Screening Level – Superfund	424	4.29–10.51	[2]

T. pyriformis IGC_{50} Dataset

This dataset containing 1093 compounds was compiled by Zhu and coworkers [21] from the work of Schultz and coworkers [22, 23]. The toxicity is expressed as the 50% growth inhibitory concentration (in mmol/L) of the *T. pyriformis* organism (a protozoan ciliate) after 40 h. Calculations were made with Log (IGC_{50}, 50% Inhibition Growth Concentration) as a response variable. Dragon 5.5 descriptors [6] were used in calculations of modelability criteria. QSAR models were built using HC [14], Dragon 5.5 [6], SiRMS[15] and CDK [16] descriptors with HC [14], kNN [17], RF [18], and SVM [24] methods. The best models were built using the HC method with HC descriptors and RF with SiRMS descriptors (R^2 = 0.85 for the external validation set) and were used in this study.

Acute Oral Toxicity Rat LD_{50} Dataset

This dataset containing 7285 compounds was compiled from the ChemIDplus database [25, 26]. The oral rat LD_{50} endpoint represents the amount of the chemical (mass of the chemical per body weight of the rat), which, when orally ingested, kills half of the rats. This dataset was recently modeled by a variety of QSAR techniques [27, 28]. Calculations were performed for Log (LD_{50}) as a response variable. Modelability criteria were calculated using Dragon 5.5 [6] descriptors. QSAR models were built using HC [14], Dragon 5.5 [6], SiRMS [15], and CDK [16] descriptors with HC [14], RF [18] and SVM [24] methods. The best model was built using the RF method with Dragon 5.5 descriptors ($R^2 = 0.66$ for the external validation set) and was used in this study.

Bioaccumulation Factor (BCF) Dataset

The bioconcentration factor BCF is defined as the ratio of the chemical concentration in biota as a result of absorption via the respiratory surface to that in water at a steady state [29]. This dataset was compiled from [4] and the final dataset consisted of 613 chemicals (after removing salts, mixtures, and ambiguous compounds). The modeled endpoint was the Log (BCF). Modelability criteria were calculated using Dragon 5.5 [6] descriptors. QSAR models were built using HC [14], Dragon 5.5 [6], SiRMS [15], and CDK [16] descriptors with HC [14], kNN [17], RF [11] and SVM [24] methods. The best model was built using the RF method with Dragon 5.5 descriptors [unpublished results] and was used in this study.

7.2.2 *Modelability Criteria*

The similarity search procedure is a kNN approach without a variable selection, i.e., it is carried out in the entire descriptor space, with no optimization of the descriptor set. Furthermore, the number of nearest neighbors is carried out like in the standard kNN developed in our laboratory [17]. As in the regular kNN, leave-one-out (*LOO*) and leave-group-out (*LGO*) cross-validation (*CV*) were used. *LOO-CV* gives rise to MODI_q^2 modelability criterion. In this work, MODI_ssR^2 modelability criterion was defined by the fivefold LGO (leave 20%-out) CV. Each dataset was divided into five equal parts. One part, in turn, was used as the prediction set, and the union of the remaining four parts was used as the "modeling" set. One most active and one most inactive compound and three compounds with activities uniformly distributed within the entire range of activities were always included in the modeling set. Both LOO-CV and LGO-CV were carried out with 1, 3, and 5 nearest neighbors.

In general, the Euclidean distances in the descriptor space between a compound and each of its k nearest neighbors are not the same. Thus, the neighbor with the smaller distance from a given compound is given a higher weight in calculating the predicted activity as follows:

$$\hat{y}_i = \frac{\sum_{j=1}^{k} y_j w_{ij}}{\sum_{j=1}^{k} w_{ij}} \tag{7.1}$$

where y_j is the observed activity value for nearest neighbor j, $\hat{y}_i$ is the predicted activity value for compound i; weights w_{ij} are defined as

$$w_{ij} = \left(1 + \frac{d_{ij}^2}{\sum_{j'=1}^{k} d_{ij'}^2} \right)^{-1} \tag{7.2}$$

d_{ij} are the Euclidean distances between compound i and each of its k nearest neighbors. For binary activity, the predicted value is rounded to the nearest integer or to a pre-determined threshold [30].

The Applicability Domain (AD) was defined by the Z-cutoff method employing the average distance between k nearest neighbors d_{av} and its standard deviation s in the modeling set. The threshold distance D_{thresh} (Eq. 7.3) is defined as

$$D_{thresh} = d_{av} + Zs \tag{7.3}$$

where Z is defined by the user (1 in this study). Compounds outside the AD were not predicted.

QSAR models obtained for 34 datasets of the training set were characterized by the criteria described in our previous publications (see, for example [30]). Leave 20%-out external cross-validation was used. First, models with high cross-validated q^2 values were selected for each tree in the RF or each kNN model. Then the models with sufficiently high q^2 values (higher than 0.6) were used for consensus prediction [31, 32] of compounds of the corresponding test sets. Then, predictions of all five prediction sets were combined and the coefficient of determination between predicted and observed activities, R^2, along with coefficients of determination R_0^2 and $R'_0{}^2$ and slopes k and k' for trend lines through the origin for predicted vs. observed and observed vs. predicted activities (see [30]) were calculated.

Similar criteria were used herein for comparison of MODI_ssR^2 and QSAR_R^2 values of QSAR models. Predictions of all five prediction sets were combined and the "cross-validated" q^2 and coefficient of determination between predicted and observed activities, R^2, along with coefficients of determination R_0^2 and $R'_0{}^2$ and slopes k and k' for trend lines through the origin for predicted vs. observed and observed vs. predicted activities were calculated. The main difference between the approach described in [30] was that, for modelability criteria, no variable selection in the similarity search procedure was used.

7.3 Results and Discussion

The modelability criteria are given for training and test sets in Tables 7.1 and 7.2. The models were built using RF (Table 7.1) and kNN methodologies (Table 7.2). A high correlation between RF and kNN_R^2 values for 34 datasets (see Fig. 7.1) is characterized by $R^2 = 0.91$. Even after the exclusion of three "outlier" datasets, $R^2 = 0.76$. Here we note that 31 datasets have both RF and kNN_R^2 values between 0.50 and 0.81. In Table 7.4, modelability criteria for the training sets are presented along with the criteria described in Sect. 7.2. Calculations were repeated for 1, 3, and 5 nearest neighbors. Modelability criteria and the corresponding statistics were also obtained for the test set (Table 7.5). In Table 7.6, the correlation coefficients between R^2 values of QSAR models and modelability criteria are given. The results are given for the training, test, and training and test sets together (the entire datasets). The corresponding regression equations for entire datasets are also given in Table 7.6. Both modelability criteria MODI_q^2 and MODI_ssR^2 were obtained for $k = 1$, 3, and 5 nearest values. QSAR models for the training sets were obtained using RF and kNN methods. The best correlations were obtained for $k = 5$ nearest neighbors for both RF and kNN models built for the training sets. Correlations for k=3 nearest neighbors are almost as good as for $k = 5$. For $k = 1$, weaker results were obtained, especially for MODI_q^2 (see Table 7.6). To illustrate our results, we have built plots of QSAR_R^2 vs. MODI_q^2 and MODI_ssR^2 values (see examples in Figs. 7.2, 7.3, 7.4, and 7.5). Regression equations for training, test, and the entire datasets are also given in these figures. We emphasize high coefficients of determination between QSAR_R^2 and modelability criteria for $k = 3$ and 5 for the whole dataset

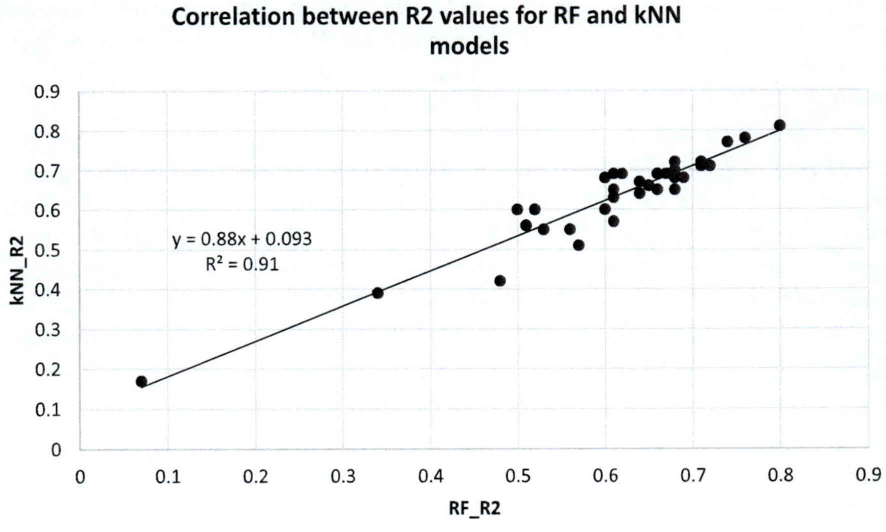

Fig. 7.1 High correlation between random forest and kNN models built for the training set

Table 7.4 Modelability criteria for the training set.

Dataset	Ncomp	MODI_q^2	MODI_ssR^2	k1	k2	R01^2	R02^2	RF_R^2	kNN_R^2
5-HT1A	1121	0.6	0.59	1	0.99	0.39	0.59	0.67	0.69
5-HT1B	438	0.57	0.58	1	0.98	0.35	0.58	0.67	0,65
5-HT1D	426	0.56	0.61	1.01	0.97	0.39	0.61	0.71	0.72
5-HT1E	87	0.1	0.19	1	0.98	−0.78	0.12	0.32	0.39
5-HT2A	1031	0.63	0.61	1	0.99	0.38	0.61	0.72	0.71
5-HT2C	791	0.53	0.52	1	0.99	0.22	0.51	0.65	0.65
5-HT3	125	0.28	0.37	1	0.97	−0.04	0.33	0.48	0.42
5-HT5	145	0.42	0.53	1.01	0.97	0.26	0.53	0.65	0.69
5-HT6	952	0.61	0.61	1	0.99	0.42	0.61	0.7	0.71
5-HT7	397	0.5	0.51	1	0.98	0.16	0.51	0.6	0.68
alpha1A	429	0.59	0.61	0.99	0.99	0.43	0.61	0.64	0.67
alpha1B	366	0.67	0.68	1	0.99	0.61	0.67	0.68	0.72
alpha2A	365	0.45	0.51	1	0.98	0.23	0.5	0.69	0.68
alpha2B	243	0.39	0.44	0.99	1	−0.07	0.44	0.51	0.56
alpha2C	284	0.56	0.56	0.99	0.99	0.32	0.56	0.68	0.7
beta1	163	0.49	0.57	1.01	0.97	0.43	0.55	0.61	0.65
beta2	202	0.52	0.58	1	0.98	0.39	0.57	0.61	0.69
D1	417	0.50	0.49	0.98	0.99	−1.65	0.2	0.68	0.68
D2	2057	0.50	0.48	0.99	1	−0.05	0.46	0.60	0.60
D3	1493	0.56	0.54	0.99	0.99	0.26	0.53	0.64	0.64
D4	639	0.47	0.48	1	0.99	0.14	0.47	0.61	0.57
D5	247	0.38	0.39	0.99	0.98	−0.17	0.38	0.57	0.51
DAT	766	0.67	0.64	0.99	0.99	0.51	0.64	0.76	0.78
H1	393	0.53	0.52	1	0.98	0.15	0.52	0.61	0.63
H2	90	0.05	0.1	0.98	1.01	−1.88	0.02	0.066	0.17
H3	874	0.50	0.55	0.99	0.99	−2.08	0.21	0.65	0.66
H4	440	0.40	0.4	1.01	0.97	−0.5	0.4	0.55	0.55
M1	331	0.45	0.5	0.98	1	0.2	0.5	0.52	0.60
M2	369	0.52	0.54	0.99	0.99	0.27	0.54	0.62	0.69
M3	378	0.55	0.58	0.99	0.99	0.37	0.58	0.68	0.69
M4	199	0.47	0.45	0.98	0.99	0.1	0.44	0.53	0.55
M5	212	0.46	0.46	0.98	1	0.15	0.45	0.50	0.60
NET	837	0.67	0.64	1.01	0.98	0.5	0.64	0.74	0.77
SRT	1093	0.73	0.72	0.99	1	0.63	0.72	0.80	0.81

and for training and test sets. The regression lines for training and test sets are close to each other. It is important to note that high correlations for the test set and for the entire dataset were obtained despite the fact that distributions of QSAR_R^2 and MODI_R^2 values are quite different for training and test sets. This also corroborates our modelability approach.

Formulas for regression lines can be used to predict QSAR modeling results qualitatively. In [2], four datasets were used to validate the modelability criteria.

Table 7.5 Modelability criteria and prediction statistics for the test sets

Dataset		$MODI\_q^2$			$MODI\_ssR^2$			$QSAR\_R^2$
		K = 1	K = 3	K = 5	K = 1	K = 3	K = 5	
1	**LC50**	0.51	0.60	0.55	0.46	0.61	0.60	0.69
2	**IGC50**	0.69	0.76	0.75	0.68	0.76	0.76	0.85
3	**LD50**	0.51	0.57	0.56	0.36	0.53	0.55	0.66
4	**BCF**	0.71	0.76	0.76	0.66	0.73	0.72	0.89
5	**RI1**	0.005	0.150	0.190	0.28	0.29	0.27	0.36
6	**RI2**	−0.14	0.07	0.12	0.18	0.095	0.16	0.43
7	**RI3**	0.24	0.42	0.44	0.34	0.46	0.36	0.46
8	**RI4**	0.18	0.38	0.37	0.38	0.41	0.44	0.53
9	**RI5**	0.041	0.340	0.360	0.18	0.35	0.37	0.51
10	**RI6**	0.18	0.19	0.19	0.22	0.26	0.3	0.37
11	**RI7**	0.18	0.39	0.42	0.37	0.4	0.32	0.58
12	**RI8**	0.12	0.16	0.19	0.21	0.23	0.22	0.39
13	**RI9**	0.28	0.086	0.110	0.18	0.1	0.19	0.29
14	**RI10**	0.18	0.41	0.42	0.37	0.49	0.44	0.48

Table 7.6 Prediction statistics for datasets included in training and test sets

QSAR method	MODI criterion	Number of NN	Training set	Test set	Both sets	
					R^2	Equation[a]
kNN	$MODI\_q^2$	1	0.76	**0.65**	**0.72**	**y = 0.58x + 0.41**
kNN	$MODI\_ssR^2$	1	0.80	**0.88**	**0.84**	**y = 0.88x + 0.19**
RF	$MODI\_q^2$	1	0.75	**0.65**	**0.72**	**y = 0.60x + 0.38**
RF	$MODI\_ssR^2$	1	0.74	**0.88**	**0.78**	**y = 0.87x + 0.18**
kNN	$MODI\_q^2$	3	0.79	**0.89**	**0.85**	**y = 0.76x + 0.26**
kNN	$MODI\_ssR^2$	3	0.88	**0.84**	**0.87**	**y = 0.84x + 0.18**
RF	$MODI\_q^2$	3	0.79	**0.89**	**0.84**	**y = 0.77x + 0.23**
RF	$MODI\_ssR^2$	3	0.82	**0.84**	**0.82**	**y = 0.83x + 0.17**
kNN	$MODI\_q^2$	5	0.86	**0.90**	**0.89**	**y = 0.81x + 0.23**
kNN	$MODI\_ssR^2$	5	0.91	**0.88**	**0.90**	**y = 0.91x + 0.16**
RF	$MODI\_q^2$	5	0.84	**0.87**	**0.90**	**y = 0.82x + 0.21**
RF	$MODI\_ssR^2$	5	0.87	**0.87**	**0.88**	**y = 0.92x + 0.14**

[a]Given equations are for estimation of $QSAR\_R^2$: x—modelability index, y—predicted $QSAR\_R^2$

It was shown that the $MODI\_ssR^2$ and $MODI\_q^2$ can be used to discriminate between modelable and non-modelable datasets. Here we demonstrate that our updated regression equations can also be used.

Predictive QSAR models were built using kNN [17] and SVM [24] with Molconn-Z [33] descriptors for a dataset containing 59 Histone Deacetylase (HDAC) inhibitors [34]. Recently, Molconn-Z descriptors became unavailable; however, almost all of them were included in Dragon 5.5 descriptors [33]. The best kNN model built for this dataset with our standard fivefold external cross-

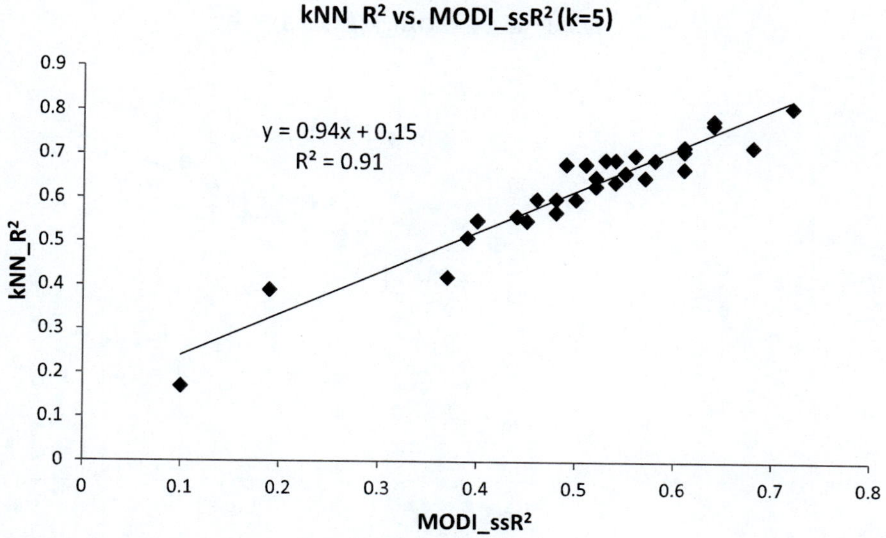

Fig. 7.2 34 datasets of the training set: correlation between the predictive power of k nearest neighbors models and MODI_ssR^2 for five nearest neighbors

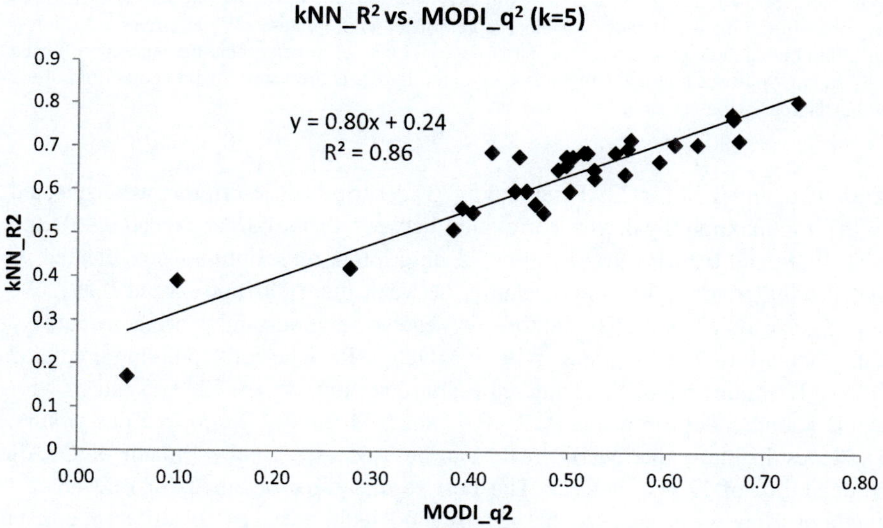

Fig. 7.3 34 datasets of the training set: correlation between the predictive power of k nearest neighbor models and MODI_q^2 for five nearest neighbors

validation protocol had $R^2 = 0.77$ [2]. MODI_ssR^2 with 1, 3 and 5 nearest neighbors were 0.60, 0.54 and 0.44 for 163 Molconn-Z [33] descriptors. The predicted QSAR_R^2 values based on regression equations (Table 7.3) were 0.72, 0.63, and 0.56, respectively. Similarly, MODI_ssR^2 with 1, 3, and 5 nearest neighbors were

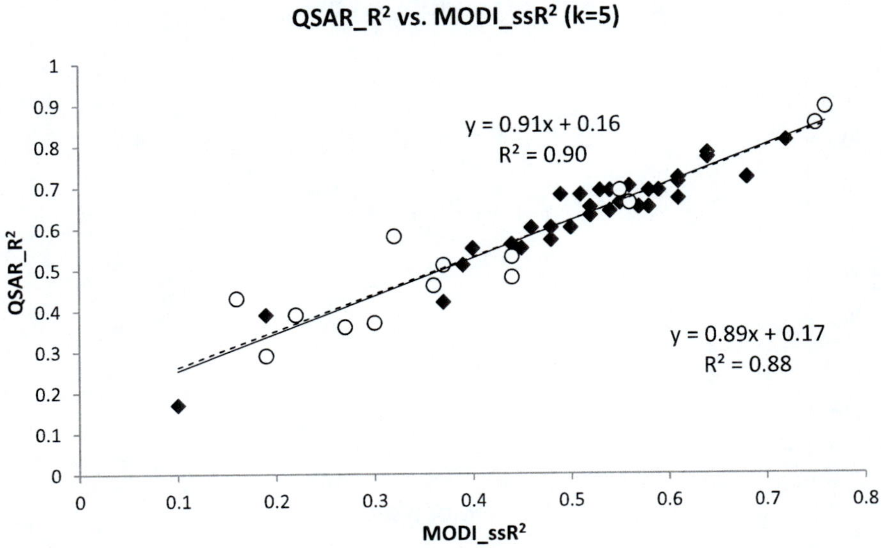

Fig. 7.4 34 datasets of the training set and 14 datasets of the test set: correlation between the predictive power of random forest (RF) models and MODI_ssR^2 for five nearest neighbors. Models were built for 34 training set datasets using RF. Models for 14 test set datasets were built as described in the main text. All datapoints represent the entire dataset. White circles are the test set. Solid line is the regression for the entire dataset. Dotted line represents the regression for the test set. On top of the plot is the regression equation for the entire dataset, while on bottom of the plot is the regression equation for the test set

0.60, 0.56, and 0.34 for 229 Dragon 5.5 [33] descriptors (descriptors were selected so that the maximum pairwise correlation between them did not exceed 0.95), and 0.71, 0.49, and 0.38 for 409 Dragon 5.5 descriptors (descriptors were selected so that the maximum pairwise correlation between them did not exceed 0.99). We hypothesize that the higher number of nearest neighbors may bring in activity cliffs that will reduce the predictivity of MODI_ssR^2. The corresponding predicted QSAR_R^2 values calculated using regression equations were 0.72, 0.65 and 0.47 for 229 Dragon 5.5 descriptors and 0.81, 0.59 and 0.51 for 409 Dragon 5.5 descriptors. These results show that MODI_ssR^2 can be used as a good estimator of QSAR modelability of HDAC dataset. The best results were obtained for one nearest neighbor. Here we emphasize that ideally modelability criteria for different counts of nearest neighbors should be calculated and we found that MODI_ss$R^2 = 0.72$. It is possible to consider the dataset with 409 descriptors and one nearest neighbor as a model already gives sufficiently high R^2 of 0.72. Interestingly, it is lower than our best estimate of QSAR_$R^2 = 0.81$ with the highest QSAR_$R^2 = 0.87$ obtained for 9 compounds of the external test set in our earlier publication [34].

The Maximum Recommended Therapeutic Dose (MRTD) database (1180 compounds) was compiled from different sources. In total, 1220 pharmaceuticals are listed in *Martindale* [35]: *The Extra Pharmacopoeia* (1973, 1983, and 1993) and

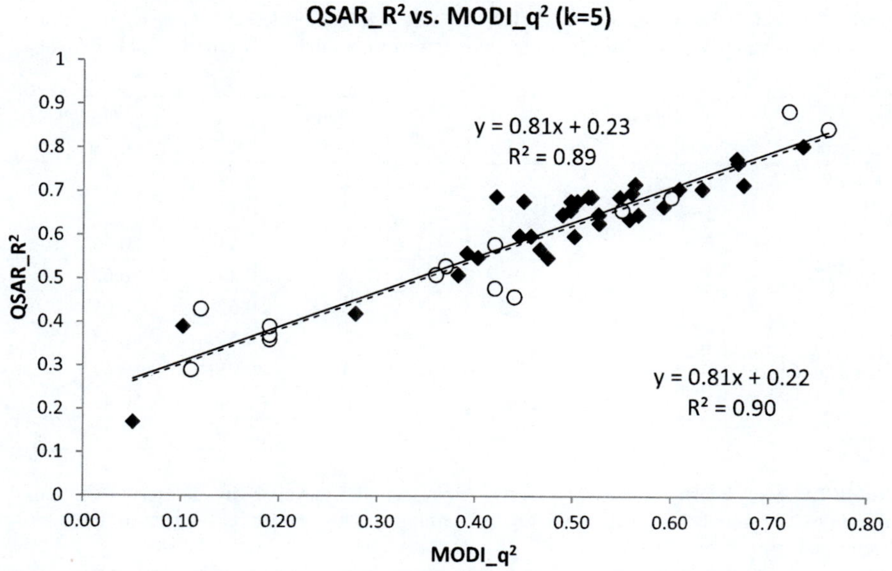

Fig. 7.5 34 datasets of the training set and 14 datasets of the test set: correlation between the predictive power of k Nearest Neighbors models and MODI_q2 for five nearest neighbors. Models were built for 34 training set datasets using RF. Models for 14 test set datasets were built as described in the main text. All datapoints represent the entire dataset. White circles are the test set. Solid line is the regression for the entire dataset. Dotted line represents the regression for the test set. On top of the plot is the regression equation for the entire dataset, while on bottom of the plot is the regression equation for the test set

The Physicians' Desk Reference (1995 and 1999) [13]. Modelability criteria were calculated using Dragon 5.5 descriptors [6]. QSAR models were built using SVM [24] and RF [11] methods with Dragon 5.5 [6], CDK [16], and SiRMS [15] descriptors. The best QSAR model was built using RF and SiRMS descriptors ($R^2 = 0.81$) [Unpublished results].

For the Alpha-1A adrenergic receptor agonists and antagonists dataset (415 compounds), we built QSAR models using SVM [24] and RF [11] methods with Dragon 5.5 [6], CDK [16], and SiRMS [15] descriptors. Modelability criteria were calculated using Dragon 5.5 [6] descriptors. The best QSAR model was built using RF and SiRMS descriptors ($R^2 = 0.81$) [Unpublished results].

For the bioavailability dataset of 1110 compounds [36], QSAR models were built using Dragon 5.5 [6] and SiRMS [15] descriptors, and for each set of descriptors, PLS [37] and RF [18]methods were used. Modelability criteria were built using Dragon 5.5 [6] descriptors. Many attempts were unsuccessfully undertaken for this dataset to build a predictive QSAR model; the highest R^2 was lower than 0.5 [38] (unpublished results).

Results of modelability studies for these datasets are presented in Tables 7.7 and 7.8. In Table 7.7, predicted QSAR_R^2 values were obtained using the corresponding

Table 7.7 Modelability studies of MRTD, Alpha1a and Bioavailability datasets. Regression equations for models built with RF for the training sets vs. modelability criteria (Table 7.6) were used

Dataset	Number of nearest neighbors	MODI_ssR^2	Estimated QSAR R^2	MODI_q^2	Estimated QSAR R^2
MRTD	1	0.36	0.49	0.19	0.49
	3	0.45	0.54	0.40	0.54
	5	0.42	0.53	0.41	0.55
Alpha1A	1	0.47	0.59	0.49	0.62
	3	0.55	0.63	0.52	0.63
	5	0.56	0.66	0.52	0.64
Bioavailability	1	0.26	0.41	0.051	0.41
	3	0.29	0.41	0.24	0.42
	5	0.22	0.34	0.26	0.42

Table 7.8 Modelability studies of MRTD, Alpha1a and Bioavailability datasets. Regression equations for models built with kNN for the training sets vs. modelability criteria (Table 7.6) were used

Dataset	Number of nearest neighbors	MODI_ssR^2	Estimated QSAR R^2	MODI_q^2	Estimated QSAR R^2
MRTD	1	0.36	0.51	0.19	0.52
	3	0.45	0.56	0.40	0.56
	5	0.42	0.54	0.41	0.56
Alpha1A	1	0.47	0.60	0.49	0.64
	3	0.55	0.64	0.52	0.66
	5	0.56	0.67	0.52	0.65
Bioavailability	1	0.26	0.42	0.051	0.44
	3	0.29	0.42	0.24	0.44
	5	0.22	0.36	0.26	0.44

regressions in Table 7.6 (models for the training set were built using RF). In Table 7.8, predicted QSAR_R^2 values obtained using the corresponding regression equations in Table 7.6 (models for the training set were built using kNN). As one can see, the modelability criteria for the Bioavailability dataset are much lower than those for the two other datasets. Almost all predicted QSAR_R^2 values for the Alpha-1A dataset are higher than 0.6, the standard threshold on a QSAR model to be considered as predictive [30]. Of course, true modeling can improve the predictive power of such models. At the same time, QSAR_R^2 for MRTD dataset is lower than 0.6; however, most of them are close to 0.55, which is only slightly less than 0.6. In this case, a more sophisticated QSAR modeling technique might be able to improve the predictivity of QSAR models developed for this dataset. On the other hand, for the Bioavailability dataset, predicted QSAR_R^2 values are significantly lower, and we can conclude that this dataset is probably not modelable. We conclude that no additional attempts to build QSAR models for this dataset with chemical descriptors are necessary.

From these results, we can observe that predicted QSAR_R^2 values are close for both RF and kNN methods. As was demonstrated above, RF_R^2 and kNN_R^2 values have a high correlation (see Fig. 7.1), so this result was expected. Besides, we can find that predicted QSAR_R^2 values are also similar for MODI_ssR^2 and MODI_q^2 values. In most cases, MODI_ssR^2 values are slightly higher than MODI_q^2 values. At the same time, the regressions QSAR_R^2 vs. MODI_ssR^2 have a smaller intercept than QSAR_R^2 vs. MODI_q^2. Thus, we recommend using MODI_ssR^2 rather than MODI_q^2. Also, we found that for one nearest neighbor, several MODI_q^2 values are negative. Consequently, we can conclude that MODI_ssR^2 is slightly better than MODI_q^2. A similar conclusion was made for category QSAR [2], where we have shown that MODI_ssCCR was slightly better than MODI_CCR.

7.4 Conclusions

In this book chapter, we describe two new modelability criteria to estimate the modelability of the continuous dataset. These criteria provide estimation if a dataset is modelable or not, i.e., if predictive QSAR models can be developed for that dataset. The two new metrics introduced here are based on the *k* nearest neighbors' approach. Their representative points consider compounds in the entire descriptor space. Both criteria were calculated using 1, 3, and 5 nearest neighbors. One of the criteria, MODI_q^2, is the q^2 value for leave-one-out cross-validation. Another index, MODI_ssR^2, is calculated using our standard fivefold external cross-validation procedure. The training set included 34 GPCR datasets, for which QSAR models were built using Dragon 5.5 descriptors along with RF and kNN algorithms. The test set contained 14 different datasets. Our results are highlighted below.

1. The modelability index has shown to be a reliable, easy, and fast metric to assess if, for a particular dataset, predictive QSAR models can be successfully developed. Here, we showed that MODI_ssR^2 provides a slightly better estimation than MODI_q^2.
2. Since the R^2 values for QSAR models built by different QSAR methods, for example, kNN and RF, correlate, it is possible to regress RF_R^2 values rather than kNN_R^2 against modelability criteria. It was demonstrated that kNN_R^2 gives slightly better results in terms of the intercept and values of MODI_q^2, which for $k = 1$ nearest neighbor can be negative. Consequently, if RF_R^2 is used, the number of nearest neighbors should preferably be 3 and 5. However, remember the results for HDAC inhibitors demonstrating that a different number of nearest neighbors should be used in modelability studies.
3. The calculated MODI_q^2 or MODI_ssR^2 for a dataset have shown to be informative of the lowest predictive bound of a QSAR model.

4. Regression equations for QSAR_R^2 vs. modelability criteria can be used for the prediction of prediction power of QSAR models in terms of R^2, which can be built for this dataset using chemical descriptors.
5. Other modelability criteria can be developed by using other fast and reliable QSAR approaches, like RF. This conclusion comes directly from a high correlation between kNN_R^2 and RF_R^2 values for QSAR models built with the same descriptors.

Finally, the modelability criteria can discriminate between QSAR models for which it is possible and not possible to build predictive QSAR models. In some cases, when a modelability criterion is extremely high, no additional QSAR modeling might be necessary. The modelability approach can dramatically decrease the time and resources used to build predictive QSAR models, especially to attempt to build QSAR models for "tough" datasets.

References

1. Golbraikh A, Muratov E, Fourches D, Tropsha A (2014) J Chem Inf Model 54:1
2. Golbraikh A, Fourches D, Sedykh A et al (2014) Modelability criteria: statistical characteristics estimating feasibility to build predictive QSAR models for a dataset. In: Leszcynski J, Shukla M (eds) Practical aspects of computational chemistry III. Springer, New York, pp 187–230
3. Wignall JA, Muratov E, Sedykh A et al (2018) Environ Health Perspect 126:057008
4. Martin TM, Harten P, Young DM et al (2012) J Chem Inf Model 52(10):2570–2578. https://doi.org/10.1021/ci300338w
5. (2012) DAI/B 73-0
6. Talete SRL (2007) Dragon v. 5.5
7. Gaulton A, Bellis LJ, Bento AP et al (2012) Nucleic Acids Res 40:D1100
8. Gaulton A, Hersey A, Nowotka M et al (2017) Nucleic Acids Res 45:D945
9. NIMH Psychoactive Drug Screening Program (PDSP)
10. Fourches D, Muratov E, Tropsha A (2010) J Chem Inf Model 50:1189
11. Polishchuk PG, Muratov EN, Artemenko AG et al (2009) J Chem Inf Model 49:2481
12. EPA (2008) ECOTOX database
13. Toxicology UE-NC for C
14. Martin TM, Harten P, Venkatapathy R et al (2008) Toxicol Mech Methods 18:251
15. Kuz'min VE, Artemenko AG, Muratov EN (2008) J Comput Aided Mol Des 22:403
16. Steinbeck C, Han Y, Kuhn S et al (2003) J Chem Inf Comput Sci 43:493
17. Zheng W, Tropsha A (2000) J Chem Inf Comput Sci 40:185
18. Breiman L (2001) Mach Learn 45:5
19. R: classification and regression with random forest. http://rss.acs.unt.edu/Rdoc/library/randomForest/html/randomForest.html. Accessed 6 Aug 2013
20. R Core Team R: a language and environment for statistical computing
21. Zhu H, Tropsha A, Fourches D et al (2008) J Chem Inf Model 48:766
22. Schultz TW, Netzeva TI (2004) Development and evaluation of QSARs for ecotoxic endpoints: The benzene response-surface model for Tetrahymena toxicity. In: Cronin MTD, Livingstone DJ (eds) Modeling environmental fate and toxicity. CRC, Boca Raton, FL, pp 265–284
23. Schultz TW (1997) Tetratox Methods 7(4): 289–309. https://doi.org/10.1080/105172397243079
24. Vapnik VN (1995) The nature of statistical learning theory. Springer, New York

25. NLM (2011) ChemIDplus: a TOXNET database
26. EPA user's guide for T.E.S.T. (Toxicity Estimation Software Tool) a program to estimate toxicity from molecular structure
27. Zhu H, Martin TM, Ye L et al (2009) Chem Res Toxicol 22:1913
28. Zhu H, Ye L, Richard A et al (2009) Environ Health Perspect 117:1257
29. Hamelink JL (1977) Current bioconcentration test methods and theory. ASTM. http://www.astm.org/DIGITAL_LIBRARY/STP/PAGES/STP32397S.htm. Accessed 6 Aug 2013
30. Tropsha A, Golbraikh A (2010) Predictive quantitative structure–activity relationships modeling: development and validation of QSAR models. In: Faulon J-L, Bender A (eds) Handbook of chemoinformatics algorithms. Chapman & Hall / CRC, London, UK, pp 213–233
31. Kovatcheva A, Golbraikh A, Oloff S et al (2005) SAR QSAR Environ Res 16:93
32. de Cerqueira Lima P, Golbraikh A, Oloff S et al (2006) J Chem Inf Model 46:1245
33. Hall L, Kellogg G, Haney D (2002) Molconn-Z, eduSoft. In: LC Ashl
34. Tang H, Wang XS, Huang X-P et al (2009) J Chem Inf Model 49:461
35. Brayfield A (2014) Martindale: the complete drug reference.37th edn. Pharmaceutical Press, London
36. Veber DF, Johnson SR, Cheng H-Y et al (2002) J Med Chem 45:2615
37. Wold S, Sjöström M, Eriksson L (2001) Chemom Intell Lab Syst 58:109
38. Artemenko A, Muratov E, Kuz'min V et al (2009) Antiviral Res 82:A56

Chapter 8
A Cluster Model for Interpretation of Surface-Enhanced Raman Scattering of Organic Compounds Interacting with Silver Nanoparticles

Nguyen Van Trang, Duy Quang Dao, Pham Vu Nhat, Phan Thi Thuy, and Minh Tho Nguyen

Abstract In this brief review, we present some illustrative computational results on the use of atomic clusters to model the solid substrate surfaces, aiming to interpret the phenomena associated with surface-enhanced Raman scattering (SERS) experiments. Using this technique, organic molecules undergo interactions with silver nanoparticles, producing characteristic chemical and spectroscopic enhancements. We exemplify that the small Ag_n cluster model can reproduce typical chemical enhancements that occur on vibrational modes of the molecules considered, including aromatic compounds, amino acids, pesticides, herbicides etc. Analyses are made for not only the ground state but also for the first excited S_1 state, which shows a reverse charge transfer trend. Two different mechanisms are considered to construe the effects induced by the incident light during the SERS phenomena. More importantly, the charge transfer mechanism allows selective and quantitative measurements of the molecule considered in its mixture with several other organic compounds. These results matter for the design of efficient mobile sensors that are expected to potentially detect pollutants and toxic compounds present in the environment and in food and agricultural products rapidly.

N. V. Trang
Institute for Tropical Technology, Vietnam Academy of Science and Technology (VAST), Hanoi, Vietnam

D. Q. Dao
Institute of Research and Development, Duy Tan University, Da Nang, Vietnam

P. V. Nhat
Department of Chemistry, Can Tho University, Can Tho, Vietnam

P. T. Thuy
School of Natural Science Education, Vinh University, Vinh, Vietnam

M. T. Nguyen (✉)
Institute for Computational Science and Technology (ICST), Ho Chi Minh City, Vietnam
e-mail: tho.nm@icst.org.vn

J. Leszczynski, M. K. Shukla (eds.), *Practical Aspects of Computational Chemistry V*, https://doi.org/10.1007/978-3-030-83244-5_8

Abbreviations

CDD	Charge density difference
CE	Chemical enhancement
CPF	Chlorpyrifos
CT	Charge transfer
DDA	Discrete dipole approximation
DFT	Density functional theory
EM	Electromagnetic mechanism
FDTD	Finite difference time domain
HOMO	Highest occupied molecular orbital
LSPR	Local surface plasmon resonance
LUMO	Lowest unoccupied molecular orbital
MD	Molecular dynamics
SERS	Surface enhanced Raman scattering
UV	Ultraviolet–visible

8.1 Introduction

An atomic cluster is an aggregate of typically a few up to several thousands of atoms and represents a distinct form of matter lying between an atom and solid bulk. When a cluster and a molecule are both generated in the gas phase, the main qualitative difference between them lies in their inherent interatomic binding forces. Atoms in a conventional molecule are known to be bound to each other by covalent and/or ionic forces. Specific stabilizing forces in a cluster allow a polyatomic system to grow much larger by stacking more and more atoms. A notable consequence is that atomic clusters often possess unique properties that basically differ from the corresponding single atom and bulk state. Fascination for free or supported atomic clusters originates not only from their non-classical geometric and electronic structures, but also from their surprising chemical reactivities. In the non-scalable regime ranging from a few to about hundreds of atoms, the structures and properties of clusters change discontinuously with not only their size but also with their charge state. Specifically, an addition of an atom, a change of a unit in the electron number, or both actions via doping, fundamentally changes the cluster identity. Such a characteristic opens opportunities for creations of new clusters, and thus, cluster-based building blocks with well-defined or tailored properties.

With the swift development in applications of nano-sized materials [1–2] and mainly the urgent need of miniaturizing optoelectronic devices, the field of cluster science has expanded broadly in recent years, establishing a contemporary inter-disciplinary domain of research, which is constantly evolving both academic and applied interests [3–6]. Basic studies of nanomaterials are of primordial importance because they consistently reveal new properties that could lead to new prospective applications. Moreover, the basic principles of such basic studies are threefold, namely, (1) a comprehension of chemical bonding in clusters that bridges the molecular phenomena with the condensed matter behaviors, (2) an establishment of new classes of clusters formed by various elements with unprecedented properties, and (3) a search for promising applications of cluster-based assembly materials [6]. While the structural motifs and electronic properties of clusters have long been, and still are, fascinating to both chemists and physicists alike, their applications have been implemented in a myriad of modern technologies such as new laser materials, infrared to visible conversion substrates, systems involving photoredox processes for solar energy conversion, photovoltaic devices, phosphorescent sensors, electroluminescent sensors for biological applications, and catalysis for chemical reactions, etc. [3–6]. As clusters are first generated in the gas phase, they need to subsequently be deposited on a solid surface, such as a metal oxide, before undergoing interactions with the reactants or the surrounding medium, leading to catalytic effects or enhancement of spectroscopic responses. They can also be stabilized by ligands before assembled in solid state materials.

Beside such actual and exciting aspects, the concept of an atomic cluster as a form of matter also allows for theoretical approaches to be implemented. It is well known that knowledge on the critical role of specific active surfaces in a nanoparticle is of great importance, since interfaces between an active material and the surrounding media determine the efficiency of, among others, heterogeneous catalysis [3]. Theoretical modeling and simulations on nanoparticle surfaces to determine their characteristics are challenging tasks as they require specific treatments to obtain an appropriate description of the solid bulk state. Properties of material are usually dependent on its dimensionality in such a manner that the behaviors of lower-dimensional systems could either be alike, or widely different from, their bulk counterparts. In this context, preliminary modeling of the bulk system is crucial in the selection of methodologies for determining the structural, vibrational, and electronic properties of the nanostructure considered.

Of the theoretical approaches to investigate nanoscale systems, and in particular their surfaces, the atomic cluster approach is of worthy interest. In fact, a solid surface could be simulated by a small-sized atomic cluster which can be treated by reliable quantum chemical computations. A suitable model often requires a bulk crystal approach to nanosized materials by building various surfaces and by cutting the periodicity of the crystal lattice [7]. Another cluster approach is based on the crystalline structure characterized inside the core of the nano-substrate considered, surrounded by surfaces reconstructed following the surface energy requirements. Such nanostructures are created from their original bulk crystals with specified atomic positions appearing in well-defined space groups. Another plainer cluster approach proceeds through full geometry optimizations of nanoclusters of the atoms, representing the nanoparticle considered without specifying a crystal structure [8, 9].

On this basis, from a viewpoint of theoretical methodology, a question of continuing interest which remains a matter of debate, is as to whether a pure atomic cluster can be used to model a solid nanoparticle [10]. Such a simple but efficient approach could provide us with a means to complement some popular experimental methods by investigation of reduced dimension nanosystems with reliable quantum chemical computations. To simulate realistic nanostructures that can reproduce experimental data, the cluster models need to be constructed with appropriate sizes [10]. As experimental observations are usually made on particles with diameters in the range of a few nanometers, modeling based on quantum-chemical methods is highly time-consuming and cannot be applied to very large systems. It is, therefore, imperative to construct small but fitting cluster models. In this chapter, we would advocate for the use of small silver clusters as models in the assignment and interpretation of the spectra obtained from surface-enhanced Raman scattering (SERS) experiments for organic compounds. The following sections present a summary of the SERS technique, followed by the use of some silver clusters selected to model the coinage metal substrates and surfaces in the experiments involving molecules such as aromatic compounds, amino acids, pesticides, herbicides, etc.

8.2 Surface-Enhanced Raman Scattering (SERS) Experiment

8.2.1 General Aspects

In 1974, largely through the pioneering work of Fleischmann and coworkers, the first surface-enhanced Raman scattering (SERS) spectra were discovered, which produced a Raman spectrum that was a millionfold more intense than expected from molecules adsorbed on specially prepared silver [11]. In recent years, due to its simple operation and high sensitivity, the SERS spectroscopy has become a powerful analytical tool for monitoring food safety as well as toxicity in the environment in detecting their fingerprint characteristics. The combination of Raman spectroscopy and nanotechnology, which are the two most popular classical and modern techniques, leads to the functioning of SERS.

Raman spectroscopy is renowned to be able to provide distinct Raman spectral data, and thereby structural information, of any specific molecule or analyte [12]. However, this method has one major drawback due to the small cross-section of Raman scattering, which often leads to low spectral resolution. Consequently, the analyte should be present of high concentration. However, a typical phenomenon was discovered for the analyte near the raw metal surface. The SERS results show that the spectral signal can be enhanced by a factor up to 10^{10}–10^{15} due to a strong local field arising from the plasmon excitation combined with direct chemical interactions between the molecule and the metal surface. Such a technique, also known as SERS, has the potential to be exploited as a single molecule spectroscopy. In other words, SERS is a powerful oscillometric spectrometric technique that allows an overly sensitive structure of a low concentration analyte to be detected based on the amplification of electromagnetic fields generated by the excitation of localized surface plasmons. Although the SERS process has been applied for several decades for chemical analysis purposes, the detail of its molecular mechanism remains far from being fully understood.

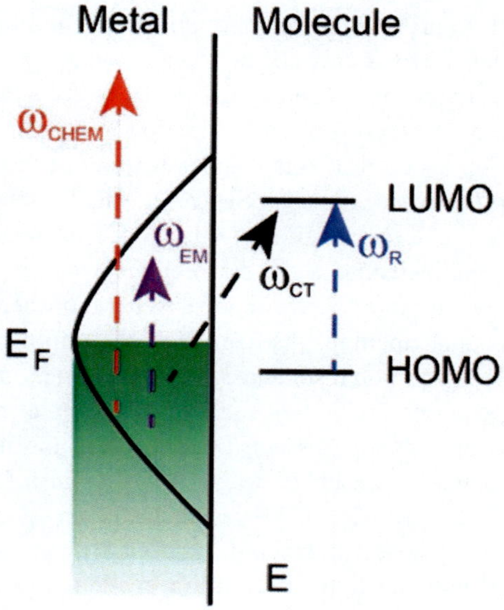

Fig. 8.1 Illustration of four different enhancement mechanisms in a SERS experiment: CHEM, chemical enhancement; CT, charge transfer; R, resonance; EM, electromagnetic mechanism

It has widely been accepted that four main mechanisms are responsible for the SERS phenomenon. Figure 8.1 summarizes these mechanisms along with the involvement of electronic configurations of the metal surface and interacting molecule. The mechanisms illustrated in Fig. 8.1 include (1) an enhancement due to non-resonant interactions between the surface and the adsorbate (CHEM), (2) a molecular resonance mechanism (resonance) where the incident beam is in resonance with a molecular excitation, (3) a charge transfer (CT) mechanism due to a resonance of the coming beam through an excitation from the metal to the adsorbate, and (4) an electromagnetic mechanism (EM) due to a very strong local field when the excitation wavelength is in resonance with the plasmon excitations in the metallic nanoparticle [13].

The mechanisms shown in Fig. 8.1 are not independent of each other. Often, a combination happens to create a total enhancement. The predominance of a certain mechanism depends on the wavelengths used in the experiment, and the specific properties of both the adsorbate and metal surfaces. In other words, these mechanisms cannot really be separated from each other according to experiment or theory. However, certain limits can be set in such a way that one mechanism will play a dominant role over the other. Reported results show that the EM is usually considered as the dominant contribution to the observed SERS signal due to the fact that local field enhancement arising from the plasmon excitation is often very large in comparison with the enhancement factor of other mechanisms. The electrodynamic enhancement mechanism is caused by the strong fields, and it is commonly simulated that the SERS intensities is enhanced by a factor corresponding to the fourth-power of the electric field enhancement, $|E_{loc}|^4$, which will be discussed later in detail. Because the EM does not include chemical interactions and charge transfer

(CT) process between both molecule and metallic surface, it cannot always clarify many SERS enhancement phenomenon such as the relative SERS signal intensity changes, presence of new peaks or frequency shifting, even with examination of the adsorbed molecule direction and polarization orientation of the EM field (cf. Fig. 8.1). For a better understanding of the SERS enhancement phenomena, other processes occurred in SERS (cf. Fig. 8.1) need to be considered.

As described in Fig. 8.1, the CT, resonance, and CHEM mechanisms are often combined into a general chemical enhancement mechanism (CE) because it is not possible to view a clear separation between them. In an EM mechanism, the enhancement results from the light amplification by the excitation of localized surface plasmon resonances. This phenomenon tends to occur locally at the energy gap positions, crevices or sharp features of plasmonic materials (such as gold, silver, and copper nanoparticles). Dependent on the material surfaces, the EM enhancement factor can reach up to ~10^{10}–10^{11}. In a CE mechanism, an enhancement factor of up to 10^3 can be reached, due to the charge transfer between the adsorbed molecule and the metallic nanoparticles, which is in resonance with the wavelength of incident light. It is still a challenge to observe a CE process occurred during a SERS experiment. Therefore, recent studies exploring the CE mechanism have been carried out by computations employing density functional theory (DFT) and time dependent-density functional theory (TD-DFT) methods.

8.2.2 Electromagnetic Mechanism of SERS (EM)

A straightforward rationale for the field enhancement in SERS can be illustrated by considering the following simple model which was first suggested by Gersten and Nitzan [14]. The Raman scattering of a molecule is influenced by the electromagnetic interaction with a polarizable source located nearby. The electromagnetic SERS enhancement is often considered as process occurred in two steps, including both local field and radiation enhancements, which can be presented as follows in Fig. 8.2 [15].

Local Field Enhancement in SERS The SERS mainly results from a localized electric field enhancement, utilizing optical resonance processes, such as those

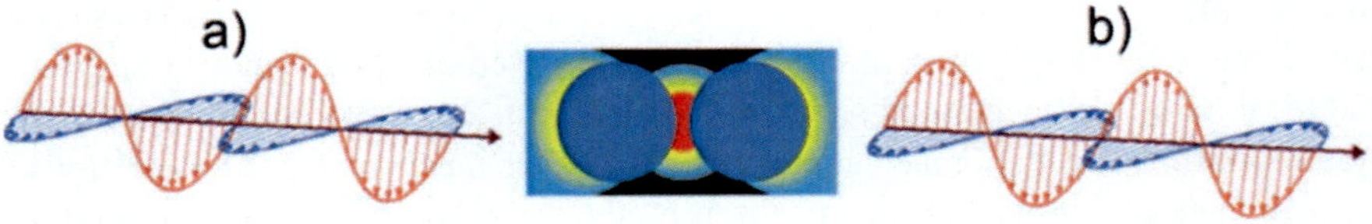

Fig. 8.2 Schematic two-step SERS enhancement mechanism: (**a**) local field enhancement and (**b**) radiation enhancement

triggered by local surface plasmon resonance (LSPR). The local electric field strength at the molecule $E_{loc}(\omega_0, r_m)$ can be enhanced as shown in Eq. (8.1):

$$E_{loc}(\omega_0, r_m) = g_1(\omega_0, r_m)\ E_0(\omega_0) \tag{8.1}$$

where $E_{loc}(\omega_0, r_m)$ is the local electric field strength at position r_m of the molecule. While the enhancement factor assigned for the incident electric field strength is $g_1(\omega_0, r_m)$, the enhanced local field, gives a stronger oscillating dipole $p_m(\omega_R, r_m)$ at the Raman scattering frequency ω_R (Eq. 8.2):

$$p_m(\omega_R, r_m) = \alpha_m^I(\omega_R, \omega_0)\ E_{loc}(\omega_0, r_m) \tag{8.2}$$

The resulting power enhancement factor at r_m is defined as follows (Eq. 8.3):

$$M_{loc}(\omega_0, r_m) = |g_1|^2 = \left|\frac{E_{loc}(\omega_0, r_m)}{M_0(\omega_0, r_m)}\right|^2 \tag{8.3}$$

Radiation Enhancement in SERS The radiation nature of $p_m(\omega_R, r_m)$ are substantially affected by the dielectric properties of its medium and their consequential optical resonance steps.

The radiation enhancement M_{Rad} can then be determined by Eq. (8.4):

$$M_{Rad}(\omega_0, r_m) = \frac{P_{Rad}}{P_0} \tag{8.4}$$

where P_{Rad} and P_0 is the capacity of the total Raman-scattered radiation with/without a nearby plasmonic nanostructure.

M_{Rad} is usually estimated using the optical reciprocity theorem (Eq. 8.5):

$$M_{Rad}(\omega_0, r_m) \approx \left|\frac{E_{loc}(\omega_0, r_m)}{M_0(\omega_0, r_m)}\right|^2 \tag{8.5}$$

The SERS enhancement factor at r_m is given by Eq. (8.6):

$$EF(\omega_0, \omega_R, r_m) \approx \frac{P_{Rad}}{P_{m,0}} = \frac{P_0}{P_{m,0}} \cdot \frac{P_{Rad}}{P_0} = \frac{P_0}{P_{m,0}} . M_{Rad}(\omega_0, r_m) \approx M_{loc}(\omega_0, r_m)\ M_{Rad}(\omega_R, r_m) \tag{8.6}$$

When the frequency of the Raman-scattered light becomes very close to that of the incident light, Eq. (8.6) can be simplified to Eq. (8.7):

$$EF(\omega_0, \omega_R, r_m) \approx \left|\frac{E_{loc}(\omega_0, r_m)}{M_0(\omega_0, r_m)}\right|^2 . \left|\frac{E_{loc}(\omega_R, r_m)}{M_0(\omega_R, r_m)}\right|^2 \approx \left|\frac{E_{loc}(\omega_0, r_m)}{M_0(\omega_0, r_m)}\right|^4 \tag{8.7}$$

Equation (8.7) is the well-known $|E|^4$-approximation for the SERS enhancement factor [15]. Despite its many approximations and simplifications, it helps us to conveniently evaluate the actual experimental SERS enhancements in a single molecule located at a position r. It is also used as a typical parameter to be evaluated for comparison with results gained from theoretical models.

The SERS method appears to be a strongly observed effect in systems capable of coupling plasmons such as electromagnetic resonances with electromagnetic plane waves, i.e., small metal characteristics and grating. Metiu and Das [16] stated on a paradigm of the technique as follows: 'Large enhancements are produced when the structure absorbs the photon and localizes it. Gratings and flat surfaces absorb the photon and 'store' the electromagnetic energy into the surface plasmon; this is delocalized in the direction parallel to the surface but localized in the perpendicular one. This increases the electromagnetic energy density near the surface. A sphere localizes the photon, by plasmon excitation, in all directions and the resulting concentration of electromagnetic energy is larger than that produced by a grating'. Presumably, the gaps between closely spaced metal features, as in aggregated colloids or cold-deposited films, concentrate the electromagnetic energy even further, producing even higher enhancements [16].

8.2.3 Chemical Enhancement Mechanism (CE)

The chemical enhancement mechanism (CE) is related to the changes in the physical and chemical abilities of a compound when adsorbed onto a metal surface, and the changes in the relative intensity and pattern of the spectrum recorded are governed by surface selection rules [17]. The CE is determined as vibrational modulations on the excitation energy and the transition dipole of the resonance. The differences in modulations on molecular orbitals due to atomic motions is frequently used to explain the enhancement arising from vibrational modes. In this regard, it is possible to quantitatively determine the chemical mechanism in SERS using electronic structure theory computations; the chemical enhancements are both system and vibrational mode dependent.

The CE has been less conclusive in the study because most SERS experiments were conducted using roughened surfaces (or particles) for which both mechanisms take part together in the same time. Indeed, some theories of CE invoke the resonance Raman scattering via a "charge transfer intermediate state". Although the formalisms of these theories vary somewhat, their key ideas are quite similar each to other. Adsorbate molecular orbitals are spread into resonances following interactions with electrons in the conduction band. While resonances of levels whose energies lie near the Fermi energy are in part filled, those that are occupied well below are fully filled (cf. Fig. 8.1). Inclusion of metallic states in chemisorption creates new ways for resonance excitation at frequencies much lower than those of intrinsic intramolecular excitations of the free adsorbate molecule. New excitations can be formed from two possibilities, namely, electrons can be changed from the filled adsorbate orbitals to unfilled metal orbitals above the Fermi level (molecule ➔ metal surface charge transfer), or metal electrons can be moved to the partly filled adsorbate affinity level (metal ➔ molecule charge transfer). The most cogent experimental proof for this picture comes from spectro-electrochemical experiments where metal-molecule charge transfer excitations show red shift upon making the electrode potential more negative (by increasing the Fermi energy), whereas molecule-metal charge transfer excitations induce blue-shift. A comprehensive theory has been developed to predict enhancement profiles, mode selectivity, overtones and combinations [18].

The most important part of the CE mechanism is a CT contribution in which a new charge transfer state appears [19]. In fact, when a molecule is adsorbed on a metal surface, its electronic states are changed upon chemisorption. The new electronic states may play a role as resonant intermediate states in Raman scattering. If the Fermi level of the metal is ranged between the highest occupied molecular orbital (HOMO) and the lowest unoccupied molecular orbital (LUMO), then CT excitations probably occur at a lower energy than essential intramolecular excitations of the adsorbate. According to Albrecht's notation [20], in the CT mechanism via Albrecht's, a term (Franck–Condon term), only the thoroughly symmetric modes are resonantly enhanced when the laser excitation is near to an allowed electronic transition and only one excited state gets involved in this progress. The resonance Raman results for vibrational mechanism that are asymmetric are usually detected when these modes are coupled with two excited states of the chromophore. The product of the symmetry models of both excited states should be equal to, or contain, the asymmetric mode. Such an action scheme has been known as the Herzberg–Teller mechanism or B-mechanism in the Albrecht's notation [20].

It is a difficult task to characterize the chemical enhancement from the EM mechanism via a visual theoretical approach. The charge difference density (CDD) in a 3D cubic image can best be used for such a visualization, which can visually reveal the photoinduced CT between the metal and the molecule, or the intra-cluster excitation at resonant electronic transitions. The CT between metal and molecule provides us with an evidence of a CT mechanism, whereas an intramolecular CT or a redistribution establishes an evidence for molecular resonance, and finally, the intra-cluster charge redistribution gives an evidence for the EM mechanism [20].

8.2.4 *Use of Silver Nanoparticles in SERS Applications*

As stated above, the SERS experiment has been proven to be a highly sensitive technique for the detection of molecules in very low concentrations, yielding a very rich set of structural information. The interactions between the adsorbed molecules and the surface of plasmonic nanostructures that are often the outstanding substrates such as gold (Au), silver (Ag) or copper (Cu) play an important role in the successful applications of SERS. In general, the SERS substrates formed by Au and Ag

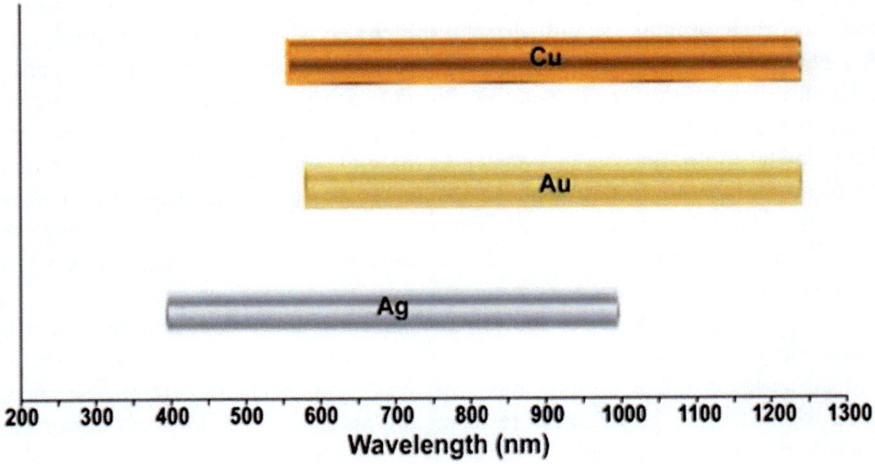

Fig. 8.3 Approximate wavelength ranges in which Ag, Au, and Cu nanoparticles are well characterized and established to be used in SERS experiments. Printed with permission from Ref. [21]. Copyright 2012, Elsevier

particles are more popular than Cu because they are air-stable materials, whereas Cu is more reactive. Let us define a localized surface plasmon resonance (*LSPR*) as a collective oscillation of electrons at the interface of the metallic substrate, which could be formed through the electron-magnetic interaction of the metal with the light at a particular wavelength. Thus, all the coinage metals exhibit suitable LSPRs that cover most of the visible and near-infrared wavelength range in which most Raman measurements happen, also making them suitable to use (Fig. 8.3) [21].

It has been shown that the Raman intensity of pyridine on an Ag electrode with the presence of Cl ions can be adjusted by varying the potential across the electrode. This phenomenon was explained by the fact that the pyridine molecule is adsorbed on the Ag surface inducing an enhancement due to a high concentration at the Ag surface. Subsequently, Van Duyne and co-workers [22] recorded that with pyridine adsorbed on rough Ag surface, Raman signals can be enhanced by 10^5–10^6 times. As for an explanation for the experiment, these authors used a tetrahedral Ag_{20} cluster model for representing the silver surface [22]. Such a surface model was confirmed through a detailed comparison of the SERS spectrum computed using quantum chemical methods with experiment. The tetrahedral Ag_{20} cluster emerges as the largest size so far used for SERS calculations found in the literature. More importantly, the tetrahedral Ag_{20} cluster provides us with a model nanoparticle structure that exhibits a well-defined fragment of a (111) surface, as well as edges and vertices that can be adopted to model coordinately unsaturated sites on the particle. As the tetrahedral Ag_{20} cluster turns out to be a reasonable reference structure for the determination of the chemical contribution to SERS enhancements, it is the most popular cluster size chosen for SERS calculations. In the following sections, we focus on the use of silver clusters to simulate nano-substrates in the interpretation of SERS experiments.

8.3 Geometrical and Electronic Structures of Silver Clusters

It is well known that although the solid bulk behaviors of the coinage metals, Cu, Ag and Au, are nearly similar to each other, some important differences emerge in the form of their small size clusters. Due to the strong surface and typical relativistic effects, the most stable structures of gold clusters tend to adopt planar and tubular structures, whereas three-dimensional (3D) structures are more popular for Cu clusters [23]. The behavior of silver clusters lies between those of Cu and Au counterparts, that is, planar structures are preferred for small Ag clusters, whereas 3D Ag structures appear slightly earlier than the Au cluster as the cluster size increases. Li and co-workers [24] found a significantly large energy gap and electron affinity for the cluster Au_{20}, based on photoelectron spectroscopy data. Density functional theory (DFT) calculations demonstrated that the most stable isomer of Au_{20} clearly possesses a tetrahedral T_d structure [24].

In general, these results may be elucidated in terms of a competition between the effects caused by an electronic order and an atomic order. For magic clusters- cluster containing 13, 38, 55 atoms, etc. a closed packing of atoms, or an atomic order, can be dominant. Accordingly, icosahedron or cuboctahedron are usually determined as the lowest-energy isomer for magic coinage metal clusters. On the other hand, based on the electron shell theory, 20 valence electrons form a closed electron shell, and consequently, an electronic order plays an important role in determining structural and electronic characteristics of 20-atom clusters. Based on the above discussion, the ground state of both Ag_{20} and Cu_{20} clusters likely results from a competition between both effects.

Recent computational studies apparently confirm that the most stable forms of Ag_{20} are energetically degenerate [25–53]. Quantum chemical results point out that while Cu_{20} adopts a compact C_s ground state structure, a pair of T_d and C_s isomers having quasi-degenerate energy content emerge for the Ag_{20} ground state. As a matter of fact, a compact C_s structure is found for the most stable Cu_{20} isomer whereas a T_d structure is located for the lowest-energy structure of Au_{20}. The energy difference of 0.6 eV between both compact C_s and T_d structures of Cu_{20} is significantly large [38]. Earlier computations pointed out that the energy difference between both lowest-lying isomers of Ag_{20}, being the high symmetry T_d and a low symmetry compact C_s structure, amounts to only 0.06 eV, indicating that their energy levels are practically degenerate [35]. It is well known that *d* electrons and *s*–*d* hybridization constitute an essential factor influencing on the shape of cluster structure. *d* electron populations are slightly smaller for small Au clusters as compared to those of corresponding Ag clusters. Moreover, as mentioned above, stronger surface and relativistic effects in Au clusters appear to lead to some basic differences between Ag_{20} and Au_{20}. Apart from these isomers, other C_{2v} or lower

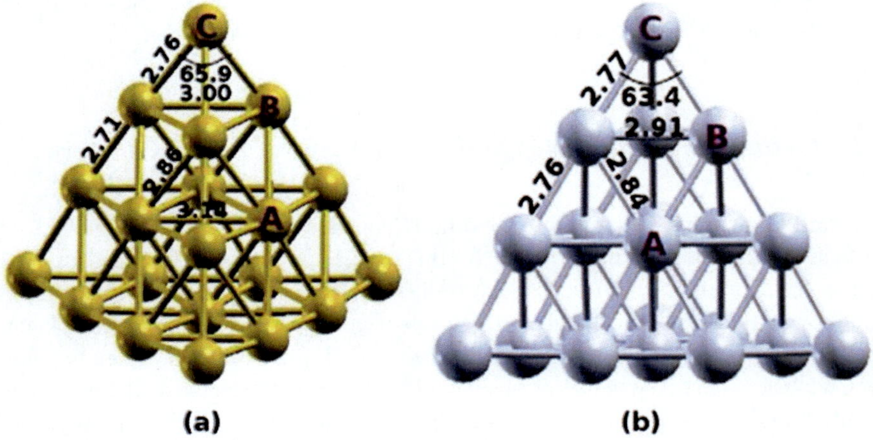

Fig. 8.4 Optimized structures and some selected binding distances of (**a**) $^1A_1$ state of T_d Au_{20} cluster, and (**b**) $^1A_1$ state of T_d Ag_{20} (PBE96 /SDD-RECP). Printed with permission from Ref. [30]. Copyright 2017, Elsevier

symmetry isomers of Ag_{20} were computed to be substantially less stable than both T_d and C_s structures [35].

The tetrahedral Ag_{20} is a relaxed moiety of the face-centered cubic (fcc) lattice of bulk silver. Figure 8.4 gives some bond distances for both tetrahedral Ag_{20} and Au_{20} clusters. The typical Ag-Ag bond lengths determined by DFT computations (using the functional PBE96) amount to 2.84, 2.77 and 2.91 Å (Fig. 8.4b) [30]. Similar equilibrium bond lengths were obtained using different DFT functionals with some small variations that could be due to the basis sets employed with *1*s-*4*p frozen core for Ag atoms. Similar to Au_{20}, three unique sites for the T_d Ag_{20} comprise of the central atoms along the planes of the faces of the pyramid, position denoted as A in Fig. 8.4b, the vertices of the pyramid (position C) and the atoms along the edges of the pyramid (position B).

Figure 8.5 displays various local energy minima obtained for Ag_{20} by Nhat et al. [25] who used the LC-BLYP functional in conjunction with a cc-pVDZ-PP basis set. As expected, the tetrahedral structure **Ag_{20}_2** displayed in Fig. 8.5 was again reported as the most stable form of Ag_{20} by several DFT calculations using different functionals including the PW91, B3LYP, TPSS and M06 [31–33, 37]. Nevertheless, Nhat et al. [25, 27] pointed out that the lower symmetry isomer **Ag_{20}_1** is the lowest-lying isomer rather than the higher symmetry T_d **Ag_{20}_2**. In fact, the **Ag_{20}_1** can be generated by adding extra Ag atoms to an icosahedral core. Recent study using the M06 functional [26] appeared to support the findings obtained earlier using the LC-B3LYP and PBE functionals, but with a negligible energy difference between both isomeric forms.

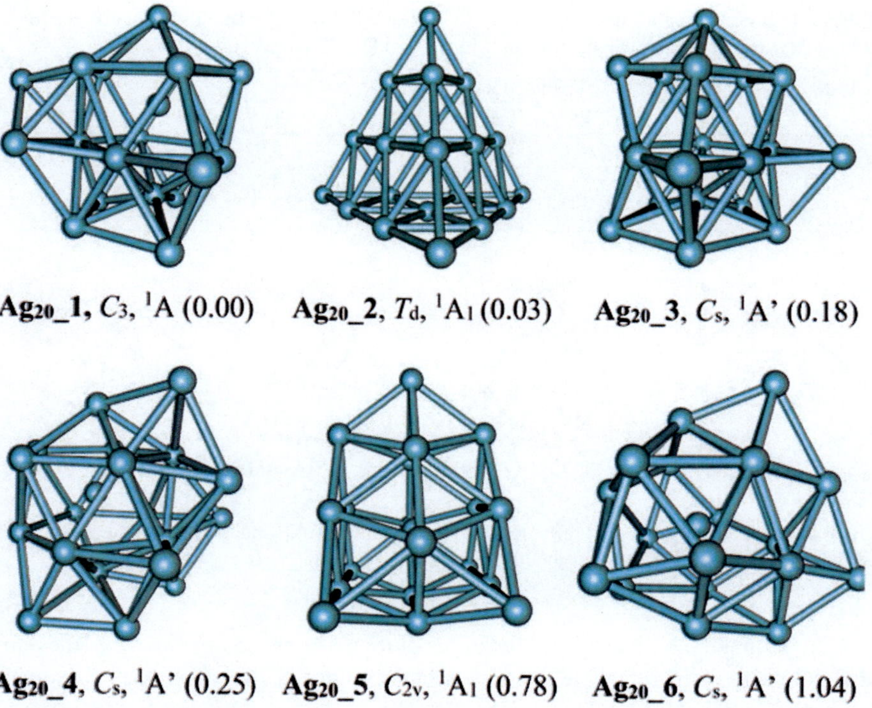

Fig. 8.5 Lower-lying structures optimized for Ag_{20} (LC-BLYP/cc-pVDZ-PP), along with symmetry point group, electronic state and relative energy (eV) with respect to the lowest-lying isomer **Ag$_{20}$_1**. Printed with permission from Ref. [25]. Copyright 2019, Wiley

In agreement with DFT calculations mentioned above, tight-binding molecular dynamics (TBMD) introduced in atomistic simulations [53] were also used to locate a stable form of Ag_{20} which is structurally similar to **Ag$_{20}$-1** in Fig. 8.5 [25]. Remarkably, the T_d **Ag$_{20}$_2** was computed to be only ~0.03 eV higher in energy than **Ag$_{20}$_1** by this MD simulation (Fig. 8.5). A larger energy gap was found by another method (~0.28 eV when using the PW91/cc-pVDZ-PP) [27]. Also formed by adding extra Ag atoms to the icosahedron Ag_{13}, only two isomers **Ag$_{20}$-3** and **Ag$_{20}$_4** are found to be less stable by 0.2–0.3 eV. Remaining isomers **Ag$_{20}$_5** and **Ag$_{20}$_6** seen in Fig. 8.5 are much less stable, being located at 0.8–1.0 eV above **Ag$_{20}$-1**. Results obtained using a variety of DFT functionals including the BP86, BPW91, PBE0, PBE and TPSS are presented in Table 8.1. Within the current error margins expected for DFT methods, being $\pm$ 0.2 eV on energetic parameters, the lowest-energy structures of Ag_{20} can be predicted to be the energetically degenerate isomers C_3 **Ag$_{20}$_1** and T_d **Ag$_{20}$_2,** even also the C_s **Ag$_{20}$_3**. Such a pattern basically differs from those of both isovalent coinage metal systems Cu_{20} and Au_{20} as illustrated in Fig. 8.6. In fact, while Cu_{20} prefers the C_3 isomer, Au_{20} clearly adopts the T_d form.

Table 8.1 Lower-lying isomeric structures of Ag_{20} (cf. Fig. 8.5) and their relative energies computed using different DFT functionals with a cc-pVDZ-PP basis set

Cluster	Relative energy (eV) Functional						
	LC-BLYP	PW91	BP86	BPW91	TPSS	PBE0	PBE
Ag_{20}_1	0.0	0.0	0.0	0.0	0.0	0.0	0.0
Ag_{20}_2	0.03	0.3	0.2	0.3	0.7	0.3	0.5
Ag_{20}_3	0.2	0.1	0.2	0.1	0.1	0.1	0.1

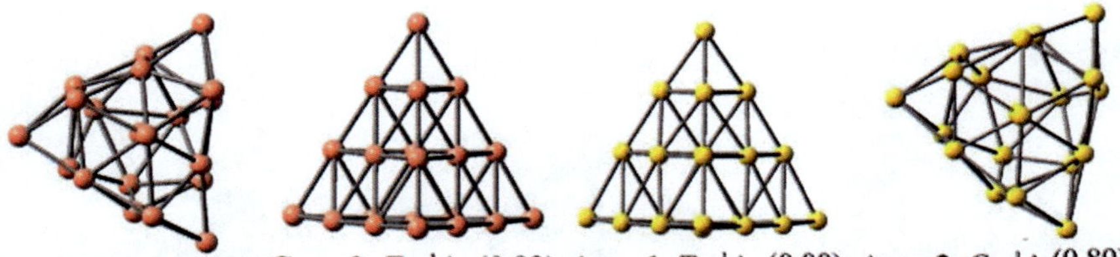

Cu_{20}_1, C_3, ^1A (0.00) **Cu_{20}_2**, T_d, $^1A_1$ (0.93) **Au_{20}_1**, T_d, $^1A_1$ (0.00) **Au_{20}_2**, C_3, ^1A (0.89)

Fig. 8.6 Lower-energy C_3 and T_d structures located for Cu_{20} and Au_{20} (eV). Printed with permission from Ref. [27]. Copyright 2018, Springer

Concerning the typical spectral features of lower-lying isomers of Ag_{20}, they are also not quite distinguishable from each other. Figure 8.7 displays the vibrational spectra of some lower-lying isomers for Ag_{20} cluster. The calculated infrared spectrum of the lowest-energy isomer **Ag_{20}_1** contains various low-frequency modes in the range of 80–150 cm^{-1} as well as stronger peaks centered at 70 and 170 cm^{-1}. However, the spectra of either **Ag_{20}_2** or **Ag_{20}_3** appear to be similar to each other, making it hard to identify them solely based on vibrational signatures. The high symmetry tetrahedron **Ag_{20}_6** presents **a** much simpler vibrational IR spectrum, defined by bands located at ~70 and 156 cm^{-1}.

The experimental absorption spectrum of Ag_{20} was recorded and interpreted [27, 28]. Accordingly, the spectrum recorded in argon is composed of a broad peak located at 3.7 eV (335 nm) and a much less intense band centered at 4.0 eV (310 nm). Figure 8.8 displays the calculated absorption spectra of a few lowest-energy isomers of Ag_{20} using density functional theory methods (TD-DFT B3LYP/ccpVDZ-PP). The lowest-energy isomer **Ag_{20}_1** turns out to be characterized by an intense band centered at ~3.8 eV (325 nm) obtained from four close lines at 3.75, 3.82, 3.93 and 3.99 eV [27]. Previous TD-DFT calculations using the CAM-B3LYP functional [45] identified corresponding lines at 3.77, 3.81, 3.89 and 3.97 eV. Overall, Fig. 8.8 shows that the calculated spectra for other isomers also reproduce the shape of the experimental spectrum. In this regard, each of the two lowest-lying isomers can equally be used as a model for the Ag_{20} cluster.

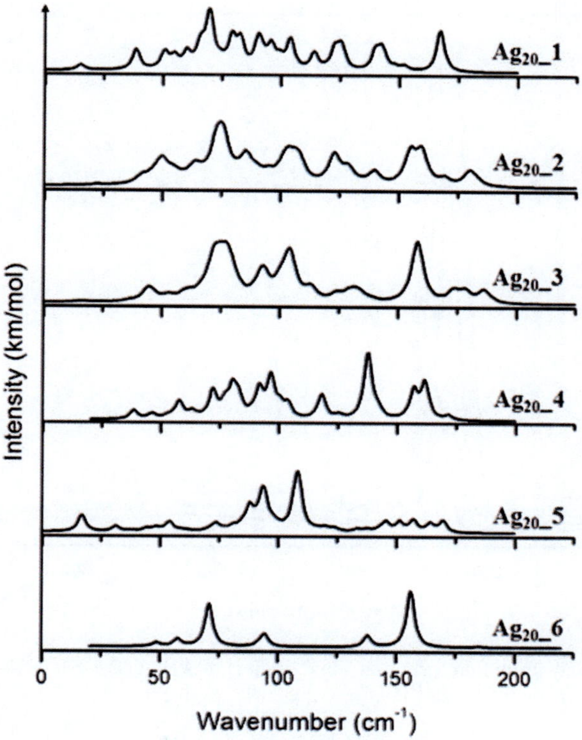

Fig. 8.7 Predicted IR spectra of lower-lying structures of Ag_{20}. Printed with permission from Ref. [27]. Copyright 2018, Springer

8.4 Detection of Single Molecules Using SERS Technique

Since the first observation of single molecule sensitivity under SERS phenomenon [48, 49], its high applicability has attracted much attention from scientists. The single molecule character of the detected SERS spectra was demonstrated by applying Poisson data. In fact, the concentration of the single molecules is ~100 times smaller in comparison to that of the aggregated silver nanoparticles, and the SERS spectra are determined by the Poisson model, relating to the number of molecules within the probed volume [12]. Most earlier studies on the chemical SERS enhancement contribution were conducted with pyridine, an aromatic molecule, which becomes a kind of classical test molecule for SERS experiment. The chemical enhancement factor could be determined by comparison of the signal of the Raman spectra of pyridine and its complexes with several types of silver particles.

Jensen et al. studied the size-dependence of the enhanced Raman scattering of pyridine adsorbed on a series of silver Ag_n (n = 2–8, 20) clusters using the TD-DFT method (Fig. 8.9) [31, 50]. Both the appearance of the SERS spectral features and the resulting enhancement are significantly dependent on the cluster size. The

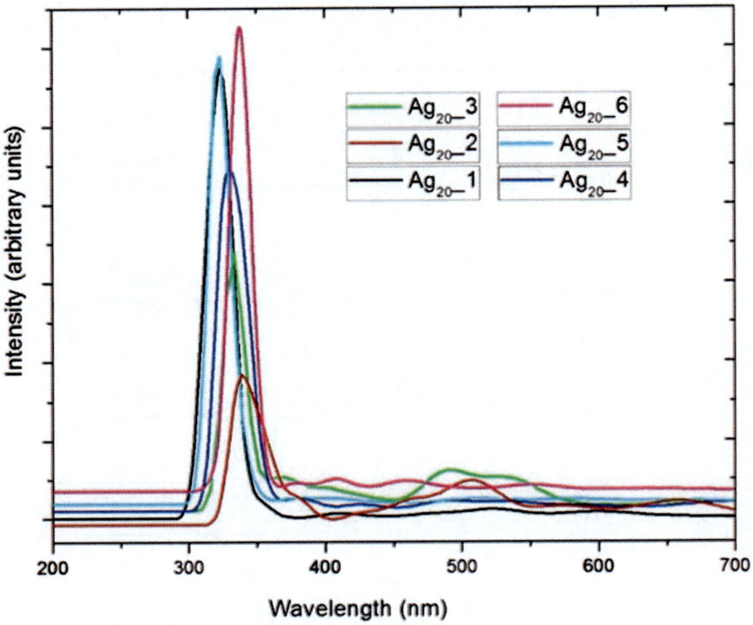

Fig. 8.8 Calculated absorption spectra of some isomers of Ag_{20} (TD-DFT B3LYP/ccpVDZ-PP). Printed with permission from Ref. [27]. Copyright 2018, Springer

total enhancements for the complexes were calculated to be at a factor of 10^3–10^4 stronger. The strongest enhancement was found for the Ag_2-Py complex. Due to the fact that such an enhancement trend depends not only on the distance connecting the molecule to the center of the metal cluster but also on the resonance polarizability, the dominance of the electromagnetic enhancement can also be argued for. In other words, the enhancement mechanism caused by the small silver clusters appears close to that of an electromagnetic mechanism for larger nanoparticles.

To better understand the contribution of a chemical mechanism to the enhancements observed in SERS, Franzen et al. [51] calculated the dimensionless excited state displacements of pyridine on Ag clusters. The Pyr-$(Ag)_n$ clusters with n = 2, 4, 8, 14, 20 were regarded as supermolecules that have absorption bands due to excitations of both Ag_n and Pyr moieties, as well as a possible charge transfer bands. Enhancements of 10^3–10^6 were determined for all complexes, which is quite close to the overall experimental SERS effect found for the Ag-pyridine system, indicating that resonance Raman is a major contributor to the effect.

Schatz and co-workers [31] showed a detailed study of the enhanced Raman scattering of a pyridine-tetrahedral Ag_{20} model system using TD-DFT computations. Similar to several earlier results, the interaction between pyridine and a silver cluster was found to be formed through the nitrogen atom in a perpendicular fashion (Fig. 8.10). The computed bond distance between the N atom and the closest silver

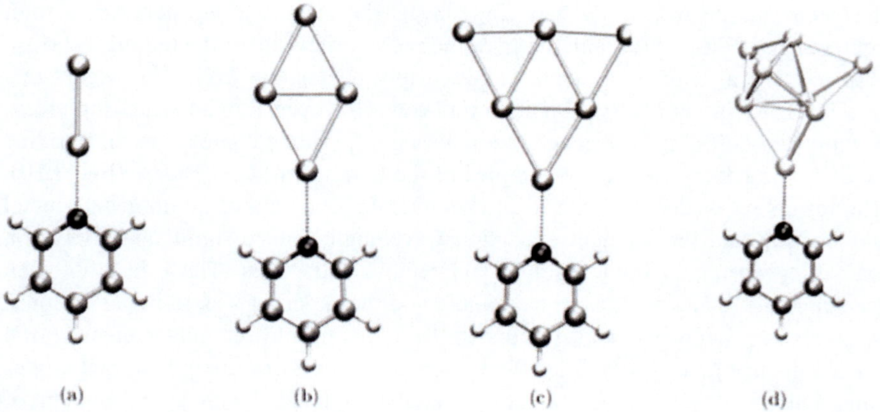

Fig. 8.9 Optimized geometries for Ag_n-Py (n = 2–8) complexes. Printed with permission from Ref. [50]. Copyright 2007, The American Chemical Society

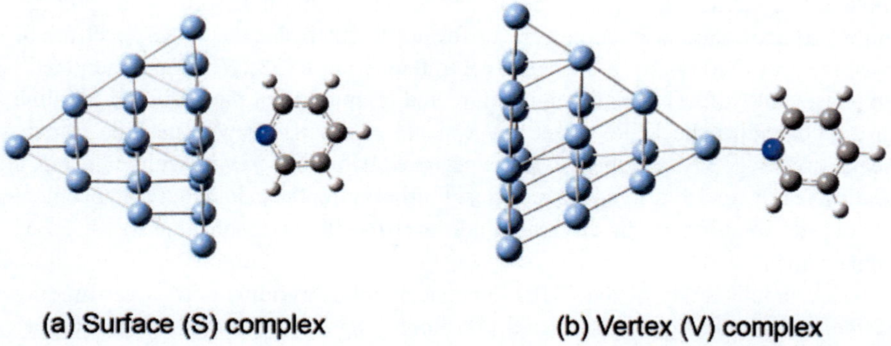

Fig. 8.10 Configurations of two pyridine-Ag_{20} complexes having C_s symmetry: (**a**) S-complex and (**b**) V-complex. Printed with permission from Ref. [31]. Copyright 2006, The American Chemical Society

atom is 2.66 and 2.46 Å for the S-complex (Fig. 8.10a) and V-complex (Fig. 8.10b), respectively.

Modeling the normal and enhanced Raman scattering of a metal-molecule-metal model on the atomic scale helps us to understand the identity of the observed SM-SERS enhancements. Due to the limitations of computational requirements of high accuracy methods, a time-dependent density functional theory (TD-DFT) method based on a short-time approximation for the Raman cross-section was used to determine both on and off-resonance Raman scattering, and this method was used to approach the absorption and Raman scattering of a pyridine-Ag_{20} model system [52]. Using this approach, a remarkable enhancement of the Raman activity due to the binding of pyridine to the Ag_{20} cluster was observed. This is a convenient method for the detection of pyridine molecules in the gas phase, which is known as a difficult target for detection without complexation to a metal surface. For example, the electronic spectrum of the bare Ag_{20} cluster shows a triple-degenerate excitation centered at ~3.4 eV. After binding pyridine at the vertex site of the tetrahedral Ag_{20} (Fig. 8.10b), a split of the peak from a triply degenerate excitation occurs due to a lowering in symmetry of both complexes. However, Raman scattering cross-sections are still almost degenerate for the Ag_{20}-pyridine complex when pyridine binds to the vertex site, i.e., V-complex of the tetrahedral Ag_{20} cluster (Fig. 8.10). The results show that both the local chemical environment of the molecule-metal binding site and the incident excitation wavelength induce significant effects on Raman spectrum, in both absolute and relative intensities. The effects of each mechanism enhancement were examined, and the ordering was found as follows: static chemical enhancements (factor of 10) < charge-transfer enhancements (10^3) < EM enhancements (10^5). The effect of enhancements of principal normal modes can be justified by their vibrational motion and the local chemical environment of the molecule. This gives us a simple but clear picture showing key details of the enhancements.

To further probe the enhancement caused by the CHEM mechanism, ab initio molecular dynamics simulations combined with a Fourier transform of the polarizability autocorrelation function were performed to determine the SERS spectrum of pyridine adsorbed on Ag_{20} cluster at room temperature [53]. This approach could emphasize the effects of both temperature and orientation of the adsorbed pyridine in the change of the SERS spectrum. Calculated results showed that the Raman peaks are in good agreement with experimental findings, even without using a scaling factor for vibrational frequencies. Furthermore, the calculated enhancement factor was found to be in the magnitude of 10^1–10^3, no doubt due to a CHEM enhancement.

A systematic study of the CHEM enhancement of pyridine derivatives binding to the tetrahedral Ag_{20} was carried out employing TD-DFT methods by simply changing the substituents on the pyridine ring in both para and meta positions [54]. This approach allowed the direct chemical interactions between the pyridine ring and the metal cluster to be probed. Results showed that the enhancement was not increased, because a larger amount of charge is actually transferred from the pyridine ring to the cluster. Instead, the energy difference between the HOMO of the metal cluster and the LUMO of the molecule seems to play a crucial role in the CHEM enhancement. A molecule giving a significant increase of this HOMO-LUMO gap tends to bring in a stronger CHEM enhancement, which is in good agreement with experimental findings.

Sun et al. investigated the chemical enhancement of SERS via CT from Ag_{20} to pyridine on resonance excitation, and the EM enhancement of SERS via intra-cluster charge redistribution on an electronic intra-cluster collective oscillation excitation [55]. From the ultraviolet to the infrared region, enhancement spectra were constructed to classify different incident excitation wavelength regions for the main chemical and EM enhancements for pyridine binding to Ag_{20} cluster in the S-complex (Fig. 8.10). The results were confirmed by the formal fragmented experimental and theoretical SERS studies of pyridine at various laser frequencies.

Silver substrates, stimulated by their multiple roles in catalysis, photography, and various theoretical and experimental studies, have been heavily investigated for the last several decades [21, 56–59]. These materials have larger potentials for clinical and therapeutic treatments due to their antibacterial and antifungal properties. Furthermore, the optical abilities of silver nanoparticles at the scale of nanometers, are also found to be superior in comparison with other transition metals [60, 61]. Therefore, silver nanoparticles and clusters have been widely applied in the field of sensors, biosensors and biomedical diagnostics [62–64]. There have been many studies that are developing techniques using sub-nanometer silver clusters interacting with biomolecules such as amino acids, peptides or nucleotides for selective detections of heavy metal ions, thiols, peptides, herbicides and nucleic acids . . . [30, 48, 51, 65, 66].

The thiol-containing cysteine amino acid can be found at many proteins, and it operates as a link to anchor these proteins to inorganic derivatives. Therefore, the interaction of cysteine with silver nanoparticles is considered as a typical illustration for adsorption of biomolecules and functionalization of metal surfaces. And so, let us consider in some detail the interactions of cysteine with a small silver cluster.

It is important to note that not only the binding site but also the binding energy are modified in various phases. In both gaseous and highly acidic solutions, the cysteine prefers to bond to silver clusters through the N-atom of the amine group. However, the S-atom of the thiolate group is the most dynamically favorable site for the bonding due to the cysteine mostly exists in deprotonated forms, in aqueous environments [67]. To illustrate this point, Fig. 8.11 shows a normal Raman spectrum of deprotonated cysteine and its SERS spectrum on Ag_{10} cluster.

According to various research [68–70], the Raman spectra of amino acids has many important peaks in the range above 3000 cm^{-1} and below 1700 cm^{-1}. Particularly, some low-intensity peaks at 1665–1585 cm^{-1}, 1605–1555 cm^{-1}, 1425–1393 cm^{-1} and 1340–1315 cm^{-1} correspond to NH_2 bending vibrations, C−C stretching peaks, and CH/CH^2 deformations, respectively. Other weak and broad bands located above 3400 cm^{-1} are assigned to symmetric and asymmetric N−H stretching vibrations. The highly overlapped bands at 3000–3200 cm^{-1} which resulted from C–H stretching modes, is the most characterized peak of an organic Raman spectrum. For Raman spectrum of cysteine, there are some strong bands near 700 and 2600 cm^{-1} that correspond to the C–S and S–H stretching model, respectively.

Binding energies (E_b) of cysteine to small Ag clusters were calculated to increase in the following ordering: aqueous solution (about –22 to –30 kcal/mol) < vacuum < acidic solution (about –10 to –16 kcal/mol). Among the small clusters, Ag- has a low affinity with cysteine, while the trimer radical Ag_3 has a large Eb. The frontier orbital's results also indicate that both forward donation (HOMO ➔ LUMO of the Cys → Ag_n process) and backward donation (Ag_n → Cys) give to the formation of intermolecular bonds in silver clusters-cysteine complexes in a balanced manner. These complexes are stabilized by both forward and backward pathways, indicating that the metal clusters play a role as either an electron acceptor or an electron donor. Hence, there are strong overlaps between HOMO and LUMO of Ag_n-cysteine

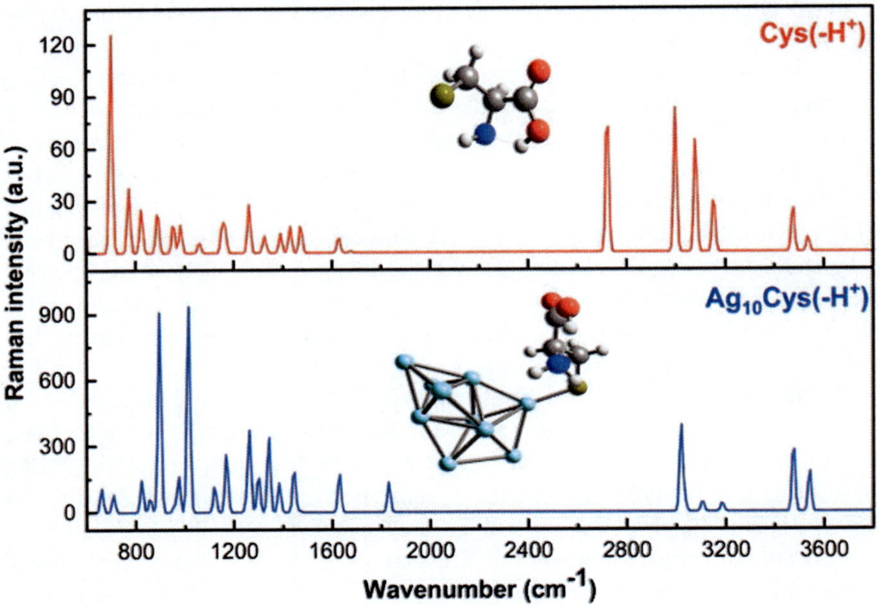

Fig. 8.11 Normal Raman spectrum of deprotonated cysteine and its SERS spectrum on an Ag_{10} cluster. Simulations were performed in aqueous solution using the LP-BLYP functional with the cc-pVTZ-PP for Ag and cc-PVTZ basis set for other atoms. Printed with permission from Ref. [67]. Copyright 2021, The American Chemical Society

complexes to form an σ-type bonding orbital. Moreover, further π-type bonding orbitals can be formed by the cysteine LUMO and the Ag_2 HOMOs.

In the computed UV-Vis spectra (Fig. 8.12), silver clusters-cysteine complexes are distinguished by high absorbance peaks at longer wavelengths than the corresponding bare species. Most of the absorption spectra of Ag_n•Cys(-H+) complexes have some characteristic intense signals located at the wavelength above 400 nm, except for complexes of cluster Ag_7 and Ag_8 sizes (cf. Fig. 8.12). Weaker transitions can be determined in both the higher and lower energy regions. For example, beside the highest peak at ~400 nm, the UV-Vis spectrum of Ag_2•Cys(-H+) has a less intense absorption band located in the range of 330–360 nm. In aqueous solution, the most important peaks are markedly blue-shifted as compared to those predicted in vacuum, except for Ag_8•Cys(-H+) and Ag_{10}•Cys(-H+). Experimental results show that there is a broad UV-Vis absorbance peak located at ~400 nm when cysteine binds covalently to silver nanoparticles [71]. Therefore, silver clusters mainly contribute to the absorption spectra of Ag_n•Cys complexes since these electronic transitions do not appear in the visible range of free cysteine. The UV spectrum of cysteine in aqueous phases has an absorption maximum at ~200 nm, relating to the combined $n \rightarrow \pi^*$ and $\pi \rightarrow \pi^*$ transitions of the carboxylic chromophore [72].

Figure 8.11 also plots the SERS spectra of the deprotonated cysteine bonded to the most stable isomer of the Ag_{10} cluster which represents the surface of a

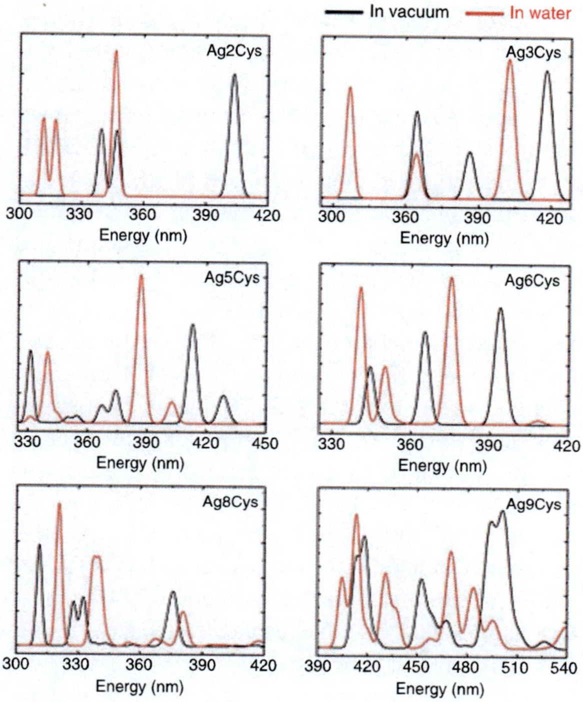

Fig. 8.12 TD-DFT absorption spectra for Ag_n•Cys(-H^+) complexes with n = 2–10 (LC-BLYP). Printed with permission from Ref. [67]. Copyright 2021, The American Chemical Society

silver nanostructure [67]. A comparison of these spectra produces some insights into important changes in band intensities due to the SERS effect. First, the cysteine SERS spectra do not have the S–H band near 2600 cm^{-1} due to the deprotonation of the thiol group at the metal surface, which is nearly the same as the experimental SERS spectrum of cysteine on silver nanoparticles [73]. Notably, there is a large enhancement of vibrational peaks due to the NH^2 bending and stretching models that are significantly small in the normal Raman spectrum of cysteine. Previously, some signals in the SERS spectra at 900–1000 cm^{-1} region due to the C–COO stretching and CH_2 bending models, were substantially enhanced [74]. However, recent calculated results [67] validated that these peaks are mostly due to the N−H bending vibrations directly oriented to the silver surface.

Furthermore, the experimental SERS spectrum of cysteine in a silver colloidal solution is also defined by the enhancement of the signals above 3100 cm^{-1}. This can be explained based on the assumption that there is an interaction between the amine head and the metal surface [74]. Quantum chemical computations [67] proved that the weak Ag–H–N bonding plays an essential role in the SERS enhancement of both N−H bending and stretching vibrations, more than an Ag−NH_2 covalent bond. Finally, there is a significant increase of the peak related to C−S stretching vibration in the energy of near 700 cm^{-1} [67], indicating that the C−S bond corresponds to the adsorption of cysteine on silver surfaces. This is demonstrated by Raman spectroscopy when the enhancement of the NH_2 vibration is significantly larger for this peak. However, this band is distinctive from that of other amino acids that do not contain a thiol functional group. Whereas the SERS chemical enhancement is accompanied with an electron transfer, their eventual correlation cannot clearly be determined. Odd-numbered Ag clusters yield more enhanced SERS peaks than even-numbered one, which can be explained by the radical ability of the former. Overall, such a SERS signature correlated with a C–S model is significantly differentiated from many other amino acids and can be used as an ideal index for a selective detection and identification of the cysteine derivatives.

Various studies on both experimental and theoretical approaches were conducted to explain the quality of the binding of different biomolecules such as amino acids and DNA nucleobases to silver metal surfaces [75–77]. These interactions act as a form of molecular chemisorption rather than a formation and breaking of new chemical bonds. The anchoring bonds, which are formed by the electron lone pair of the X atoms (X = S, N, O) and the anti-bonding orbitals of the metals, play an essential role in the stability of these complexes [78]. Another significant contribution is the non-conventional O–H$\cdots$M hydrogen bonds formed by an electron transfer from the metal atom (M) to the hydrogen of X-H bonds. Recent studies also indicated that deprotonated cysteine residue tend to stabilize the fluorescent silver clusters formed in the protein templates [79].

Notwithstanding the significance of nano-bio interfaces, the nature of interactions between proteins and silver surface is still a matter of debate. Although some theoretical studies were carried out to address relevant issues, most computational investigations were executed in the gas phase, and the effects of the biological environment have not been taken into account. Investigation on the binding of M^+ ions and some small clusters M_n (M = Au, Ag, n = 2, 3) with several amino acids and DNA bases also excluded the solvent effects [35, 80]. A recent systematic theoretical study [50] aimed to decipher the adsorption behaviors of the thiol-containing cysteine on a silver surface. Small silver clusters Ag_n in the size range of n = 2–10 were adopted as reactant models to simulate the metallic nano-surface. Molecule-surface interactions were simulated not only in vacuum but also in aqueous solution, and a chemical enhancement mechanism of the SERS phenomenon in these systems was also inspected [50, 81].

An et al. [64] recently prepared the flower-like silver nanoparticles (Ag-NPs) of dimension going from 0.3 to 0.5 μm. The 150 mg $AgNO_3$ and 100 mL HNO_3 0.1 M were added together and stirred until the complete dissolution. Another solution of 500 mg ascorbic acid in 10 mL deionized water was also prepared. The two solutions were then stirred together at room temperature for 3 h. Ag-NPs were thus obtained after washing the obtained products in several hours and centrifuging at 4000 rpm [64]. The obtained Ag-NPs with the surface properties of flower tips favors the adsorption of the small molecules on its rough surface. The normal Raman and SERS spectra were recorded by a Raman spectrometer using 633 excited lasers. This apparatus allowed the spectral range from 2000 to 200 cm^{-1} to be recorded The Ag-NPs obtained above were used as SERS substrate to analyze the melamine

($C_6H_6N_6$) that was illegally added into the milk to increase the apparent protein content. The experimental normal Raman spectrum shows two intense peaks located at 676 and 983 cm^{-1} corresponding to ring breathing vibration and C-N-C bending vibration, respectively. And the typical peak at 676 cm^{-1} is shifted into 684 cm^{-1}, and the one at 983 cm^{-1} is shifted into 980 cm^{-1} in the experimental SERS spectra of melamine. Overall, the melamine-Ag_n cluster complexes are shown to be a good model for SERS modeling in reproducing the CE mechanism via charge transfer (CT) process.

In fact, An et al. [64] evaluated the interactions of the melamine molecule with different Ag_n cluster models with $n = 4$, 8, 10 and 20 atoms. The most stable interaction configurations found to consist in the perpendicular interaction of melamine at its ring-N atom onto one Ag atom located at the corner site of the cluster. The experimental vs. calculated values of two typical peaks on SERS spectra are in good agreement with each other. The experimental peak observed at 676 cm^{-1} is also found on the calculated SERS spectra at 688 and 685 cm^{-1} by using the Ag_4, Ag_8, Ag_{10}, and Ag_{20} models, respectively. Similarly, the experimental peak centered at 983 cm^{-1} is reproduced at 1003, 996, 998 and 983 cm^{-1} using the Ag_n clusters.

Furthermore, the cluster sizes exert a noticeable influence on the intensities of the Raman and SERS signals. As seen in Fig. 8.13, while the intensity of the 685 cm^{-1} peak in the normal Raman spectra is about 1.4 (Fig. 8.13a), the intensity of this peak increases 2.1 times (intensity of 2.9) by using the Ag_4 cluster model (Fig. 8.13b). And when using the Ag_8 and Ag_{10} clusters, the intensity of 685 cm^{-1} peak also increases up to 2.4 times (intensity of 3.5) and 2.6 times (intensity of 3.7) (Fig. 8.13c, d), respectively. Finally, in the case of melamine–Ag_{20} cluster, the peak intensity is saturated at 3.6 (increase by 2.5 times, Fig. 8.13e). Overall, these results again demonstrate that the pyramidal Ag_{20} cluster is a good model to reproduce spectroscopic properties of the absorbed molecule.

The SERS technique is also used to detect pesticides that have played an important role in modern agriculture. Besides their obvious beneficial actions, the intensive use of pesticides did and still cause much pollution to the environment. The over-abundant amounts of pesticides accumulated in the soil, water, plants, crops, food and beverage, etc. also severely affect human health because of their toxicity. A pesticide monitoring method using reliable and reproducible detection methods is, therefore, an essential demand. Although many chromatographic methods such as LC/GC-MS, HPLC or TLC can detect the ultralow concentration of pesticides, the SERS has emerged as a powerful technique in the analysis of pesticide residues owing to its several fundamental conveniences such as ultrasensitive and on-site capacity with low-cost detection, faster turnover, simpler protocols, and in situ samplings.

A recent theoretical study by Ngo et al. [63] provides further understanding on the effects of chemical enhancement (CE) mechanism and the influence of interaction configurations (or geometries) of chlorpyrifos (CPF, which is a pesticide) with nanoparticles on the SERS spectra. Interaction of the CPF molecule with Ag nanoparticles was considered not only by using the Ag_{20} cluster model but also

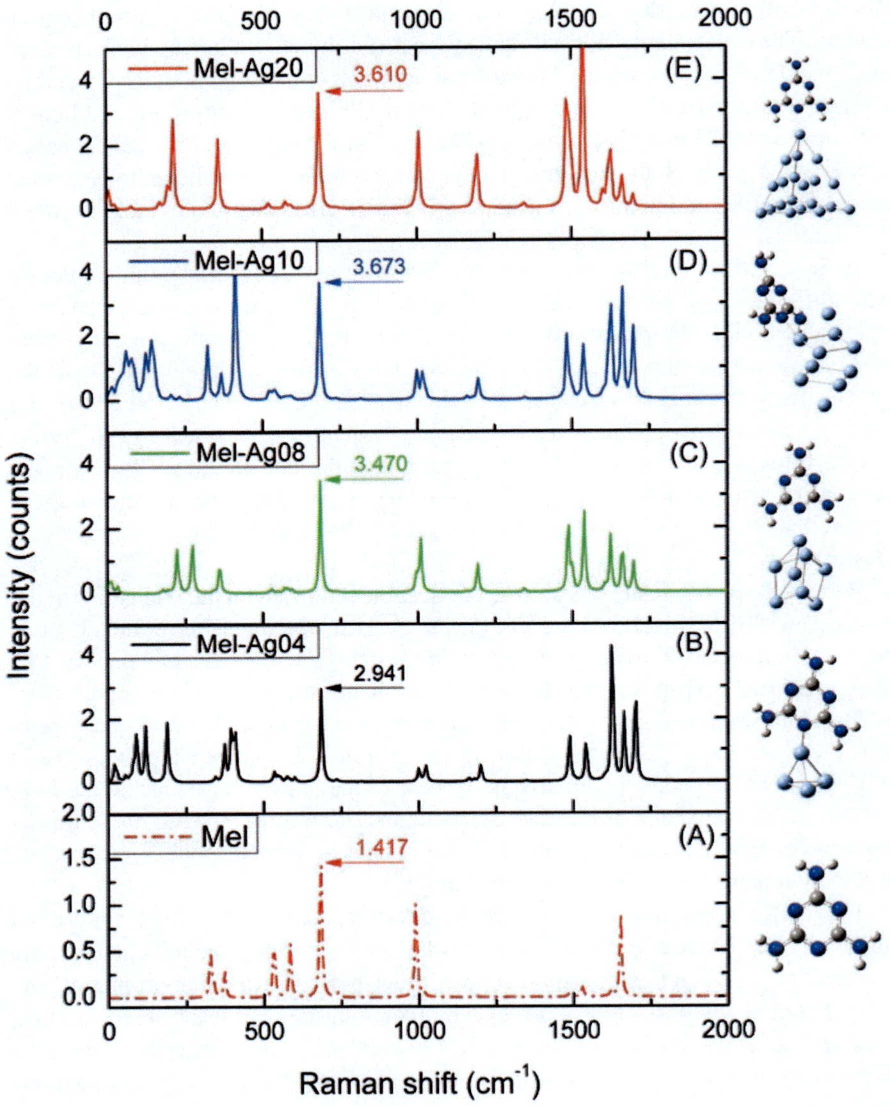

Fig. 8.13 Calculated normal Raman spectra of melamine (**a**) and SERS spectra of the most stable melamine-silver complexes: melamine–Ag_4 (**b**); melamine–Ag_8 (**c**); melamine–Ag_{10} (**d**) and melamine–Ag_{20} (**e**) and their optimized structures. Calculations were carried out using the B3LYP functional with the LanL2DZ basis set for Ag atoms and the 6-31G(d) for C, N, and H atoms. Printed with permission from Ref. [64]. Copyright 2016, Elsevier

by computations using the extended Ag(111) solid surface model with periodic boundary conditions and plane-wave pseudopotential.

DFT calculations showed that both interactions between one Ag atom on the top of tetrahedral cluster with S4 atom, and between the cluster surface with the

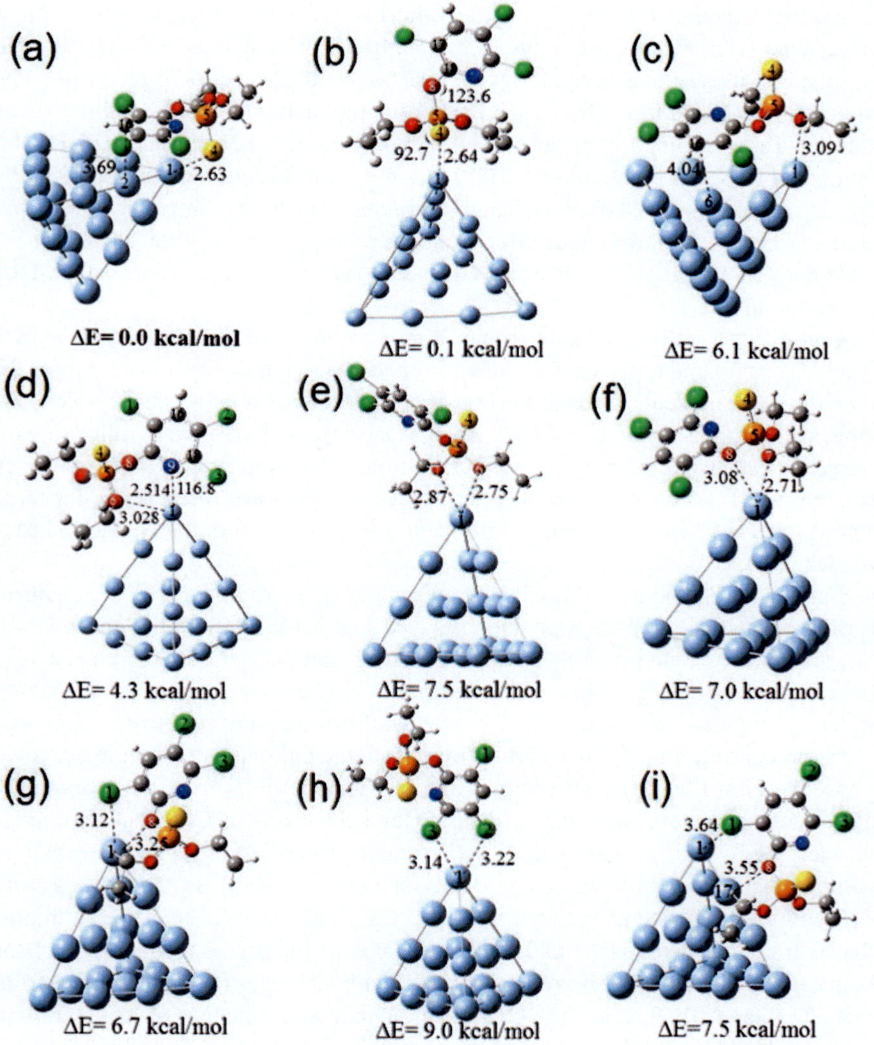

Fig. 8.14 Optimized structures, relative energies ΔE) of all possible interaction configurations between chlorpyrifos (CPF) and Ag_{20} cluster. Bond lengths are given in Å and bond angles indegrees. Printed with permission from Ref. [63]. Copyright 2020, The American Chemical Society

pyridine ring of CPF lead to the most stable configuration of the CPF–Ag20 complex (Fig. 8.14a). Another stable configuration with relative energy (ΔE) being only 0.1 kcal/mol is also recognized in which only the S4 atom interacts with the Ag atom at the top of the Ag_{20} cluster (Fig. 8.14b). The bond distance S4-Ag amounts to ~2.6 Å.

Both interaction modes have quite different geometric shapes, even with a small energy difference of only ~0.1 kcal/mol. For that reason, first-principle simulations using the extended Ag(111) surface with the periodic-repeated slabs model were needed to further probe further the interactions of CPF with the silver surface. To determine the dispersion correction between the CPF and AgNPs, the optimized Becke88 functional (optB88) was used together with the nonlocal vdW-DF correlation (optB88-vdW exchange-correlation function). Since the Ag-NPs is characterized by a quite rough surface, considerable defect sites (like adatom sites) and under-coordinated sites (like step edges sites) are expected to be present besides the plane surface.

Consequently, all three nanoparticle surface models result in remarkably similar results as compared to the ones obtained from the Ag_{20} tetrahedral cluster model. In fact, the CPF molecule binds to the Ag(111) surface via both the bond between S4 atom with surface Ag atom and the van der Waals interaction of the pyridine ring on terrace Ag sites and the bond length of the surface Ag atom and S4 atom is ~2.7 Å (cf. Fig. 8.15). Such a results deduced from calculations using the periodic-repeated slabs model is in good agreement with the those obtained from the T_d Ag_{20} cluster model.

Furthermore, Ngo et al. [63] also analyzed the charge transfer (CT) mechanism for the most stable CPF–Ag_{20} complexes (see Fig. 8.14a) in both the ground state S_0 and the first excited state S_1 based on TD-DFT method. The HOMO and LUMO populations and charge density difference (CDD) maps were determined for both states. Computed results for the S_0 state confirm the molecule-to-metal charge transfer occurred from filled adsorbate orbitals to the unfilled metal region occupied above the Fermi-level. Fermi energies of silver nanoparticles from the vacuum are also visualized, along with those for the π and π^* levels of CPF. The difference in energy between the Fermi level and frontier orbital levels of the adsorbate is comparable in terms of frequency to the incident light [82]. In the excited state S_1, with the influence of incident light, the CT flow occurs in the reverse trends, that is from the metal to the CPF ligand. In such a situation, a resonance happens between the incident photon frequency (ω_{inc}) with a charge-transfer transition from metal to ligand. To elucidate the CT transition under the influence of incident light, optical UV absorption spectra were simulated using TD-DFT method [63]. While the HOMO (becoming a SOMO) in the S_1 state is primarily located at the Ag_{20} moiety, the LUMO (becoming a SOMO) is mainly centered at the pyridine moiety of CPF ligand. These results imply that electron densities are transferred from the Ag_{20} cluster to the CPF ligand following incident light. This result is in good agreement with the CDD map determined at the first excited S_1 state. It is important to note that a laser of 532 nm used in these experimental conditions caused the electronic excitation to occur at the 532 nm (2.33 eV) which is in resonance with the charge transfer transition of the Fermi level to the π^* level of CPF.

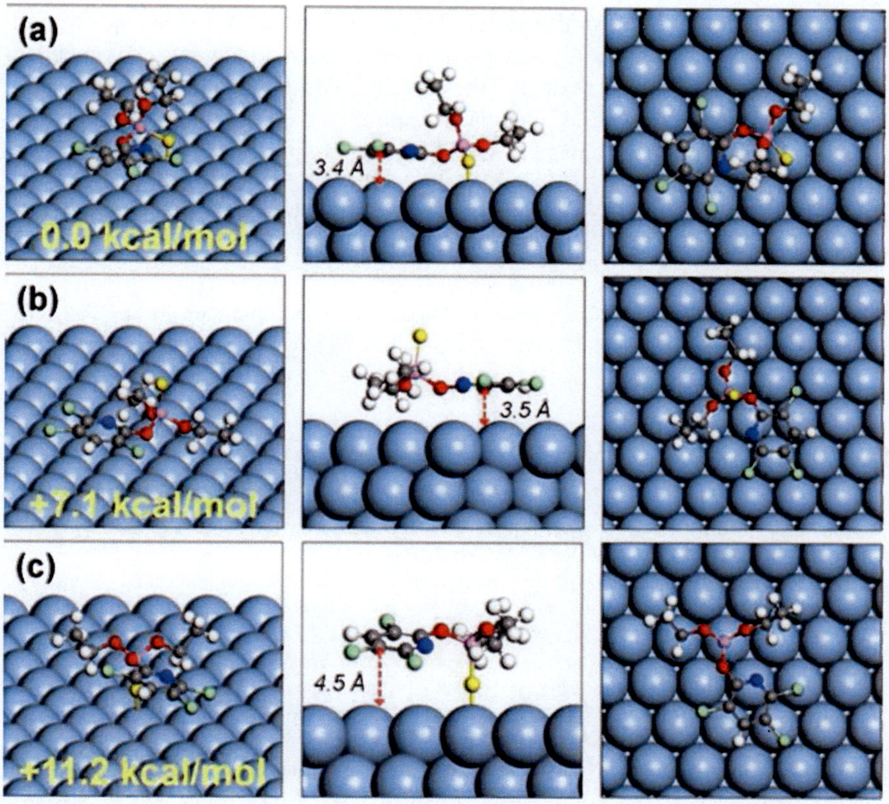

Fig. 8.15 Adsorption structure and binding energies relative of CPF on Ag(111) surface in perspective view (left), side view (center) and top view (right) for each isomer. Printed with permission from Ref. [63]. Copyright 2020, The American Chemical Society

8.5 Concluding Remarks

In this chapter, we outlined the development of a theoretical treatment using quantum chemical computations for the surface-enhanced Raman scattering (SERS) experiments. The latter constitutes a powerful vibrational spectrometric technique which allows overly sensitive detection of analytes at very low concentrations. The current state of knowledge concerning the mechanism of SERS, and some recent computational studies in which small silver clusters were used as models for substrate materials were described. Our opinion on the effects of both electromagnetic (EM) and chemical (CE) enhancement mechanisms on the SERS spectra were based on the results obtained from small cluster models such as the Ag_{10} and Ag_{20} cluster. Some structural issues such as the most stable structure of the Ag_{20} cluster as well as the molecular interactions of silver clusters with organic molecules such as pyridine and pesticides were discussed in detail. A silver cluster as small as Ag_{10}, and in particular the tetrahedral Ag_{20} cluster, have been shown to have quite good behavior in representing the Ag nanoparticle surfaces, and they are recommended to be used in further studies related to the adsorption of the organic molecules onto silver nanoparticles. The cluster model can naturally be extended to the simulation of nano-surfaces of other coinage metals including copper and gold. These results are also of importance for the design of efficient mobile sensors that are expected to rapidly detect toxic compounds present in the environment and in food and agricultural products.

Acknowledgements This work was funded by VinGroup (Vietnam) and supported by VinGroup Innovation Foundation (VinIF) under project code VINIF.2020.DA21.

References

1. Rao CNR, Müller A, Cheetham AK (2004) The chemistry of nanomaterials: synthesis, properties and applications. Wiley VCH, Weinheim
2. Daniel MV, Astruc D (2004) Gold nanoparticles: assembly, supramolecular chemistry, quantum-size-related properties, and applications toward biology, catalysis and nanotechnology. Chem Rev 104:293–351
3. Heiz U, Landman U (2007) Nanocatalysis. Springer, New York
4. Diez I, Ras RHA (2011) Fluorescent silver nanoclusters. Nanoscale 3:1963–1974
5. Sungmoon C, Dickson RM, Yu J (2012) Developing luminescent silver nanodots for biological applications. Chem Soc Rev 41:1867–1898
6. Nguyen MT, Kiran B (2017) Clusters. Structure, bonding and reactivity. Springer, New York
7. Martsinovich N, Jones DR, Troisi A (2010) Electronic structure of TiO_2 surfaces and effect of molecular adsorbates using different DFT implementations. J Phys Chem C 114:22659–22271
8. Qu ZW, Kroes GJ (2006) Theoretical study of the electronic structure and stability of titanium dioxide clusters $(TiO_2)_n$ with n = 1–9. J Phys Chem B 110:8998–9011
9. Gałyńska M, Persson P (2013) Emerging polymorphism in nanostructured TiO_2: quantum chemical comparison of anatase, rutile, and brookite clusters. Int J Quantum Chem 113:2611–1622
10. Makowska-Janusik M, Kassiba, AH (2017) In: Leszczynski J (ed) Handbook of computational chemistry. Chapter 23, Springer, New York
11. Fleischmann M, Hendra PJ, McQuillan AJ (1974) Raman spectra of pyridine adsorbed at a silver electrode. Chem Phys Lett 26:163–166
12. Cialla D, März A, Böhme R, Theil F, Weber K, Schmitt M, Popp J (2012) Surface-enhanced Raman spectroscopy (SERS): progress and trends. Anal Bioanal Chem 403:27–54
13. Jensen L, Aikens CM, Schatz GC (2008) Electronic structure methods for studying surface-enhanced Raman scattering. Chem Soc Rev 37:1061–1073
14. Gersten J, Nitzan A (1980) Electromagnetic theory of enhanced Raman scattering by molecules adsorbed on rough surfaces. J Chem Phys 73:3023–3037
15. Ding SY, You EM, Tian ZQ, Moskovits M (2017) Electromagnetic theories of surface-enhanced Raman spectroscopy. Chem Soc Rev 46:4042–4076
16. Metiu H, Das P (1984) The electromagnetic theory of surface enhanced spectroscopy. Annu Rev Phys Chem 35:507–536
17. Sharafdini R, Mohammadpour M, Ramazani S, Jamshidi Z (2020) Theoretical simulation of surface-enhanced resonance Raman spectroscopy of cytosine and its tautomers. J Raman Spectrosc 51:55–65

18. Campion A, Ivanecky III JE, Child CM, Foster M (1995) On the mechanism of chemical enhancement in surface-enhanced Raman scattering. J Am Chem Soc 117:11807–11808
19. Mohammadpour M, Khodabandeh MH, Visscher L, Jamshidi Z (2017) Elucidation of charge-transfer SERS selection rules by considering the excited state properties and the role of electrode potential. Phys Chem Chem Phys 19:7833–7843
20. Xia L, Chen M, Zhao X, Zhang Z, Xia J, Xu H, Sun M (2014) Visualized method of chemical enhancement mechanism on SERS and TERS. J Raman Spectrosc 45:533–540
21. Sharma B, Frontiera RR, Henry AI, Ringe E, Van Duyne RP (2012) SERS: materials, applications, and the future. Mater Today 15(1-2):16–25
22. Jeanmaire DL, Van Duyne RP (1977) Surface Raman spectroelectrochemistry: Part I. heterocyclic, aromatic, and aliphatic amines adsorbed on the anodized silver electrode. J Electroanal Chem Interfacial Electrochem 84:1–20
23. Wang J, Wang G, Zhao J (2003) Structures and electronic properties of Cu_{20}, Ag_{20}, and Au_{20} clusters with density functional method. Chem Phys Lett 380:716–720
24. Li J, Li X, Zhai HJ, Wang LS (2003) Au_{20}: a tetrahedral cluster. Science 299:864–867
25. Nhat PV, Si NT, Nguyen MT (2019) Comment on theoretical investigations on geometrical and electronic structures of silver clusters. J Comput Chem 40:1990–1993
26. McKee ML, Samokhvalov A (2017) Density functional study of neutral and charged silver clusters Ag_n with $n = 2$–22. Evolution of properties and structure. J Phys Chem A 121:5018–5028
27. Nhat PV, Si NT, Nguyen MT (2018) Elucidation of the molecular and electronic structures of some magic silver clusters Ag_n ($n = 8, 18, 20$). J Mol Model 24:209–221
28. Baishya K, Idrobo JC, Öğüt S, Yang M, Jackson K, Jellinek J (2008) Optical absorption spectra of intermediate-size silver clusters from first principles. Phys Rev B 78(7):075439
29. Zhao J, Luo Y, Wang G (2001) Tight-binding study of structural and electronic properties of silver clusters. Eur Phys J D 14:309–316
30. De H, Paul A, Datta A (2017) Theoretical study of Au_4 thymine, Au_{20} and Ag_{20} uracil and thymine complexes for surface enhanced Raman scattering. Comput Theor Chem 1111:1–13
31. Zhao L, Jensen L, Schatz GC (2006) Pyridine-Ag_{20} cluster: a model system for studying surface-enhanced Raman scattering. J Am Chem Soc 128:2911–2919
32. Tsuneda T (2019) Theoretical investigations on geometrical and electronic structures of silver clusters. J Comput Chem 40:206–211
33. Alonso JA (2000) Electronic and atomic structure, and magnetism of transition metal clusters. Chem Rev 100:637–678
34. Liao MS, Watts JD, Huang MJ (2014) Theoretical comparative study of oxygen adsorption on neutral and anionic Ag_n and Au_n clusters ($n = 2$–25). J Phys Chem C 118:21911–21927
35. Sych TS, Reveguk ZV, Pomogaev VA, Buglak AA, Reveguk AA, Ramazanov RR, Kononov AI (2018) Fluorescent silver clusters on protein templates: understanding their structure. J Phys Chem C 122(51):29549–29558
36. Fernández EM, Soler JM, Garzón IL, Balbás LC (2004) Trends in the structure and bonding of noble metal clusters. Phys Rev B 70:165403
37. Tian D, Zhang H, Zhao J (2007) Structure and structural evolution of Ag_n ($n = 3$–22) clusters using a genetic algorithm and density functional theory method. Solid State Commun 144:174–179
38. Woodruff DP (2007) Atomic clusters: from gas phase to deposited. Elsevier, Amsterdam
39. Nhat PV, Si NT, Leszczynski J, Nguyen MT (2017) Another look at structure of gold clusters Au_n from perspective of phenomenological shell model. Chem Phys 493:140–160
40. Pyykko P (1988) Relativistic effects in structural chemistry. Chem Rev 88:563–594
41. Schwerdtfeger P, Dolg M, Schwarz WE, Bowmaker GA, Boyd PD (1989) Relativistic effects in gold chemistry. I Diatomic gold compounds. J Chem Phys 91:1762–1774
42. Johansson MP, Pyykkö P (2004) The importance of being tetrahedral: the cadmium pyramids Cd_n $n = 4, 10, 20, 35$ and 56. Phys Chem Chem Phys 6:2907–2909
43. Jackson KA (1993) First-principles study of the structural and electronic properties of Cu clusters. Phys Rev B 47:9715

44. Taylor KJ, Pettiette-Hall CL, Cheshnovsky O, Smalley RE (1992) Ultraviolet photoelectron spectra of coinage metal clusters. J Chem Phys 96:3319–3329
45. Rabilloud F (2013) UV-visible absorption spectra of metallic clusters from TD-DFT calculations. Eur Phys J D 67:18
46. Alameddin G, Hunter J, Cameron D, Kappes MM (1992) Electronic and geometric structure in silver clusters. Chem Phys Lett 192:122–128
47. Jackschath C, Rabin I, Schulze W (1992) Electron impact ionization of silver clusters Ag_n, n $\leqq$ 36. Zeitschrift für Physik D 22:517–520
48. Kneipp K, Wang Y, Kneipp H, Perelman LT, Itzkan I, Dasari RR, Feld MS (1997) Single molecule detection using surface-enhanced Raman scattering (SERS). Phys Rev Lett 78:1667
49. Nie S, Emory SR (1997) Probing single molecules and single nanoparticles by surface-enhanced Raman scattering. Science 275:1102–1106
50. Jensen L, Zhao LL, Schatz GC (2007) Size-dependence of the enhanced Raman scattering of pyridine adsorbed on Ag_n ($n = 2-8, 20$) Clusters. J Phys Chem C 111:4756–4764
51. Gaff J, Franzen S (2012) Resonance Raman enhancement of pyridine on Ag clusters. Chem Phys 397:34–41
52. Jensen L, Zhao LL, Autschbach J, Schatz GC (2005) Theory and method for calculating resonance Raman scattering from resonance polarizability derivatives. J Chem Phys 123:174110
53. Su JP, Lee YT, Lu SY, Lin JS (2013) Chemical mechanism of surface-enhanced Raman scattering spectrum of pyridine adsorbed on Ag cluster: ab initio molecular dynamics approach. J Comput Chem 34:2806–2815
54. Morton SM, Jensen L (2009) Understanding the molecule−surface chemical coupling in SERS. J Am Chem Soc 131:4090–4098
55. Sun M, Liu S, Chen M, Xu H (2009) Direct visual evidence for the chemical mechanism of surface-enhanced resonance Raman scattering via charge transfer. J Raman Spectrosc 40:137–143
56. Goodacre R, Graham D, Faulds K (2018) Recent developments in quantitative SERS: moving towards absolute quantification. Trends Anal Chem 102:359–368
57. Zhang Y, Zhao S, Zheng J, He L (2017) Surface-enhanced Raman spectroscopy (SERS) combined techniques for high-performance detection and characterization. Trends Anal Chem 90:1–13
58. Camden JP, Dieringer JA, Wang Y, Masiello DJ, Marks LD, Schatz GC, Van Duyne RP (2008) Probing the structure of single-molecule surface-enhanced Raman scattering hot spots. J Am Chem Soc 130:12616–12617
59. Langer J, Jimenez de Aberasturi D, Aizpurua J, Alvarez-Puebla RA, Auguié B, Baumberg JJ, Choo J (2019) Present and future of surface-enhanced Raman scattering. ACS Nano 14:28–117
60. Xu ML, Gao Y, Han XX, Zhao B (2017) Detection of pesticide residues in food using surface-enhanced Raman spectroscopy: a review. J Agric Food Chem 65:6719–6726
61. Cañamares MV, Pozzi F, Lombardi JR (2019) Raman, SERS and DFT analysis of the main alkaloids contained in Syrian rue. J Phys Chem C 123:9262–9271
62. Yao G, Zhai Z, Zhong J, Huang Q (2017) DFT and SERS study of 15N Full-labeled adenine adsorption on silver and gold surfaces. J Phys Chem C 121:9869–9878
63. Ngo TC, Trinh QT, An NTT, Tri NN, Trung NT, Truong DH, Huy BT, Nguyen MT, Dao DQ (2020) SERS spectra of the pesticide chlorpyrifos adsorbed on silver nanosurface: the Ag_{20} cluster model. J Phys Chem C 124:21702–21716
64. An NTT, Dao DQ, Nam PC, Huy BT, Tran HN (2016) Surface enhanced Raman scattering of melamine on silver substrate: an experimental and DFT study. Spectrochim Acta A 169:230–237
65. Ivanov E, Ando RA, Corio P (2016) Solid-liquid-liquid extraction as an approach to the sensitive detection of a hydrophobic pollutant through surface-enhanced Raman spectroscopy. Vib Spectrosc 87:116–122
66. Klimeš J, Bowler DR, Michaelides A (2011) Van der Waals density functionals applied to solids. Phys Rev B 83:195131

67. Nhat PV, Si NT, Tien NT, Nguyen MT (2021) Theoretical study of the binding of the thiol-containing cysteine amino acid to the silver surface using a cluster model. J Phys Chem A 125:3244–3256
68. Colthup N (2012) Introduction to infrared and Raman spectroscopy. Elsevier, Holland
69. Podstawka E, Ozaki Y, Proniewicz LM (2004) Surface-enhanced Raman spectroscopy investigation of amino acids and their homodipeptides adsorbed on colloidal silver. Part I. Appl Spectrosc 58:570–580
70. Jenkins AL, Larsen RA, Williams TB (2005) Characterization of amino acids using Raman spectroscopy. Spectrochim Acta A 61:1585–1594
71. Jing C, Fang Y (2007) Experimental (SERS) and theoretical (DFT) studies on the adsorption behaviors of L-cysteine on gold/silver nanoparticles. Chem Phys 332:27–32
72. Fleming DG, Finnerty JJ, Campos-Vallette M, Célis F, Aliaga AE, Fredes C, Koch R (2009) Experimental and theoretical Raman and surface-enhanced Raman scattering study of cysteine. J Raman Spectrosc 40:632–638
73. Toh HS, Batchelor-McAuley C, Tschulik K, Compton RG (2014) Chemical interactions between silver nanoparticles and thiols: a comparison of mercaptohexanol against cysteine. Sci China Chem 57:1199–1210
74. Nikolić VD, Ilić DP, Nikolić LB, Stanković MZ, Stanojević LP, Savić IM, Savić IM (2012) The synthesis and structure characterization of deoxyalliin and alliin. Adv Technol 1:38–46
75. Otto C, Van den Tweel TJJ, De Mul FFM, Greve J (1986) Surface-enhanced Raman spectroscopy of DNA bases. J Raman Spectrosc 17:289–298
76. Sundaram J, Park B, Kwon Y, Lawrence KC (2013) Surface enhanced Raman scattering (SERS) with biopolymer encapsulated silver nanosubstrates for rapid detection of foodborne pathogens. Int J Food Microbiol 167:67–73
77. Cialla D, Pollok S, Steinbrücker C, Weber K, Popp J (2014) SERS-based detection of biomolecules. Nanophotonics 3:383–411
78. Pakiari AH, Jamshidi Z (2007) Interaction of amino acids with gold and silver clusters. J Phys Chem A 111:4391–4396
79. Buglak AA, Ramazanov RR, Kononov AI (2019) Silver cluster–amino acid interactions: a quantum-chemical study. Amino Acids 51:855–864
80. Buglak AA, Kononov AI (2020) Comparative study of gold and silver interactions with amino acids and nucleobases. RSC Adv 10:34149–34160
81. Miyamoto M, Hada M (2020) Surface-enhanced Raman scattering of M_2–pyrazine–M_2 (M = Cu, Ag, Au): analysis by natural perturbation orbitals and density functional theory functional dependence. J Comput Chem 41:1628–1637
82. Brolo AG, Irish DE, Smith BD (1997) Applications of surface enhanced Raman scattering to the study of metal-adsorbate interactions. J Mol Struct 405:29–44

Index

A
Ab initio molecular dynamics (AIMD), 2, 19, 139, 149, 164
Adaptive Intermolecular REBO (AIREBO), 5, 6, 8, 19, 20
Adaptive multiple time step (A-MTS), 45
Adenosine, 142, 143
Adiabatic electron affinity (AEA), 137–141, 145
A_{g20} cluster, 264, 268, 271, 272, 277, 279–282
Alkaline hydrolysis, v, vi, 216–231
All-atom level, 37
ANTA, 185, 193
Arid soils, v, 172, 184
Artificial intelligence, 172, 203–207
Assisted Model Building with Energy Refinement (AMBER), 3, 11, 57, 105
Associative memory,Water mediated, Structure and Energy Model (AWSEM), 54–56
Atomic forces, 182, 206
Atomistic MD simulation, 115, 118

B
Basis sets, 72, 76, 80, 91, 92, 97, 105, 106, 111, 118, 173, 174, 176, 180, 193, 266, 268, 274, 278
Benzene, 86, 105–111
Binding energies, 80, 82, 84, 85, 88, 89, 92, 96, 98, 102–106, 109–112, 176, 177, 181, 184–186, 188–191, 194, 195, 273, 281
Bioaccumulation factor (BCF), 242, 246
Boltzmann integrals, 37
Boltzmann probabilities, 48, 198–200
Born-Mayer, 78
Born-Oppenheimer, 35, 36

C
Canonical Monte Carlo, 43, 44, 46
Cation-π complex, 106–110
Cation-π interaction, 73, 105–106, 108, 109, 118
CG dynamics, 33
Charge density, 174, 184, 186, 187, 189, 197, 280
Charge density difference (CDD), 184, 186–187, 189, 197, 263, 280
Charge transfer (CT), 72, 74, 84, 93–95, 118, 172, 186, 259, 260, 262, 263, 270, 272, 277, 280
Charge transfer intermediate state, 262
Charge transfer to solvents (CTTS), 72, 74, 85, 89, 93–95, 118
CHARMM force field, 115
Chebyshev Interaction Model for Efficient Simulation (ChIMES), 15, 16, 20
Chemical development kit (CDK), 236, 241, 242, 249
Chemical enhancement (CE), 259, 260, 262–263, 269, 272, 276, 277, 281
ChemIDplus, 241
Chlorpyrifos (CPF), 277, 279–281
Classical MD, 116
Classical molecular dynamics, 78
Clays, 194
Coarse-grained approach, v, 13, 33, 60

J. Leszczynski, M. K. Shukla (eds.), *Practical Aspects of Computational Chemistry V*, https://doi.org/10.1007/978-3-030-83244-5

Computational chemistry, v, 1, 4, 134, 171–207
Contaminants, v, 171–207
Corundum, 184
Coulomb-electrostatic, 36, 56
CRYSOL program, 51
Crystal orbital overlap population, 178
Crystal orbital overlap population (COOP), 178, 191, 196
Cysteine, 49, 273–276

D

DeepCDpred, 53
Density functional theory (DFT), v, 6, 7, 17–19, 32, 72, 73, 76, 80–82, 84, 92, 106, 109, 111–113, 116, 118, 135, 138, 139, 141, 145, 146, 149, 172–184, 188, 192, 200–202, 204–207, 260, 265–268, 278
Density functional tight binding (DFTB), vi, 17, 18, 20, 172, 200–202, 205–207
Density of states (DOS), 47, 177–178, 184, 195, 196
Diffusion equation method, 47
Dihydrocytosine, 127
Dihydrothymine, 127
Dipole-dipole, 73, 76, 84, 112, 116, 118
Dipole-induced dipole, 73, 118
Dissociative electron attachment (DEA), 127, 129, 130, 132, 134, 136, 138–141, 145–147, 149, 150, 153–156, 159, 160, 163–165
Distance scaling method (DSM), 47
DNAN, 185, 190, 193–195, 197–200
DNT, 185, 193–195
Drude model, 75, 76, 79

E

Electromagnetic mechanism (EM), 259–263, 270, 272, 281
Electron paramagnetic resonance (EPR), 48
Electron traps, 145
Empirical valence bond (EVB), 9–11, 13, 19, 20
Enzymatic catalysis, 32
Equation of motion, 1
Exchange correlation functional, 111, 174, 184
Exchange-repulsion, 76, 95

F

Fatty acid dimers, 85
Force matching, 15, 40
FOX7, 185, 186, 188, 189

G

Gay-Berne potentials, 37
Geometry optimization, 257
Graphical processing units (GPUs), 33
GROMOS, 3
Guanosine, 142, 143

H

Halide-water clusters, 74, 89, 93–95, 118
Hammett equation, 74
Hammett parameters, 111
Hammett substituent constants, 72
Hellmann-Feynman theorem, 182–184
Henry's Law, 197
Herzberg-Teller mechanism, 263
Hexahydro-1,3,5-trinitro-1,3,5-triazine (RDX), 215–231
Hierarchical clustering (HC), 236, 241, 242
HiRe-RNA, 56–57
Histone deacetylase inhibitors (HADC), 246, 248, 251
Hybrid functionals, 179
Hydrated electron, v, 125–165
Hydrogen bonding, 6, 55, 56, 73, 85–88, 96, 101–104, 115, 116, 118
Hydrogen-deuterium exchange, 48, 51–52
Hypoxia, v, 126–129, 151, 152, 154, 165
Hypoxia-activated prodrugs, 152, 154
Hypoxia-selective cytotoxins (HSCs), 152
Hypoxic cells, 126, 128, 151, 156

I

Inhibition constant (K_i), 178, 179
π-π interaction, 74, 109–114, 118
Interaction energies, 4, 37, 74–81, 83, 84, 106–109, 115, 116, 184

K

Kaolinite, 194, 195
Kubo cluster-cumulant functions, 32, 37

L

LAMMPS, 7, 55
Lennard-Jones potential, 3, 38, 77, 78
LEWIS, 15, 20, 172, 178, 184, 189, 190, 195–197, 199, 207
Lewis acidity, 172, 178, 184, 190, 196–197, 207
Liquid chromatography, 49
Liquid-crystal simulations, 37
London dispersion forces, 73

M

Machine learning (ML), 18, 19, 53, 203–207
MARTINI force field, 37, 38, 42, 56
Maximum recommended therapeutic dose (MRTD), 248, 250
Metal complexes, 157–165
Metallic nanoparticles, 128, 157–165, 259, 260
Metal oxides, 257
3-methylindole, 106–108, 110
Modelability, v, vi, 233–252
Modelability index, 235, 236, 246, 251
Modified nucleosides, 128–134
Møller-Plesset (MP), 72, 73, 83, 105
Monte Carlo, 36, 41–44, 46, 47, 52, 56, 162
Multiscale modeling, 33
Multi-state empirical valence bond (MS-EVB), 11–13, 19, 20
Multi-state empirical valence model, 11
Munitions, v, 171, 172, 183–185, 190, 192–200, 207

N

NARES-2P, 42, 43, 48, 58, 59
Nearest neighbors, 204, 235, 242–244, 247–251
4-nitro-2,4-diazabutanal, 218, 230
Non-covalent interactions, v, 71–118
Nonlocal van der Waals functionals, 181
Normoxic cells, 126
Nosé-Hoover, 45
Nosé -Poincare, 45
NQ, 185, 186, 188–190, 197
NTO, 185–188, 190, 193–195, 197
Nuclear magnetic resonance (NMR), 43, 48, 52–53, 85, 115
Nucleic acids, 32, 33, 42, 43, 55–58, 78, 129

O

Octahydro-1,3,5,7-tetranitro-1,3,5,7-tetrazocine (HMX), 215–231
Optimized Potential for Efficient protein structure Prediction (OPEP), 54, 56–57
Optimized Potential for Liquid Simulations (OPLS), 3
Oxygen mimetics, 151–157, 165

P

Pepsi-SAXS, 51
Photoinduced CT, 263
Physics-based potentials, 32–41, 43, 45–48, 53–61
P-methylphenol, 106–108, 110
Potential of mean force (PMF), 32–39, 58, 60
Protein-folding simulations, 34, 56
Pyridine, 264, 269–272, 279–281

Q

QM/MM approach, 3, 33, 164
QSAR modeling, 245, 250
Quadrupole-dipole, 73, 76, 105, 107, 118
Quadrupole-quadrupole, 73, 76, 118

R

Radiosensitizers, v, 125–165
Radiotherapy, 126–135, 151, 159, 164, 165
Reaction mechanism, 13, 19, 150, 216, 231
Reactive empirical bond order (REBO), 4–6, 19, 20
Reactive force-field, v, 1–20, 200, 201
ReaxFF, 4, 6–9, 13, 19, 20
ReaxFF-lg, 8, 20
Reduction chemistry, 33, 190, 192, 193
RexPoN, 14

S

SASSIM method, 51
Schematic of 2,4,6-trinitrotoluene (TNT), 184–188, 190, 191, 193–196, 198, 199
Second-order perturbation theory, 180
Self-interaction errors, 172, 178–180
Silver clusters, v, 258, 265–270, 273, 274, 276, 281
Silver nanoparticles, 255–282

Simulated kinetics, 222, 228, 231
South American Initiative for a Rapid and Accurate Hamiltonian (SIRAH), 38, 54, 57
Structure-property relationship, 72
Supramolecular assemblies, 115
Surface-enhanced raman scattering (SERS), vi, 255–282
Symmetry-adapted perturbation theory (SAPT), 72, 84, 110

T
Theoretical chemistry, 175
Thermodynamic cycle, 95, 192

U
UNIfied COarse gRaiNed model (UNICORN), 58–59
UNited RESidue (UNRES), 35, 38, 39, 41–43, 47, 48, 52, 54, 56, 58, 59, 61

V
Velocity-Verlet, 45
Vertical attachment energy (VAE), 137, 138, 140, 141
Vertical detachment energy (VDE), 137–141, 145
Vibrational feshbach resonance (VFR), 155
Vibration-rotation-tunneling (VRT), 85, 86
Virtual-bond potentials, 37

W
Weak interactions, 72–74, 84